5/85

Casebooks
in Earth Sciences
Series Editor: R. N. Ginsburg

Living Fossils

Edited by
Niles Eldredge and Steven M. Stanley

With 72 Illustrations

Springer Verlag
New York Berlin Heidelberg Tokyo

Editors

Niles Eldredge, Department of Invertebrates, The American Museum
of Natural History, Central Park West at 79th Street, New York, NY
10024/USA

Steven M. Stanley, Department of Earth and Planetary Sciences, The
Johns Hopkins University, Baltimore, MD 21218/USA

Series Editor

Robert N. Ginsburg, University of Miami, School of Marine and Atmo-
spheric Science, Fisher Island Station, Miami, FL 33139/USA

On the front cover: (above) Fig. 14.1. *Acipenser sturio* L. (p. 149);
(below) Fig. 13.1. *Polytpterus senegalus* (p. 144).

Library of Congress Cataloging in Publication Data
Main entry under title:
Living fossils.
(Casebooks in earth sciences)
Bibliography: p.
Includes index.
1. Zoology—Miscellanea. 2. Evolution. I. Eldredge,
Niles. II. Stanley, Steven M. III. Series.
QL58.L58.1984.591.3'8.84-1300.

Typeset by Bi-Comp, Inc., York, Pennsylvania
Printed and bound by Halliday Lithograph, West Hanover, Massachusetts
Printed in the United States of America.

9 8 7 6 5 4 3 2 1

ISBN 0-387-90957-5 Springer-Verlag New York Berlin Heidelberg Tokyo
ISBN 3-540-90957-5 Springer-Verlag Berlin Heidelberg New York Tokyo

Series Preface

The case history approach has an impressive record of success in a variety of disciplines. Collections of case histories, casebooks, are now widely used in all sorts of specialties other than in their familiar application to law and medicine. The case method had its formal beginning at Harvard in 1871 when Christopher Lagdell developed it as a means of teaching. It was so successful in teaching law that it was soon adopted in medical education, and the collection of cases provided the raw material for research on various diseases. Subsequently, the case history approach spread to such varied fields as business, psychology, management, and economics, and there are over 100 books in print that use this approach.

The idea for a series of *Casebooks in Earth Science* grew from my experience in organizing and editing a collection of examples of one variety of sedimentary deposits. The project began as an effort to bring some order to a large number of descriptions of these deposits that were so varied in presentation and terminology that even specialists found them difficult to compare and analyze. Thus, from the beginning, it was evident that something more than a simple collection of papers was needed. Accordingly, the nearly fifty contributors worked together with George de Vries Klein and me to establish a standard format for presenting the case histories. We clarified the terminology and some basic concepts, and when the drafts of the cases were completed we met to discuss and review them. When the collection was ready to submit to the publisher, and I was searching for an appropriate subtitle, a perceptive colleague, R. Michael Lloyd pointed out that it was a collection of case histories comparable in principle to the familiar casebooks of law and medicine. After this casebook [*Tidal Deposits,* (1975)] was published and accorded a warm reception, I realized that the same approach could well be applied to many other subjects in earth science.

It is the aim of this new series, *Casebooks in Earth Science,* to apply the discipline of compiling and organizing truly representative case histories to accomplish various objectives: establish a collection of case histories for both reference and teaching; clarify terminology and basic concepts; stimulate and facilitate synthesis and classification; and encourage the identification of new questions and new approaches. There are no restrictions on the subject matter for the casebook series save that they concern earth science. However, it is clear that the most appropriate subjects are those that are largely descriptive. Just as there are no fixed boundaries on subject matter, so is the format and approach of individual volumes open to the discretion of the editors working with their contributors. Most casebooks will of necessity be communal efforts with one or more editors working with a group of contributors. However, it is also likely that a collection of case histories could be assembled by one person drawing on a combination of personal experience and the literature.

Clearly the case history approach has been successful in a wide range of disciplines. The systematic application of this proven method to earth science subjects holds the promise of producing valuable new resources for teaching and research.

Miami, Florida
July, 1984

Robert N. Ginsburg
Series Editor

Contents

List of Contributors

Roger L. Batten
Department of Invertebrates, The American Museum of Natural History, Central Park West at 79th St., New York, NY 10024

Alan H. Cheetham
Department of Paleobiology, National Museum of Natural History, Smithsonian Institution, Washington, DC 20560

Mitchell W. Colgan
Earth Sciences Board, University of California, Santa Cruz, CA 95064

Joel Cracraft
Department of Anatomy, University of Illinois, P.O. Box 6998, Chicago, IL 60680

Eric Delson
Department of Anthropology, Lehman College, City University of New York, Bronx, NY 10468, and Department of Vertebrate Paleontology, American Museum of Natural History, New York, NY 10024

Niles Eldredge
Department of Invertebrates, The American Museum of Natural History, Central Park West at 79th St., New York, NY 10024

Robert J. Emry
Department of Paleobiology, National Museum of Natural History, Smithsonian Institution, Washington, DC 20560

Daniel C. Fisher
Museum of Paleontology, University of Michigan, Ann Arbor, MI 48109

Peter Forey
Department of Palaeontology, British Museum (Natural History), Cromwell Road, London SW7 5BD

Brian G. Gardiner
Department of Biology, Queen Elizabeth College, University of London, Campden Hill Road, London W8 7AH

Michael T. Ghiselin
Department of Invertebrate Zoology, California Academy of Sciences, Golden Gate Park, San Francisco, CA 94118

P. Humphry Greenwood
Department of Zoology, British Museum (Natural History), Cromwell Road, London SW7 5BD

Robert R. Hessler
Scripps Institution of Oceanography, University of California, San Diego, La Jolla, CA 92093

Carole S. Hickman
Department of Paleontology, University of California, Berkeley, CA 94720

Richard S. Houbrick
Department of Invertebrate Zoology (Mollusks), National Museum of Natural History, Smithsonian Institution, Washington, DC 20560

Christine Janis
Newnham College and Department of Zoology, University of Cambridge, Cambridge, England, Current address: Division of Biology and Medicine, Box G, Brown University, Providence, RI 02912

John G. Maisey
Department of Vertebrate Paleontology, The American Museum of Natural History, Central Park West at 79th St., New York, NY 10024

Eugene R. Meyer
Department of Earth and Planetary Sciences, The Johns Hopkins University, Baltimore, MD 21218

Michael Novacek
Department of Vertebrate Paleontology, The American Museum of Natural History, Central Park West at 79th St., New York, NY 10024

Colin Patterson
Department of Paleontology, British Museum (Natural History), Cromwell Road, London SW7 5BD

Alfred L. Rosenberger
Department of Anthropology, University of Illinois, Box 4384, Chicago, IL 60630

Frederick R. Schram
Department of Geology, Natural History Museum, P.O. Box 1390, San Diego, CA 92112

Hans-Peter Schultze
Museum of Natural History, University of Kansas, Lawrence, KS 66045

Jeffrey H. Schwartz
Department of Anthropology, University of Pittsburgh, Pittsburgh, PA 15260

Steven M. Stanley
Department of Earth and Planetary Sciences, The Johns Hopkins University, Baltimore, MD 21218

Ian Tattersall
Department of Anthropology, The American Museum of Natural History, Central Park West at 79th St., New York,, NY 10024

Richard W. Thorington, Jr.
Department of Vertebrate Zoology, National Museum of Natural History, Smithsonian Institution, Washington, DC 20560

Elisabeth S. Vrba
Transvaal Museum, P.O. Box 413, Pretoria 0001, South Africa

Peter Ward
Department of Geology, University of California, Davis, CA 95616

E. O. Wiley
Museum of Natural History, University of Kansas, Lawrence, KS 66045

Judith E. Winston
Department of Invertebrates, The American Museum of Natural History, Central Park West at 79th St., New York, NY 10024

Katherine E. Wolfram
Department of Vertebrate Paleontology, The American Museum of Natural History, Central Park West at 79th St., New York, NY 10024

Living Fossils: Introduction to the Casebook

Niles Eldredge and Steven M. Stanley

Department of Invertebrates, American Museum of Natural History, New York, NY 10024
Department of Earth and Planetary Science, The Johns Hopkins University, Baltimore, MD 21218

Science is a perpetual collision between ideas and observations. In evolutionary theory, the ideas we entertain usually concern process: We seek to know just how evolution actually works. But our "observations" are problematical; ranging from DNA sequences, distributions of alleles in populations, and on up through the origins and extinctions of major taxa, the "data" of evolutionary biology are of vastly different scale, quality, and, some would say, significance. But all these "data" share a common feature: All are states of biological systems that represent the outcome of the evolutionary process. All are historical entities—these particular genes, allelic frequencies, species, and phyla. And all that we think we know about them constitutes not so much raw data or observation as detailed hypotheses.

Clearly, progress in evolutionary theory depends very much on our getting as accurate a picture as possible about what the evolutionary process produces. Similar, recurrent sorts of evolutionary phenomena—crossing over, mutation, allopatric speciation,[1] phylogenetic trends—are classes of events of some consider-

able generality. These are the sorts of phenomena, these are the patterns that evolutionary theory seeks to explain.

It has been emphasized in recent years, perhaps *ad nauseam,* that evolutionary biologists must keep their notions of process divorced as much as possible from their perception of evolutionary patterns. If we assume we know all there is to know about a process at the outset, we bias our very vision of the patterns, sacrificing what little hope for objectivity we may harbor in elucidating life's history. But there is a still graver objection. We need independently tested hypotheses of pattern to test competing hypotheses of evolutionary process. If we are to improve evolutionary theory, we need as objective a view as possible of the fruits of that process.

Hence this casebook. The recent flurry of activity in paleobiology directed at patterns and processes of macroevolution abundantly underscores the critical role that reliable information plays in evaluating the relative merits of alternative theories of evolutionary mechanisms. Living fossils—the phenomenon of "arrested evolution"—is one of several patterns commonly singled out for macroevolutionary analysis (Stanley 1979; Eldredge and Cracraft 1980). The typically low diversity of such low-rate lineages improves the possibility of accurately summarizing both anatomical and taxic diversity, past and present, for each group without

---

[1] Mutation and allopatric speciation are both usually, and properly, construed as processes. But the phenomenon of mutation and the phenomenon of allopatric speciation are also *kinds of events*—hence biologic patterns that suggest the existence of a process of the same name. This duality of meaning is similar to that of the word "adaptation" (Simpson 1953:160).

much difficulty—cardinal qualities that enhance the very appeal of such taxa for close study. A recent analysis (Vrba and Eldredge 1984) tabulates some four or five major competing theories of the process(es) underlying the production of evolutionary trends. Of these, species selection itself turns out to be a class of processes, with a number of variant versions supplied by an array of different theorists. We need some criterion by which we may establish which of these ideas are the more powerful, the more accurate and apt descriptors of nature.

Thus, the goal of this book is simplicity itself: to provide anyone interested in the general phenomenon of slow evolutionary rates with enough case histories (as portrayed by a spectrum of biologists and paleontologists with a heterogeneous assemblage of theoretical backgrounds and inclinations) to decide if there is anything to this supposed phenomenon of arrested evolution, these living fossils. Is there anything about these data that requires special explanation (as, for example, Simpson 1944, 1953, thought), or are we merely dealing with the left-hand tail of a normal distribution of evolutionary rates, *all* of which fall out as various permutations of factors conspire in the "normal" course of evolutionary events? Or do we even have a good grasp of what those permutations are that do give us a spectrum of evolutionary rates?

And then there is the further goal: If there are one or two fundamental and identifiable patterns here, suggested by case after case with a haunting similarity, perhaps we can use the examples in this casebook to come to grips with competing notions of macroevolutionary processes. Some of the authors try to do precisely this. Though their primary charge was to present a brief sketch of one or more examples of living fossils, loosely following certain criteria and guidelines, some authors chose as well to evaluate the gamut of theory that they and others have proposed to explain living fossils. In a final chapter, Eldredge briefly considers this question of commonality of pattern. But the question, as in any case it must, remains wide open: This is a *source* book of information and first-order interpretation, as well as a guide to the literature, which is intended to allow readers to test their own hypotheses. This is first and foremost a casebook of information, not a book on theory. Theory guides us, tells us what is interesting about the information. But it is the information, the carefully constructed cases themselves, that is given to the reader for consideration and further use.

This book is *not* an encyclopedia. Some serendipity inevitably crept in as the papers were collected, and naturally our initial plan was to include all of the more famous examples of living fossils. But *Sphenodon* and *Lingula* are not to be found in these pages, nor have we included a single plant. Indeed, all but one example are metazoans. These gaps simply reflect our editorial inability to obtain manuscripts from qualified biologists, certainly not a conscious effort to omit some examples in favor of others. Botanists tend to feel that most evolutionary theory addresses animals rather than plants, so the lack of some good discussions of plants—on *Gingko,* for example—is particularly to be regretted.

On the bright side, the 34 cases in this book[2] reveal a great deal not only about the many-sided phenomenon of living fossils, but also of the diverse styles and approaches of actually doing comparative biology these days. The impact of cladistics is clear: Nearly all contributors discuss the phylogenetic relationships of the taxa they consider, and not only the familiar rhetoric (e.g., "synapomorphies"), but also the more critical analysis that cladistics promises is present in these contributions. Some authors stop right there, with a careful presentation of the relationships, assessment of character resemblance among fossil and living species, plus an account of the temporal and spatial diversity of the group in question. Others go further: Some assess the adaptive nature of the evolutionary transformation that the group has undergone, and others examine theoretical explanations for the pattern of arrested evolution they see. The core information, as available, is to be found in each contribution; but what each author chooses to make of living fossils is very much idiosyncratic, and the essays of this book really do reveal the diversity of style and opinion in modern systematics and paleontology.

---

[2] Each essay contains a single example, with the exception of three discussed in the single chapter by Batten. Cracraft and Delson and Rosenberger present surveys of Aves and Primates, respectively, greatly augmenting the actual number of cases in the book.

We began this project with a criterion for what constitutes a living fossil that is perhaps a bit broader than our own (Eldredge 1975, 1979; Stanley 1975, 1979) and other earlier published versions, if for no reason other than to cast the net widely. The aim was to see just how many different sorts of phenomena were hiding together under the general rubric of living fossil. As we wrote our prospective authors:

The only criterion for a group's inclusion here is that a living species must be anatomically very similar (bordering on identity) to a fossil species which occurs very early in the history of the lineage. Thus some examples will embrace 500 million years, while others will involve 100 million years or less—depending upon the time of origin of the lineage in question.

Thus sponges, as an entire phylum, cannot qualify simply because they are primitive—by such a yardstick, virtually everything is a living fossil. But, as in Vrba's (this volume) impalas, a species may be a living fossil even if its lineage arose as recently as, say, the Miocene. We went on to specify what we expected for each case:

We shall ask each contributor to prepare a short presentation succinctly summarizing the following: (1) how close is the example in terms of anatomical similarity; in other words, what are the *bona fides* of the case? What is the time scale involved?; (2) what are the phylogenetic relationships (i.e., sister group) of the lineage in question?; (3) what is the species-level diversity (a) today and (b) through geologic time (you will be allowed two or three illustrations, and this item can be summarized in a simple plot); (4) how are the ecological niches of the living species realized (i.e., are they physiological and ecological specialists or generalists)?; (5) what do the species distributions in (a) space (areal extent—especially for the living species) and (b) time (longevity—for fossils, obviously) look like?; (6) what is the nature of within-species genetic and phenotypic variation?; and (7) how, in general, does the sister group (item 2 above) compare with respect to these parameters? We realize, of course, that in many instances data will not be available for some of these questions.

By and large we got what we asked for. Some authors conclude theirs is not, after all, a good example of a living fossil. Others thought their cases fit right in with some sort of concept of living fossils. With these few guidelines in mind, we feel these contributions can be read and easily compared with a great deal of profit.

Literature

Eldredge, N. 1975. Survivors from the good old, old, old days. Natural History 81(10):52–59.

Eldredge, N. 1979. Alternative approaches to evolutionary theory. In: Schwartz, J. H., Rollins, H. B. (eds.), Models and methodologies in evolutionary theory. Bull. Carnegie Mus. Nat. Hist. 13:7–19.

Eldredge, N., Cracraft, J. 1980. Phylogenetic patterns and the evolutionary process. Method and theory in comparative biology. New York: Columbia U. Press.

Simpson, G. G. 1944. Tempo and mode in evolution. New York: Columbia U. Press.

Simpson, G. G. 1953. The major features of evolution. New York: Columbia U. Press.

Stanley, S. M. 1975. A theory of evolution above the species level. Proc. Nat. Acad. Sci. 72:646–650.

Stanley, S. M. 1979. Macroevolution: pattern and process. San Francisco: Freeman.

Vrba, E. S., Eldredge, N. In press. Individuals, hierarchies and processes: towards a more complete evolutionary theory. Paleobiology.

1
Evolutionary Stasis in the Elephant-Shrew, *Rhynchocyon*

Michael Novacek

Department of Vertebrate Paleontology, American Museum of Natural History, New York, NY 10024

Introduction

One of the most bizarre groups to turn up in the zoological literature is the mammalian order Rhinogradentia (Stümpke 1967, published posthumously). This group of 150 outrageously endowed species—informally known as *snouters*—was apparently confined to the obscure South Sea islands of Hi-yi-yi. I say "was" in the acknowledgment that, quite tragically (and conveniently?), snouters, islands, and Stümpke were erased in a nuclear test accident two decades ago.

Some dour mammalogists may actually regard Stümpke's monograph as a work of pure and silly fantasy. Perhaps the skeptics might be disarmed by a reference to the surviving counterparts of the rhinogrades: the order Macroscelidea, or the elephant-shrews. These mouse to rabbit-sized insectivores (in eating habits, not affinities) are wholly confined at present and (by all indications) in the past to the African continent. With their long, flexible snouts, stilt-like limbs, and odd, sometimes cursorial gait, elephant-shrews inspire immediate comparison with the creatures of Hi-yi-yi (Fig. 1).

Peculiar appearance does not, however, establish a species' reputation as a living fossil. Indeed, elephant-shrews seem unlikely candidates for such a title when evaluated in the company of many of the groups treated in this volume. About 25 species (10 of them extinct) of macroscelideans have been described (Patterson 1965; Corbet and Hanks 1968; Kingdon 1974; Butler 1978). Such diversification certainly outstrips that understood for pangolins, aardvarks, and the like. (Note that the tree-shrews, a popular archetype for a primitive "placental" mammal, also make a fair showing of diversity; they comprise 5 Recent genera, 16 Recent species, and a few newly discovered fossils; see Walker 1968; Luckett 1980; Jacobs 1980; and Tattersall, this volume).

Why then, should elephant-shrews warrant attention in a study of living fossils? One reason pertains to the concept that these animals represent a distinct ordinal-level taxon, and, by inference, a taxon of great antiquity in the history of eutherian mammals. Such a concept did not always hold. Elephant-shrews were traditionally regarded as a subcategory of the ill-defined order Insectivora (Gregory 1910; Simpson 1945). Butler (1956), however, broke with tradition in recognizing macroscelideans as an order of mammals wholly separate from the "true" Insectivora, a distinction supported by several more recent studies (Patterson 1965; McKenna 1975; Novacek 1980). Major shifts in taxonomic rank do not, in themselves, always denote major shifts in taxonomic thinking. In this case, however, revision was accompanied by an emphasis on the remote divergence time for Macroscelidea, a divergence time much earlier than that suggested by the first occurrence of

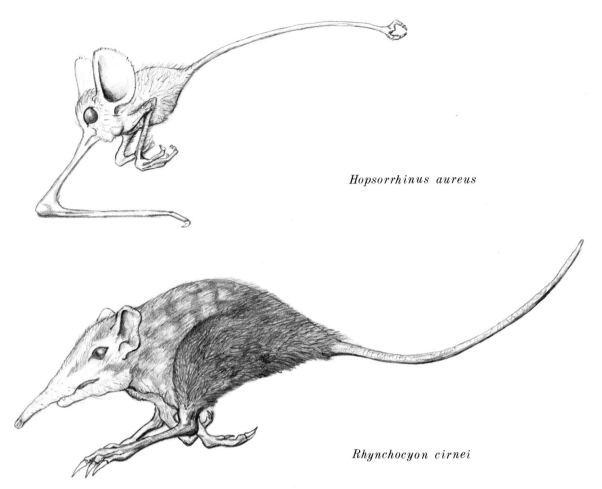

Hopsorrhinus aureus

Rhynchocyon cirnei

Fig. 1. Above, a snouter (after Stümpke 1967:43). Below, an elephant-shrew at running speed (after Kingdon 1974). If pursued, *Rhynchocyon chrysopygus,* "ran away in a gait that was very similar to the stotting of gazelle, or . . . ran away in its typical, swift, cursorial bounding gait. While stotting or bounding, it hammered the leaf litter loudly with its rear feet, producing a very characteristic 'crunch, crunch, crunch' sound as it fled" (Rathbun 1979:51).

elephant-shrews in the fossil record (McKenna 1975; Table 2 in Novacek 1982). Hence, this group aptly illustrates how a restructuring of higher level phylogeny can alter our impressions of evolutionary rates. Compared with most higher level mammal groups of similar age (e.g., artiodactyls, rodents, perissodactyls, cetaceans, etc.), macroscelideans are sluggish with respect to the tempo of "taxification."

But a more compelling reason for considering the Macroscelidea in the context of living fossils derives from the geochronologic endurance of one of its members. The genus *Rhynchocyon* is known from fossils that date back to the early Miocene of east Africa (Butler and Hopwood 1957; Butler 1969). The minimum range of 20 million years suggested by this record may seem unremarkable to those accustomed to the vast stretches of time represented by horseshoe crabs and garfish. Nevertheless, only a handful of Recent mammalian genera compare with *Rhynchocyon* in longevity. This evolutionary stasis is not, however, mirrored by the history of *Rhynchocyon*'s nearest living relatives, the diverse Macroscelidinae. As discussed below, a close comparison of these sister-groups reveals evidence bearing on theories that relate variance in evolutionary tempos to certain ecological, behavioral, and physiological factors (Eldredge 1979; Stanley 1979).

Characters, Cladograms, Fossil Occurrence, and Phylogeny

The principal taxonomic study of Recent macroscelideans is that of Corbet and Hanks (1968). The fossils have been treated most comprehensively by Patterson (1965), whose coverage has since been emended by descriptions of additional taxa reviewed in Butler (1978). Rathbun (1979) presented a phylogenetic tree for macroscelidean genera based, as he acknowledged, primarily on Patterson's (1965) analysis. The tree was given without explicit

character information, but its basic framework, with some notable exceptions, is corroborated by distributions of shared, special features taken from the diagnoses of Patterson (1965), Corbet and Hanks (1968), Butler and Hopwood (1957), and Butler and Greenwood (1976). This information is abstracted in Figs. 2 and 3 and Table 1 herein. The figures should be regarded only as an interpretive summary of published work; first-hand analysis is highly warranted, as indicated by the lack of resolution at certain nodes in the cladogram (Fig. 2). In the discussion to follow, I have retained the formal category names recognized in the literature. Obviously, the acceptance of both the ordinal status of Macroscelidea and a cladogram resembling that shown in Fig. 2 would require adjustments in the current classification.

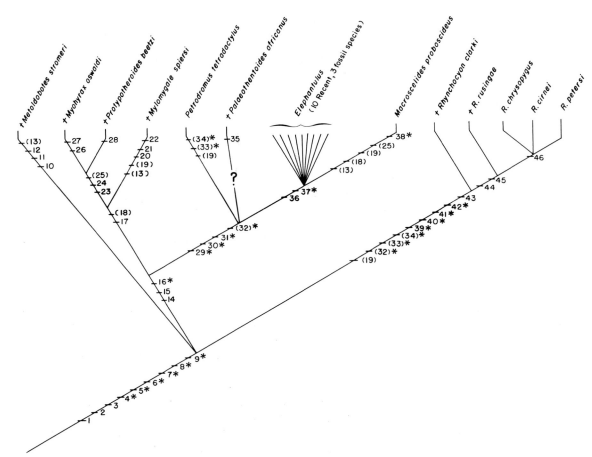

Fig. 2. A cladogram for Macroscelidea. Numbers refer to characters listed in Table 1. Parentheses indicate characters that occur more than once in the cladogram and are interpreted as convergent events under this scheme. Asterisks indicate characters of anatomical regions not represented in all fossil taxa in the cladogram.

Table 1. Derived characters for cladistic relationships shown in Fig. 2.[a]

---

1. M_3 and M^3, when present, small or vestigial.
2. Posterior teeth low-crowned, cusps and connecting lophs worn early to cutting ridges.
3. P^4 large, molariform.
4. Compound auditory "bulla," with variable contributions from ecto-, entotympanic, squamosal, petrosal, alisphenoid, basisphenoid, and pterygoid.
5. Jaw condyle set well above level of cheek teeth.
6. Coronoid process with small, hooklike dorsal edge.
7. Distal limb elements elongate, closely appressed, or fused.
8. Pollex and hallux small.
9. Proboscis long and flexible.
10. Jaw symphysis extensive, terminates below P_3.
11. I_3 large and procumbent.
12. I_3 with deep lingual groove.
13. Anterior teeth small and crowded.
14. Cheek teeth prismatic, with pronounced lingual reentrant folds.
15. Cheek teeth slightly hypsodont, greater emphasis of cutting crests.
16. Upper canine small, premolariform.
17. Mandible with strongly convex ventral border.
18. Cheek teeth strongly hypsodont.
19. M_3 absent.
20. P_4–M_2 tall, columnar, strongly compressed anteroposteriorly.
21. P_4–M_2 with very deep lingual and labial reentrant folds, crown pattern "rodent-like."
22. Alveolar border concave below P_4–M_2.
23. I^{1-2} enlarged, procumbent, with enamel restricted to labial faces.
24. Mandible below posterior cheek teeth very deep, robust.
25. Fossettes, fossettids on crowns of upper and lower cheek teeth, respectively.
26. Cement in fossettes of upper cheek teeth.
27. M^3 single rooted.
28. Fossettids deep, extending to base of P_3–M_2.
29. Distal ulna rudimentary, fused with radius.
30. Ilio-sacral fusion involving first and second vertebrae.
31. Middle section of the palate with large vacuities.
32. M^3 absent.
33. Lower incisors with expanded, bilobed tips.
34. Hallux absent.
35. P_4 with labial swelling between trigonid and talonid and posteriorly positioned metaconid.
36. Posterior edge of palate highly fenestrated.
37. Sagittal supratemporal crests very weak or absent.
38. Auditory bulla grossly inflated.
39. Facial part of skull broad and flattened.
40. Upper incisors vestigial or absent.
41. Feet digitigrade.
42. Pollex absent.
43. Anterior edge of coronoid process more gently sloping.
44. Larger body size.
45. Protostylid on P_2, P_3, DP_3.
46. Metastylid on DP_4, M_1.

---

[a] For discussions of characters, see Appendix.

As Figs. 2 and 3 indicate, the Macroscelidinae, which includes *Petrodromus, Macroscelides, Palaeothentoides,* and the diverse *Elephantulus,* accounts for most of the species within the order. *Elephantulus* and *Macroscelides* have fossil records that begin in the Late Pliocene (Butler and Greenwood 1976). *Palaeothentoides* (Stromer 1932; Patterson 1965)

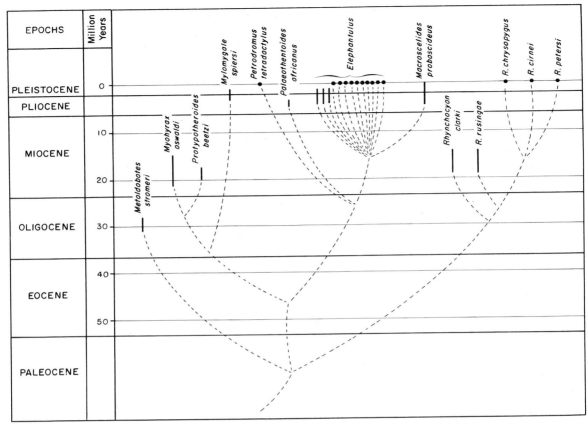

Fig. 3. Phylogenetic tree for Macroscelidea. Vertical bars indicate observed geochronologic ranges based on fossil occurrence. Solid circles represent extant species with no fossil record. Dashed lines represent speculative phylogeny that conforms with the *relative* splitting sequence given in the cladogram.

is known only from the somewhat earlier Pliocene Klein Zee fauna. *Petrodromus,* a genus very specialized in limb structure, has no fossil record.

Macroscelidines can be recognized by a set of synapomorphies in the limbs, sacral vertebrae, palate, and dentition (characters 29–32, Table 1, and Figs. 2, 4, 5, and 6). Resolution within this group, however, is far from satisfactory. Patterson (1965:305) remarked that *Palaeothentoides* combines characters of *Elephantulus* and *Macroscelides,* but his description of this fossil leads one only to the conclusion that it is an aberrant member of the Macroscelidinae. Most problematic is the lack of defining characters for *Elephantulus.* The impression given by the available work is that this genus is paraphyletic, and the problem of resolution among its many alleged members is extremely challenging. Broom (1937) recognized a

subset of *Elephantulus* species based on specializations of the upper second premolar (P^2), and, as Butler (1978) noted, this early concept has some merit. Corbet and Hanks (1968) derived a distance phenogram from 31 characters that, in part, corroborate Broom's (1937) original analysis. Their characters are, however, largely ambiguous in polarity, and their results have not been scrutinized from a different taxonomic perspective. Unfortunately, Corbet and Hanks (1968:66) "distinguished" *Elephantulus* only by the lack of specializations observed for *Macroscelides* and *Petrodromus.* The problem is illustrated by the "shaving-brush" pattern depicted in the cladogram (Fig. 2).

Butler (1978) has argued for a much older radiation of the Macroscelidinae than a literal reading of the fossil record indicates. This view finds support in Patterson's (1965) tentative reference of the Oligocene *Metoldobotes* (Fig. 7)

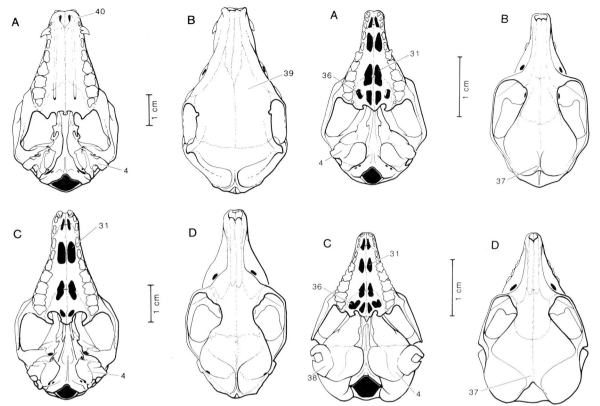

Fig. 4. Diagrams of (A) ventral and (B) dorsal views of skulls of *Rhynchocyon chrysopygus,* AMNH (American Museum of Natural History) 187234; and (C) ventral and (D) dorsal views of skull of *Petrodromus tetradactylus,* AMNH 83775. Numbers indicate traits listed and described in Table 1.

Fig. 5. Diagrams of (A) ventral and (B) dorsal views of skull of *Elephantulus brachyrhynchus,* AMNH 115696; and (C) ventral and (D) dorsal views of skull of *Macroscelides proboscideus.* Numbers indicate traits listed and described in Table 1.

to this subfamily. Butler (1978:58) denied, however, the relevance of this evidence by claiming that *Metoldobotes* is probably "an extinct offshoot from an Eocene macroscelidid stock, indicative of a radiation that had already taken place by early Oligocene time." As the cladogram (Fig. 2) shows, there is no obvious relationship between *Metoldobotes* and a particular subgroup of macroscelideans. The genus is, nonetheless, easily recognized by a number of unique features. It represents an isolated branch stemming from the base of the elephant-shrew tree (Fig. 3).

The extinct family Myohyracinae, which customarily includes only *Myohyrax* and *Protypotheroides,* is characterized by dental specializations that indicate a herbivorous diet (Table 1 and Fig. 7). Accordingly, myohyracines were at

one time regarded as hyracoids (Andrews 1914; Stromer 1926; Whiteworth 1954). Patterson (1965) convincingly argued for the transfer of this subfamily to the Macroscelidea (see also comments of Butler 1978). Although Patterson and other authors recognized the Myohyracinae as an isolated lineage, a concept also depicted in Rathbun's (1979) tree, a few dental characters (characters 14, 15, and 16 in Table 1, and Fig. 2) indicate a possible close relationship with Macroscelidinae. As Butler (1978) noted, the "prismatic" crown pattern on the cheek teeth in macroscelidines and myohyracines clearly departs from the condition in *Rhynchocyon* (Figs. 6 and 7).

Patterson (1965) recognized the subfamily Mylomygalinae based on a single genus. *Mylomygale* does, however, share some of the den-

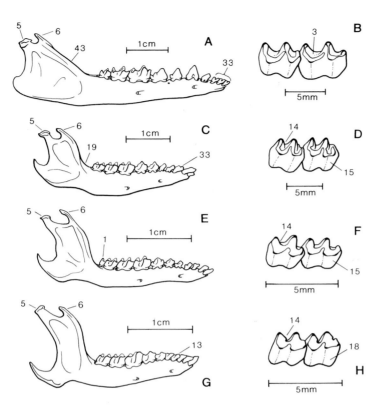

Fig. 6. Lateral views of right lower jaw (A, C, E, G) and right P_4, M_1 (B, D, F, H) for *Rhynchocyon chrysopygus*, AMNH 187234 (A, B); *Petrodromus tetradactylus*, AMNH 83775 (C, D); *Elephantulus brachyrhynchus*, AMNH 115696 (E, F); and *Macroscelides proboscideus* (G, H). Numbers indicate characters listed and described in Table 1.

tal specializations of the Myohyracinae; it departs from the latter only in showing unique, remarkably rodent-like features of the crown. A possible monophyletic grouping for these taxa is suggested by the common possession of strongly hypsodont cheek teeth and a distinct, convex curvature of the ventral border of the mandible (characters 17 and 18, Table 1, and Fig. 2). Hypsodonty also occurs in *Macroscelides* (Fig. 2), but not to the extent observed in *Mylomygale* and myohyracines (Fig. 7).

In contrast to the foregoing groups, *Rhynchocyon*, as noted previously, shows an interesting lack of diversity, both in species numbers and morphological range. The genus is recognized as the sole member of the subfamily Rhychocyoninae in published classifications. The two Miocene species described from the Rusinga fauna of Kenya have been set apart from the extant species by their "slightly more primitive characters" (Butler 1978:57). Differences between fossil and Recent forms pertain to details of the naso-facial region of the skull and the posterior premolars and molars (character 46, Table 1, Fig. 2, and Butler and Hopwood 1957;

Butler 1969). These differences led Butler (1978) to suggest that the Miocene species might be separated from the Recent taxa at the subgeneric level. His suggestion is, however, contradicted by his statement (1978:59) that *R. rusingae* is closer to living species than *R. clarki,* in its larger size and in having an extra cuspule on the back of the premolars (characters 44, 45, Table 1, and Fig. 2), and thus "may be near the direct ancestry of the two living species of *Rhynchocyon.*" These remarks imply—contrary to establishment of a fossil subgenus—the existence of a monophyletic group comprising the living species and *R. rusingae* and excluding *R. clarki.*

Differences among the Recent species of *Rhynchocyon* are, if anything, less apparent than those distinguishing the fossils. Corbet and Hanks (1968) recognized three different species based on variable color patterns of the rump area, although Butler (1978), as indicated by the above quote, recognizes only two. Moreover, Kingdon (1974) considered the variation in color pattern too minor to justify the recognition of more than one species. These differ-

Fig. 7. (A) Lateral view of right lower jaw; and (B) occlusal view of right P³–M³ of *Myohyrax oswaldi;* (C) occlusal view of left P₄, M₁ of *Mylomygale spiersi;* (D) occlusal view of P₃₋₄, M₁₋₂ of *Metoldobotes stromeri.* Numbers indicate characters listed and described in Table 1. (All figures redrawn from Patterson 1965.)

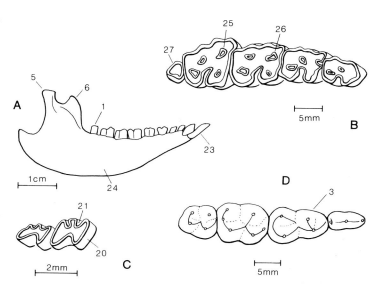

ences of opinion suggest, despite the general endorsement for the classification of Corbet and Hanks (e.g., Rathbun 1979), that the morphological range shown by the Recent species is very narrow. To sum, the comparative evidence derived from both fossil and Recent forms of *Rhynchocyon* provides a compelling case of stasis in mammals.

There is no clear evidence for the occurrence of either fossil or Recent macroscelideans outside of Africa. A few suggestions for elephant-shrews from other continents have been rejected. Most noteworthy of these cases is Filhol's (1892) description of *Pseudorhyncocyon* from the Eocene of western Europe as an early macroscelidean. Butler and Hopwood (1957) argued convincingly against this reference, and their conclusion is supported by Sigé's (1974) study of more complete dentitions and jaws of *Pseudorhyncocyon.* It is probable, as Sigé suggests, that this genus is a very specialized member of the archaic leptictids.

The relationships of macroscelideans with other mammalian orders is a matter of great uncertainty. The order is characterized by a large number of very distinctive, derived features (characters 1–9, Table 1, and Fig. 2). These do not, however, offer definite clues on sister-groups. The once venerable idea that these animals were closely related to tupaiids in the Menotyphla (Haeckel 1866; Gregory 1910) has been virtually demolished by a considerable number of studies (Butler 1972; Patterson 1965;

McKenna 1975; Luckett 1980; Novacek 1980, and others cited therein). Menotyphla is, under any of its previous characterizations, a classic example of paraphyly. McKenna (1975) suggested relationships between macroscelideans and more archaic eutherians from the Cretaceous of Asia and, perhaps remotely, with lagomorphs (rabbits and pikas). Szalay (1977) associated elephant-shrews more intimately with lagomorphs based on features of the tarsus. Elsewhere, I have disputed some of the arguments for homology presented by Szalay for this relationship (Novacek 1978). I have, however, noted some provisional evidence for the placement of macroscelideans somewhere within a clade that also contains lagomorphs, rodents, the extinct anagalids, and zalambdalestids (Novacek 1982). This association echoes the ideas first proposed by McKenna (1975) but involves a somewhat different morphological character set than that emphasized by the latter author.

Given the uncertainty over the affinities of the macroscelideans, one may justifiably ask on what basis one can recognize the derived characters shown in the cladogram for the Macroscelidea. Reference to an alternate number of possible outgroups, such as those noted above, as well as insectivorans, edentates, and marsupials, constitute the evidence for the derived status of those characters listed in Table 1. I have tried to clarify the reasons behind these assessments in the appendix at the end of this chapter. Unfortunately, the lack of resolution among the higher level groups of eutherian mammals does not allow a very straightforward analysis of characters based on outgroup comparisons.

Most of the 46 characters in Table 1 appear only once in the cladogram shown in Fig. 2, but there are some notable exceptions. The highest degree of homoplasy is demonstrated by features involving loss of the posteriormost molars (characters 19 and 32, Table 1, and Fig. 2). Loss of M^3 and the (not always coincident) loss of M_3 several times in the cladogram is an outcome of analysis developed from the assumption of parsimony. It is clear, nonetheless, that loss of posterior molars has occurred with some frequency in various mammalian groups.

From a taxonomic perspective, the most obvious case of conflicting data comes from the derived similarity in several traits between *Macroscelides* and *Mylomygale*. Evidence for the monophyly of the Macroscelidinae (characters 29–32) leads one to the conclusion that the similarity between *Macroscelides* and *Mylomygale* is the result of convergence. One might tritely caution that *Mylomygale* is represented only by teeth and jaws, and more complete material might place it differently in the cladogram. As is commonly the case, many fossil taxa of this group are not represented by all the anatomical regions considered in constructing the cladogram.

Evolutionary Rates in Elephant-Shrews

Darwin (1859:105–108) recognized living fossils as members of clades that demonstrate remarkably slow evolution. Subsequently, the matter of what constitutes slow evolution has been subject to diverse opinion. Several recent reviews characterize living fossils as fulfilling three criteria: (1) They are living members of a group of marked longevity in the geologic record, (2) they demonstrate very little morphologic divergence from earlier occurring members of the same clade, and (3) they belong to a clade that shows very low taxonomic diversity through most, or all, of its known history (Eldredge 1979; Stanley 1979). Intuitively, one

might expect some correlation between criteria 2 and 3. A variety of cases suggests, however, that morphological "diversity" can be high but taxonomic diversity low (e.g., Cambrian echinoderms, Sprinkle 1976). The converse relationship is perhaps even more frequent. Most taxonomists are doubtless aware of certain groups within their area of interest that are species rich but morphologically conservative, an observation that provides no end of frustration to the tyro armed with a field guide and confronted, for example, with the numerous species of deer mice or little brown bats.

Despite these cases, there is more than intuitive reason for arguing that morphologic and taxic rates are often strongly correlated. Stanley (1979:122–127), for example, marshalled evidence that, in more than a dozen clades, very sluggish morphologic change accompanied very low rates of speciation. Such bradytelic (slowly evolving) lineages are best interpreted, according to Stanley (1979), under a punctuational model of evolution, because this model, rather than a gradualistic one, predicts that low diversity over time yields little morphologic evolution owing to the lack of opportunity for speciation. In a similar fashion, Eldredge (1979) proposed that the predictable relationship between morphologic and speciation rates demonstrated the value of a taxic approach (one emphasizing speciation events) to explanations of evolutionary history.

One problem with such generalizations is that it is often difficult to discriminate between the evidence for species diversity and morphologic divergence. Thus, Schopf et al. (1975) claimed that the high rates of taxonomic diversification attributed to mammals merely reflect our sensitivity to a sample rich in morphological characters (and their terminology). A rejoinder to this claim is the observation that many lower level taxa of mammals are recognized, as in numerous other groups, by a very restricted portion of the total anatomy. Accordingly, the characterizations of such taxa are not any richer than those of less complex groups (Eldredge 1976; Novacek and Norell 1982). It is acknowledged, nonetheless, that the question of morphologic complexity and taxonomic recognition has not been investigated empirically enough to address some of the issues raised by Schopf et al. (1975).

One way, then, to monitor the degrees of morphologic change, as opposed to the degree of taxonomic diversification, is to consider the richness of characterizations for members of the group in question. In a cladistic context, this amounts to an inspection of the number of special features accumulated at each node in the cladogram. Contradictory characters, interpreted as instances of convergence or evolutionary reversal, would further augment a measure of overall morphologic change. As Fig. 2 demonstrates, appreciable morphologic divergence is represented by macroscelidean clades. A number of specializations derived from distinctly different anatomical systems can be recruited at various nodes of the cladogram (terminal taxa are, however, primarily identified by details of tooth anatomy; see discussion above). Moreover, a notable range of morphology is represented by comparisons among the three major subtaxa: the myohyracines, macroscelidines, and rhynchocyonines. This pattern suggests that, despite their relatively moderate taxic diversity, elephant-shrews as a group hardly fit our notion of a bradytelic lineage.

Such an impression does not, however, hold for at least one major branch of the elephant-shrew tree. The monogeneric status of the Rhynchocyoninae clearly reflects the lack of morphologic range within this taxon. As noted above, the differences among living species and their Miocene relatives are trivial enough to throw into question the standard classification of this group (Kingdon 1974; and comments above). Hence, the label of "living fossil" seems highly appropriate for extant species of *Rhynchocyon,* with respect to the above-noted criteria for longevity, morphologic stagnation, and low rates of speciation.

Some Correlates

This example of stasis in *Rhynchocyon* might be considered in light of theories that seek to explain the variance in rates of evolution. Such theories commonly address the question: What factors predispose clades to be either bradytelic or tachytelic? More traditional explanations are couched in phyletic, or transformational, terms. A transformational theory for the occurrence of bradytely, for example, might invoke low ge-

netic variability, weak directional selection (e.g., Simpson 1953), or a long-term stability in habitat and adaptation. Such interpretations have been criticized at length by Eldredge (1979), Stanley (1979), and Gould (1980), authors who explicitly favor a taxic approach to the study of evolutionary history.

Working for this taxic perspective, Eldredge (1979) fashioned a theory for the occurrence of bradytelic lineages, based on the central premise that there is a significant, perhaps causal, connection between morphological change and rates of speciation (see also Stanley 1979). This connection, he further argued, is initiated and maintained by a correlative set of biological factors. Bradytelic lineages tend to have a majority of eurytopic species; namely, species that are generalists with respect to habitat, diet, physiological tolerances, and related parameters. Moreover, studies by Fryer and Iles (1969) and Jackson (1974) suggest that eurytopes do not usually occur sympatrically with closely related species. In this way allopatry impedes any divergence in specialization that might arise in response to interactions of these species. Eldredge (1979:14) ties these variables together with the following statement: "It would thus appear that speciation rates within eurytopic lineages are automatically dampened by their ecological strategy, and as a corollary morphological change will be retarded." In contrast, the stenotopic syndrome (very select habitat, dietary preference, narrow physiological tolerance, etc.) can be linked with marked sympatry, high species diversity, and strong morphological divergence (Fryer and Iles 1969; Eldredge 1979).

How well do these generalizations hold for a comparison of rhynchocyonines and their nearest living relatives, the macroscelidines? The answer is, quite well, given the available information. Fortunately, the relevant data are better than one might expect for these exotic mammals, largely as a result of the systematic review of Corbet and Hanks (1968) and the superb analysis of life-history traits and social behavior by Rathbun (1979). The latter study is focused primarily on two species, *Rhynchocyon chrysopygus* and *Elephantulus rufescens.* Rathbun, nevertheless, convincingly demonstrates, with a summary of more anecdotal work, that similar patterns hold for related species in either subfamily. As is shown in Table 2, a series of differences in biological traits between rhynchocyonines and macroscelidines are strongly, but not perfectly, consistent with the above generalizations. The only ambiguous element of this comparison pertains to habitat preference. All three species of *Rhynchocyon* inhabit thick, riverine bush or closed canopy forests (Corbet and Hanks 1968; Rathbun 1979). Species of *Elephantulus,* by contrast, show a diversity of habitats ranging from savannah woodland to subdesert, but many of these species appear to prefer a given habitat (rocky outcrops, scrub, grassland, or open woodland) encompassed within this range. One might thus infer that species of *Rhynchocyon* are more generalized within the limits prescribed by their more mesic environment. It is unclear, however, whether this habitat range is any broader than that shown by a given species of *Elephantulus.*

Behavioral differences between rhyncho-

Table 2. Differences in species diversity, morphologic range and various biological parameters in Rhynchocyoninae and Macroscelidinae.

Traits	Rhynchocyoninae	Macroscelidinae
No. genera, +[a]	1	4
No. species, +	5	16
Morphologic range, +	Species distinguished by minor differences in body size, teeth, pelage patterns	Marked differences between many species (Fig. 2)
Habitat, ?[b]	Thick bush, closed canopy forest, mesic conditions	Variable, from savannah woodlands to subdesert; with different species showing preferential microhabitats
Distribution, +	Distinctly allopatric for all three living species	Many species broadly sympatric, some syntopic
Activity[c], +	Diurnal, no definite activity peaks; 2.7% intraspecific interaction out of total activity budget	Polycyclic activity, with activity peaks at dawn and dusk; 13% intraspecific activity (males)
Trails, +	Not usually constructed	Complex trails, focus of much activity
Nests, −[d]	Constructed from leaves, twigs, and rootlets	Not known for any species; some species use burrows
Grooming[c], +	Absent	Elaborate
Territorial aggression[c], +	Present, generalized; expressed by "chasing."	Present, more elaborate; ritualized sequence of "foot-drumming," "mechanical walk," chasing, fighting
Marking[c], +	Scent-marking, no static-optic-scent-marking	Scent-marking, use of dung piles and urine for purpose of static-optic-scent-making of territories
Foraging[c], +	Uniformly active interrupted with by brief periods of digging; 79% total activity	Foraging activity from a few seconds to 5 minutes interrupted by trail cleaning and other activities; 30% total activity, females, 13%, males
Diet[c], +	Leaf-litter invertebrates in approximate proportion to availability, except for abundant millipedes (probably unpalatable); some plant material	Primarily ants and termites, strongly selected over other abundant invertebrates; very little plant material; some species more generalized (e.g., *Petrodromus tetradactylus*)

[a] +: Conformity between observed differences and rate control model discussed in text.
[b] ?: Ambiguity.
[c] Differences in grooming, aggression, marking, foraging, and diet are based primarily on Rathbun's (1979) observations of *Rhynchocyon chrysopygus* and *Elephantulus rufescens.* Both subfamilies share a number of specialized life-history traits (e.g., monogamy) not listed in table.
[d] −: Contradiction to model discussed in text.

cyonines and macroscelidines can be more readily applied to the eurytopy–stenotopy contrast. The stenotopy syndrome is much more evident in *Elephantulus rufescens* and in *Rhynchocyon chrysopygus* with regard to activity patterns, social interactions, territorial display, foraging behavior, and diet (Table 2). Admittedly, these characterizations hinge on the premise that more involved social interactions and narrower feeding strategies indicate more specialized conditions. As Rathbun (1979) notes, this suite of behavioral traits is generally found in other, less thoroughly documented species of macroscelidines, although *Macroscelides* departs from species of *Elephantulus* in aspects of behavior and dietary preference. One exception to the correlates of a eurytopic tendency in *R. chrysopygus* is the nest-building activity in this species, a behavior evidently absent in *E. rufescens*. This exception does not drastically affect the overall pattern. Ecological traits form mosaics analogous to the networks of primitive and derived morphological traits. Nevertheless, either system allows for generalizations about organisms and taxa.

One final and very striking contrast between rhynchocyonines and macroscelidines is revealed by distributional data. Several species of *Elephantulus* are broadly sympatric, although they are rarely recorded in close proximity within the same habitat. Corbet and Hanks (1968), who exhaustively reviewed this distributional information, suggested that sympatric species of *Elephantulus* are more likely isolated by habitat preference than by differential exploitation of the same habitat. If their statement is accurate, it denotes a narrower habitat preference for species of this genus than the range data indicate (Table 2). The relatively more complex social behavior and feeding strategies of *E. rufescens* suggest, however, that a combination of behavioral displacement and habitat preference may be in operation for sympatric species of *Elephantulus*. Perhaps relevant here is the observation that *Macroscelides proboscideus* is broadly sympatric and syntopic with *Elephantulus rupestris* (Corbet and Hanks 1968), but that the two species apparently differ in diet. Most species of *Elephantulus* eat mainly ants and termites, whereas *Macroscelides* feeds on a variety of invertebrates and plant material (Brown 1964; Sauer 1973; Rathbun 1979). Per-

haps this divergence in diet can be related to the obvious differences in tooth morphology (Table 1 and Figs. 2 and 5), although such a parallel has not been rigorously examined.

In strong contrast to this pattern, all three living species of *Rhynchocyon* are distinctly allopatric in distribution. There is a definite 30-km range gap between *R. chrysopygus* and *R. petersi,* the coastal species of eastern Kenya and Tanzania (Corbet and Hanks 1968: 64). Likewise, neither inland nor coastal populations of *R. cirnei* overlap with ranges of the other two species. Rathbun (1979) hypothesized that central populations of *R. cirnei* were isolated by the development of the East Africa Highlands and Rift Valley. These tectonic events occurred as much as 16 million years ago (Andrews and Van Couvering 1975), suggesting a long-term separation of central and coastal populations that is certainly belied by the minor morphological differences between these populations (analysis here is complicated by various taxonomic views; see comments under characters 45 and 46 in the Appendix). To account, then, for Rathbun's plausible biogeographic scenario and the conservative variation within *Rhynchocyon*, one must conclude that the character divergence among various subgroups of this genus was remarkably sluggish.

Conclusions

In summary, comparisons between the two extant sister-groups of elephant-shrews display a correlation web of biological traits and evolutionary tempos predicted by Eldredge (1979), Stanley (1979), and others. Rhynchocyonines, for example, not only strongly exemplify the expected correlation between low rates of speciation and conservative morphological change, they also show the eurytopic tendencies—as inferred from behavioral and ecological strategies—demonstrated for other bradytelic groups. Moreover, Jackson's (1974) prediction that closely related eurytopic species will tend to have allopatric distributions is clearly borne out in the case of *Rhynchocyon*.

What general implications can be found in this consistency between theory and observation? Clearly, a hypothesis linking morphological change, speciation rates, and a variety of ecological parameters is difficult to frame in other than a narrative form. Such a hypothesis might be abandoned in the face of a "better" explanation of the data, but it is not susceptible to falsification by a single contradictory observation (see also Eldredge 1979:15) because theories proposed for rate controls, like many other biological theories, are not meant to be sacrosanct. They merely claim that species rates and morphologic change are *usually* correlated, bradytelic lineages are *usually* eurytopic, closely related eurytopes are *usually* allopatric, and so on. In this manner, general explanations of the variance in evolutionary tempos (e.g., Boucot 1975; Eldredge 1979; Stanley 1979; Johnson 1982) parallel many of the hypotheses developed by ecologists over the past three decades. Unfortunately, most evolutionary and ecological theories share a problem related to their resiliency. If one or a few contradictions are accommodated, if predictions are allowed to slacken, and if *ad hoc* hypotheses are copiously provided, it becomes uncertain what degree of contradiction is required for the eventual abandonment of a theory.

But it may be asking too much of all biological theories to attain equal levels of precision. Perhaps it is enough, in certain instances, to extract from the biological morass a network of variables that shows repeatability across widely different organisms and situations. In this context, it is surprising how closely the example discussed herein matches a pattern of relationships customarily envisioned for marine invertebrates. Thus, one might acknowledge that theories of rate controls may provide useful generalizations concerning the interplay of phylogeny, distribution, and ecology in the biotic world. One might stop short, however, of appreciating these theories as profound and inviolate explanations of process in evolution.

Appendix

The following are notes on the evaluations of characters listed in Table 1.

Character 1. The extremely small size of the last upper and lower molar is clearly a departure from the primitive eutherian conditions, where this tooth is smaller, but not markedly so, then the anterior molars. The evidence that the vestigial M_3^3 represents a derived condition is the widespread distribution of well-developed teeth at this position among other mammalian clades. The inferred trend here is further reduction (character 27) succeeded by loss of M_3 and M^3, as represented by the absence of these teeth (characters 19 and 32) in certain groups of macroscelideans. As noted in the text, the acceptance of the scheme of relationships shown in Fig. 2 requires the hypothesis that loss of these teeth is a highly convergent condition. However, the expression of the condition at a more general level—namely, the marked reduction of M_3^3—seems useful for characterizing the order as a whole.

Character 2. The gestalt of the crowns of the cheek teeth with regard to profile, ridges, and wear suggests a grinding mode of occlusion that seems derived for Eutheria. It is generally agreed that the primitive condition for the infraclass is a trenchant, sectorial tooth, characteristic of many living insectivorans (Butler 1972; Novacek 1977a) and a large number of early clades (Lillegraven 1969). Obviously, many other groups show the molar features broadly described here. If the nearest outgroup of Macroscelidea is likely to be lagomorphs or rodents, character 2 is of equivocal value for distinguishing elephant-shrews.

Character 3. The status of the "molariform" condition of P_4^4 is a problem open to much uncertainty and debate. It should be noted that eutherians primitively had five premolars, and loss of a premolar somewhere in the middle of the series occurred in the early phylogenesis of this infraclass (McKenna 1975). Thus, the teeth customarily denoted P_4^4 might be more correctly identified as P_5^5 or DP_5^5. In any case, the molariform aspect of the last premolar is found in diverse mammalian groups, but this condition differs from the simple, connate, or trenchant tooth observed in many Mesozoic eutherians, early primates, and insectivorans. Butler (1956), Novacek (1977a), and others have argued that the latter condition is more primitive for eutherians. Nevertheless, this question of polarity has not been resolved.

Character 4. Macroscelideans are unique within eutherian mammals in having an osseous auditory bulla that incorporates a large number of separate elements (Van der Klaauw 1931; Novacek 1977b; MacPhee 1981). Certain other mammals have compound bullae, but these bullae usually comprise only two or three elements (see Novacek 1977b, 1980; MacPhee 1981). It is plausible that the primitive condition in eutherians is abullate (as in shrews, tenrecids, aardvarks, and certain fossil forms), or one wherein the ectotympanic or some other bone has a small process that serves as an incipient floor of the tympanic cavity. A popular view that many bullar conditions were derived from a cartilaginous or osseous entotympanic bulla has been rejected in more recent reviews (Novacek 1977b; MacPhee 1979). The bullar condition for Macroscelidea clearly distinguishes this group. Basicranial regions for several fossil taxa are unknown, however, so whether the complex bulla unites all recognized elephant-shrew taxa is uncertain.

Characters 5 and 6. The high condyle and small, hooklike coronoid process in macroscelideans are traits that strongly differ from the typical condition in eutherians and their nearest outgroup, the marsupials. Rodents and lagomorphs share some, but not strong, similarities with elephant-shrews in these features, suggesting a basis for special relationship (McKenna 1975; Novacek 1982).

Character 7. The elongation of the distal limb elements and their partial fusion have long been noted as macroscelidean specializations, correlated with their curious, cursorial locomotion (Evans 1942; Rathbun 1979; Novacek 1980). Limb proportions among living elephant-shrews show some variation (Tables 2 and 3 in Evans 1942). It has been tempting to compare macroscelideans with lagomorphs in this regard, but the two groups share no striking specializations in details of the limb and ankle construction.

Character 8. The small pollex and hallux suggests further correlation with the cursorial locomotion of macroscelideans. There is little doubt that the small size of these elements represents a relatively derived condition within Eutheria. The hallux is absent in *Petrodromus* (character 34) and the hallux and pollex are absent (characters 34 and 42) in *Rhynchocyon*, implying a more specialized expression of the reduced condition of these elements. These traits are described in Evans (1942).

Character 9. "Elephant-shrews have a most distinctive head and skull with very large eyes and exceptionally long proboscis. As the nose is constantly being thrust into soft soil or crevices, and as the nostrils are at the tip of the proboscis, there is a need both for flexibility and also for a stout tube that cannot be sealed easily by pressure while the animal is probing about in the leaf-litter. In *Rhynchocyon* this tube is lined by thirty cartilaginous rings similar to those found in the larynx." (Kingdon 1974:8).

Character 10. The symphysis in *Metoldobotes* extends to P_3, and thus is longer than in all other known members of the order, in which it extends only to C or P_1 (Patterson 1965:301).

Characters 11 and 12. The size and construction of I_3 in *Metoldobotes* is unique within the order. *Myohyrax* and *Protypotheroides* have very large I^{1-2} (character 23) but small I^3. Some macroscelidines have a lingual groove on the lower incisors, but in no case is this feature as pronounced as in *Metoldobotes* (Patterson 1965:299).

Character 13. (Inferred) size reduction and crowding of the anterior teeth is a rare condition in the order. It occurs only in *Metoldobotes, Mylomygale,* and *Macroscelides.* Homology in this condition for these taxa is extremely unlikely (see Fig. 2). Such a condition is also improbable in any contender for a macroscelidean outgroup.

Characters 14 and 15. The prismatic, hypsodont cheek teeth in myohyracines and macroscelidines have not been emphasized as evidence in uniting these taxa, yet these conditions clearly differ from those in *Metoldobotes* and *Rhynchocyon* (see also Butler 1978; and remarks in text). Within the myohyracine–macroscelidine grouping, further emphasis of hypsodonty (character 18) occurs in all members of the former subfamily and in *Macroscelides.* The most specialized expression of these traits is represented in *Mylomygale spiersi,* where the cheek teeth are very tall, columnar, and compressed (character 20) and the reentrant folding is very deep (character 21). The resemblance of the cheek teeth in this species to the highly prismatic teeth of certain hystricomorph rodents is therefore quite striking (Patterson 1965:311). The trend thus advocated here involves both increased hypsodonty and increased infolding on the margins of the crown. With respect to these trends, it is very plausible that *Metoldobotes* represents the primitive condition within the group. The teeth in this genus have a low-crowned, bunodont appearance reminiscent of the primitive condition in rodents (as seen in paramyids and sciurids), hyopsodontine condylarths, or paromomyiform primates. Patterson (1965:301) compared aspects of the molar morphology in *Metoldobotes* favorably with *Rhynchocyon,* but these reside in primitive resemblance. Based on study of collections in the American Museum of Natural History, I find that *Rhynchocyon* does not approach the "prismatic" condition seen in *Petrodromus* or *Elephantulus,* which represent the most conservative members of the myohyracine–macroscelidine complex with respect to this crown pattern.

Character 16. The small premolariform upper canine is thought to represent a departure from the primitive eutherian condition, wherein, it has been argued (Butler 1972; Novacek 1977a), the canine is large and trenchant. The latter condition is common to many didelphid marsupials, erinaceomorph insectivores, and Cretaceous *Kennalestes* and *Asioryctes.* Given the uncertainty over the identity of macroscelidean relatives, the status of character 16 remains tentative.

Character 17. The strong convex bowing of the ventral border of mandible is a rare trait both within macroscelideans and many other mammalian groups. This condition serves to link *Mylomygale* with *Myohyrax* and *Protypotheroides.* The more extreme deepening of the lower jaw in the latter two genera (character 24), implies an emphasis of horizontal or propalinal occlusion, promoted by a well-developed masseter muscle complex. In this respect, there is a strong, but parallel, similarity with hyracoids, certain rodents, and many ungulate groups. In lagomorphs, the angular process is well developed and adds to the depth of the mandible, but the ventral border is not strongly bowed. From this character, and characters 14, 15, 25, 26, and 28, it is easy to see how *Myohyrax* and *Protypotheroides* were originally regarded as hyracoids (e.g., Andrews 1914; Stromer 1926).

Character 18. See comments under characters 14 and 15.

Character 19. See comments under character 1.

Characters 20 and 21. See comments under characters 14 and 15.

Character 22. The highly concave alveolar border of the mandible is a condition unique to *Mylomygale* within Macroscelidea. It is a specialization perhaps related to the occlusal function in this taxon.

Character 23. The I_{1-2}^{1-2} condition in *Myohyrax* and *Protypotheroides* departs from that in all other macroscelideans where these teeth are known. I_1^1 are larger and more procumbent in both rodents and hyracoids (I_2^2 are lost in these taxa). The Rodentia show a restriction of the enamel, to the anterior, rather than labial, surfaces of I_1^1.

Character 24. See comments under character 17.

Characters 25 and 26. The development of distinct fossettes and fossettids on *unworn* cheek teeth is observed only in *Myohyrax, Protypotheroides,* and *Macroscelides.* The formation of cement (character 26) in the fossettes of upper teeth in *Myohyrax* represents further specialization. Patterson (1965) provides a detailed discussion of these traits.

Character 27. A single-rooted M^3 seems part of the trend for reduction and loss of this tooth. See comments under character 1.

Character 28. The very deep fossettids in *Protypotheroides* are interpreted as a more extreme expression of character 25.

Character 29. The small size (or inferred reduction) of the ulna and its fusion with the radius serves to group *Petrodromus, Elephantulus,* and *Macroscelides,* and exclude *Rhynchocyon.* In the latter, as in many other mammals, the ulna is well developed and broadly separate from the radius. Clearly, the fused condition is derived for Macroscelidea. This macroscelidine synapomorphy is discussed by Evans (1942) and Corbet and Hanks (1968).

Character 30. The remarks under character 29 apply as well for the more extensive iliosacral fusion in macroscelidines.

Character 31. The presence of large palatal vacuities in macroscelidines is only tentatively cited as a derived macroscelidean trait. These openings are small or lacking in *Rhynchocyon,* ptilocercine tupaiids, dermopterans, primates, and most lipotyphlous insectivores (Novacek 1980, Table 5:79). Large palatal openings are, however, present in marsupials, rodents, lagomorphs, and early Tertiary anagalids (Simpson 1931; Evans 1942). Since the latter three groups have been suggested as possible close relatives to macroscelideans, the lack of palatal openings in *Rhynchocyon* might be plausibly interpreted as a secondary condition. At present, I choose not to accept this scenario, and I score the character 31 as a synapomorphy for macroscelidines. Note that the cranioskeletal conditions 29, 30, 31, and 32 are not known for *Palaeothentoides*. The latter is known only from a lower jaw and dentition. The reference of this genus within the family is therefore highly uncertain, as indicated in the cladogram (Fig. 2). *Palaeonthentoides* shows the strongly prismatic cheek–tooth pattern of macroscelidines and myohyracines, but not the deep, ventrally convex jaw of the latter group.

Character 32. See comments under character 1.

Character 33. Lower incisors are bilobed at their tips only in *Petrodromus* and (where known) in *Rhynchocyon*. The conditions in these two taxa are not, however, perfectly matched: Expansion and formation of lobes is less developed in *Petrodromus* (Evans 1942:95). The resemblance, at any rate, seems convergent under a parsimony scheme.

Character 34. The absence (or inferred loss) of the hallux within macroscelideans is unique to *Petrodromus* and *Rhynchocyon*. This character has been used in the classification of Corbet and Hanks (1968:66) to distinguish *Petrodromus*. Again, parsimony leads to the conclusion that loss of the hallux occurred independently in *Petrodromus* and *Rhynchocyon*. This loss has been tied to the more digitigrade, cursorial tendency in these two taxa (Evans 1942). See also comments under character 8.

Character 35. "P$_4$ [in *Palaeothentoides*] is the most distinctive tooth of the series. None of the living forms has a labial re-entrant between trigonid and talonid nearly filled by a swelling, and in none is the metaconid so far posterointernal to the protoconid." (Patterson 1965:304).

Character 36. The ornate fenestration of the posterior palate in *Macroscelides* and *Elephan-*

tulus is an extremely rare trait in eutherian mammals. The character is discussed in Evans (1942) and Corbet and Hanks (1968). Fenestration of this cranial region is incipient in *Petrodromus,* absent in *Rhynchocyon.*

Character 37. The weak development, or absence, of supratemporal and sagittal crests is clearly a derived eutherian feature (Butler 1956) usually correlated with expansion of the braincase.

Character 38. The marked inflation of the auditory bulla in *Macroscelides* is unique within the order. This development, which gives the basicranium a *Dipodomys*-like appearance, is cited as the primary diagnostic trait of the genus in Corbet and Hanks (1968). The function of inflated auditory bullae is the subject of more speculation than understanding (Novacek 1977b).

Character 39. No macroscelidine shares the flattening and expansion of the facial region of the skull with *Rhynchocyon*. This condition is most likely a specialization for the latter that might relate to the highly developed snout muscles and very elongate proboscis. See also comments under character 9.

Character 40. Absence of upper incisors in *Rhynchocyon* is unambiguously unique and diagnostic in this taxon. Upper dentitions are, however, not well known in fossil members of this genus.

Character 41. *Rhynchocyon* is the most digitigrade of any elephant-shrew (see Evans 1942).

Character 42. See comments under character 8.

Character 43. The more gentle slope of the anterior edge of the coronoid process in *Rhynchocyon* is a clear departure from other macroscelideans. Under any of the various assumptions for outgroups, this character is clearly derived within the order.

Character 44. Large body size is a distinctive mark of *Rhynchocyon,* but is this trait primitive or derived? I opt for the latter assessment, but acknowledge its tentative stature.

Characters 45 and 46. Butler and Hopwood (1957) and Butler (1969, 1978) cited details of the premolars and M$_1$ in order to differentiate various fossil and recent species of *Rhynchocyon*. Only those distinguishing the three (?) living species and a group containing the extant species plus *R. rusingae* are listed in Table 1. Characters that might resolve the trichotomy including the three living species are not clearly evident. The distinct, golden rump patch, overlying a very thick dermal shield, distinguishes *Rhynchocyon chrysopygus* (Corbet and Hanks 1968; Kingdom 1974; Rathbun 1978), but the absence of this feature in *R. cirnei* and *R. petersi* is hardly evidence for linking these two species.

The uncertain taxonomy within this genus complicates its biogeographic analysis. *Rhynchocyon cirnei* has a wide range, and is divided into several subspecies based on distributional information, pelage color, and superficial features of the tail. The most distinct population, recognized as *R. c. stuhlmanni* is confined to central Africa (portions of Zaire, Sudan, and Uganda), and is separated from coastal *R. chrysopygus* and *R. petersi* by several hundred kilometers. Other proposed subspecies are distributed around Lake Tanganyika and Lake Malawi (*R. c. reichardi*) and in coastal parts of southern Tanzania and northern Mozambique. These subspecies are, however, well separated from populations of *R. petersi* and *R. chrysopygus* (Corbet and Hanks, Fig. 2:58). The coastal subspecies of *R. cirnei* are interpreted by Kingdon (1974:42) as "hybrid-morphs" of *R. c. reichardi* and *R. petersi*. Kingdon therefore recognizes only one species, *R. cirnei*, and he regards the *petersi* and *chrysopygus* groups as "incipient species."

If, as Rathbun (1979) argues, the central populations of *Rhynchocyon* were isolated from coastal populations by the formation of the East Africa Highlands and Rift Valley, Kingdon's (1974) arrangement would require the secondary dispersal of inland "subspecies" toward the coast, and their reproductive mixing with populations of *R. ("c.") petersi*. This hypothesis predicts that the relationships of some coastal and central populations are best expressed by a reticulate phylogeny, or that some coastal and some inland populations share a special relationship not in common with other populations. Alternatively, one might propose that *Rhynchocyon* once had an ancestral distribution throughout east Africa, and Miocene tectonic events isolated inland from coastal populations, but these populations have not diverged significantly subsequent to their isolation. This hypothesis predicts that allopatric coastal populations are more closely related to each other than they are to inland populations. It seems difficult indeed to recruit evidence bearing on the choice between these hypotheses, although the second might be preferable on parsimony grounds. In any case, this dilemma underscores the conservative phylogenesis in *Rhynchocyon*.

Acknowledgments: I thank Karl Koopman (American Museum of Natural History) for access to collections of Recent elephant-shrews. For useful comments on the manuscript, I thank Malcolm McKenna, Karl Koopman and Niles Eldredge. Ray Gooris prepared Figures 2 and 3, Lorraine Meeker, Figures 4–7. Alejandra Lora and Barbara Werscheck typed the manuscript. This study was supported by the Frick Laboratory Endowment (American Museum of Natural History).

Literature

Andrews, C. W. 1914. On the lower Miocene vertebrates from British East Africa collected by Dr. Felix Oswald. Geol. Soc. London Quart. Jour. 70:163–186.

Andrews, P., Van Couvering, J. A. H. 1975. Palaeoenvironments in the East African Miocene. Contrib. Primat. 5:62–103.

Boucot, A. J. 1975. Evolution and extinction rate controls. Amsterdam: Elsevier.

Broom, R. 1937. On some new Pleistocene mammals from limestone caves of the Transvaal. S. Afr. J. Sci. 33:750–768.

Brown, J. C. 1964. Observations on the elephant shrews (Macroscelididae) of equatorial Africa. Proc. Zool. Soc. London 143:103–119.

Butler, P. M. 1956. The skull of *Ictops* and the classification of the Insectivora. Proc. Zool. Soc. London 126:453–481.

Butler, P. M. 1969. Insectivores and bats from the Miocene of East Africa: new material, pp. 1–38. In: Leakey, L. S. B. (ed.), Fossil vertebrates of Africa, Vol. 1. New York and London: Academic.

Butler, P. M. 1972. The problem of insectivore classification, pp. 253–265. In: Joysey K. A., Kemp, T. S. (eds.), Studies in vertebrate evolution. Edinburgh: Oliver and Boyd.

Butler, P. M. 1978. Insectivora and Chiroptera, pp. 56–68. In: Maglio, V. J., Cooke H. B. S., (eds.), Evolution of African mammals. Cambridge, Massachusetts, London: Harvard U. Press.

Butler, P. M., Greenwood, M. 1976. Lower Pleistocene elephant-shrews (Macroscelididae) from Olduvai and Makapansgat. pp. 1–56. In: Savage, R. J. G., Coryndon, S. C. (eds.), Fossil vertebrates of Africa, Vol. 4. London: Academic.

Butler, P. M., Hopwood, A. T. 1957. Insectivora and Chiroptera from the Miocene rocks of Kenya Colony. Fossil mammals of Africa 13. London: Brit. Mus. (Nat. Hist.).

Corbet, G. B., Hanks, J. 1968. A revision of the elephant-shrews, family Macroscelididae. Bull. Brit. Mus. (Nat. Hist.), Zool. 16:47–111.

Darwin, C. R. 1859. On the origin of species. London: John Murray.

Eldredge, N. 1976. Differential evolutionary rates. Paleobiology 2:174–177.

Eldredge, N. 1979. Alternative approaches to evolutionary theory, pp. 7–19. In: Schwartz, J. H., Rollins, H. B. (eds.), Models and methodologies in evolutionary theory. Bull. Carnegie Mus. Nat. Hist. 13:1–105.

Evans, F. G. 1942. The osteology and relationships of the elephant-shrews (Macroscelididae). Bull. Amer. Mus. Nat. Hist. 80:85–125.

Filhol, H. 1892. Note sur un insectivore nouveau. Bull. Soc. Philom. (8)4:134.

Fryer, G., Iles, T. D. 1969. Alternative routes to evolutionary success as exhibited by African cichlid fishes of the genus *Tilapia* and the species flocks of the great lakes. Evolution 23:359–369.

Gould, S. J. 1980. Is a new and general theory of evolution emerging? Paleobiology 6:119–130.

Gregory, W. K. 1910. The orders of mammals. Bull. Amer. Mus. Nat. Hist. 27:1–524.

Haeckel, E. 1866. Generelle Morphologie der Organismen. Berlin: Reimer.

Jackson, J. B. C. 1974. Biogeographic consequences of eurytopy and stenotopy among marine bivalves and their evolutionary significance. Amer. Nat. 108:541–560.

Jacobs, L. L. 1980. Siwalik fossil tree shrews, pp. 205–216. In: Luckett, W. P. (ed.), Comparative biology and evolutionary relationships of tree shrews. New York: Plenum.

Johnson, J. G. 1982. Occurrence of phyletic gradualism and punctuated equilibria through geologic time. J. Paleont. 56:1329–1331.

Kingdon, J. 1974. East African mammals, an atlas of evolution in Africa, Vol. IIA. London: Academic.

Lillegraven, J. A. 1969. Latest Cretaceous mammals of upper part of Edmonton Formation of Alberta, Canada, and a review of marsupial–placental dichotomy in mammalian evolution. U. Kansas Paleont. Contrib. 50 (Vertebrata 12):1–122.

Luckett, W. P. 1980. The suggested evolutionary relationships and classification of tree shrews, pp. 3–31. In: Luckett, W. P. (ed.), Comparative biology and evolutionary relationships of tree shrews. New York: Plenum.

MacPhee, R. D. E. 1979. Entotympanics, ontogeny and primates. Folia primatol. 31:23–47.

MacPhee, R. D. E. 1981. Auditory regions of primates and eutherian insectivores: morphology, ontogeny, and character analysis. Contr. Primat. 18:1–282.

McKenna, M. C. 1975. Toward a phylogenetic classification of the Mammalia, pp. 21–46. In: Luckett, W. P., Szalay, F. S. (eds.), Phylogeny of the primates, a multidisciplinary approach. New York: Plenum.

Novacek, M. J. 1977a. A review of Paleocene and Eocene Leptictidae (Eutheria: Mammalia) from North America. PaleoBios 24:1–42.

Novacek, M. J. 1977b. Aspects of the problem of variation, origin, and evolution of the eutherian auditory bulla. Mam. Rev. 7:131–149.

Novacek, M. J. 1978. Evolution and relationships of the Leptictidae (Eutheria: Mammalia). Ph.D. Thesis, U. California, Berkeley.

Novacek, M. J. 1980. Cranioskeletal features in tupaiids and selected Eutheria as phylogenetic evidence, pp. 35–93. In: Luckett, W. P. (ed.), Comparative biology and evolutionary relationships of tree shrews. New York: Plenum.

Novacek, M. J. 1982. Information for molecular studies from anatomical and fossil evidence on higher eutherian phylogeny. pp. 3–41. In: Goodman, M. (ed.), Macromolecular sequences in systematic and evolutionary biology. New York: Plenum.

Novacek, M. J., Norell, M. A. 1982. Fossils, phylogeny, and taxonomic rates of evolution. Syst. Zool. 31:369–378.

Patterson, B. 1965. The fossil elephant shrews (Family Macroscelididae). Bull. Mus. Comp. Zool., Harvard U. 133(6):297–335.

Rathbun, G. B. 1978. Evolution of the rump region in the golden-rumped elephant-shrew. Bull. Carnegie Mus. Nat. Hist. 6:11–19.

Rathbun, G. B. 1979. The social structure and ecology of elephant-shrews. Fortschr. Verhalt. Zeitschr. Tierpsychol. 20:3–76.

Sauer, E. G. F. 1973. Zum Sozialverhalten der Kurzohrigen Elefantenspitzmaus, *Macroscelides proboscideus*. Z. Säugetierkd. 38:65–97.

Schopf, T. J. M., Raup, D. M., Gould, S. J., Simberloff, D. S. 1975. Genomic versus morphologic rates of evolution: influence of morphologic complexity. Paleobiology 1:63–70.

Sigé, B. 1974. *Pseudorhyncocyon cayluxi* Filhol,

1892, insectivore geant des phosphorites du Quercy. Paleovertebrata 6(1–2):33–46.

Simpson, G. G. 1931. A new insectivore from the Oligocene, Ulan Gochu horizon, of Mongolia. Amer. Mus. Novit. 505:1–22.

Simpson, G. G. 1945. The principles of classification and a classification of mammals. Bull. Amer. Mus. Nat. Hist. 85:1–350.

Simpson, G. G. 1953. The major features of evolution. New York: Columbia U.

Sprinkle, J. 1976. Classification and phylogeny of "pelmatozoan" echinoderms. Syst. Zool. 25:83–91.

Stanley, S. M. 1979. Macroevolution: pattern and process. San Francisco: Freeman.

Stromer, E. 1926. Reste land- und süsswasserbewohnender Wirbeltiere aus den Diamantfeldern Deutsch-Südwestafrikas, pp. 107–153. In: Kaiser, E. (ed.), Die Diamantenwüste Südwestafrikas, Vol. 2. Berlin: Dietrich Riemer.

Stromer, E. 1932. *Palaeothentoides africanus* nov. gen. nov. spec. ein erstes Beuteltier aus Afrika. Sitzungsber. math-nat. Abt. Bayer Akad. Wiss. 1932:177–190.

Stümpke, H. 1967. Bau and Leben der Rhinogradentia. Stuttgart: Gustav Fischer Verlag.

Szalay, F. S. 1977. Phylogenetic relationships and classification of the eutherian Mammalia, pp. 315–374. In: Hecht, M. K., Goody, P. C., Hecht, B. M. (eds.), Major patterns in vertebrate evolution. New York: Plenum.

Van der Klaauw, C. J. 1931. The auditory bulla in fossil mammals, with a general introduction to this region of the skull. Bull. Am. Mus. Nat. Hist. 62:1–352.

Walker, E. P. 1968. Mammals of the world. 2nd ed. Baltimore: Johns Hopkins Press.

Whiteworth, T. 1954. The Miocene hyracoids of East Africa. Fossil mammals of Africa. Brit. Mus. (Nat. Hist.) 7:1–58.

2
The Tree Squirrel *Sciurus* (Sciuridae, Rodentia) as a Living Fossil

Robert J. Emry and Richard W. Thorington, Jr.

Department of Paleobiology and Department of Vertebrate Zoology, National Museum of Natural History, Smithsonian Institution, Washington, DC 20560

Introduction

The familiar living squirrel, *Sciurus,* is not among the classic and often-cited examples of living fossils, although squirrels have long been recognized as being among the most primitive members of the Rodentia, the mammalian order that has exceeded all others in specific diversity. In the sense that they represent the least derived family of a very diverse order, squirrels in general might be called living fossils. The recently discovered skeleton of *Protosciurus* (perhaps the oldest squirrel fossil) shows that the earliest recognized sciurid is strikingly similar in its osteology to living *Sciurus*. In the sense that it has evolved very little from what is apparently the primitive squirrel morphotype, *Sciurus* is a living fossil.

Anatomy

The skeleton of *Protosciurus* has been compared in detail with those of extant sciurids and with other appropriate fossil rodents (Emry and Thorington 1982). Most of the details need not be repeated here; it is perhaps easier to point out the differences, which are few and mostly subtle, than to catalog the similarities. *Protosciurus* is primitive in retaining a protrogomorphous masseter; i.e., it is not sciuromorphous. But even though technically not

sciuromorphous, its chewing and gnawing apparatus is advanced in several ways compared with most protrogomorphs. For example, the masseteric fossa of the mandible terminates anteriorly beneath M_1 rather than beneath M_2, and the rostrum is relatively short. The primary functional advantage of sciuromorphy is in bringing the jaw muscles closer to the incisors, thereby increasing the force of the bite for gnawing; part of the same advantage is gained by shortening the rostrum, thus bringing the incisors closer to the jaw muscles. *Protosciurus* has a lyrate area between the parietal crests (i.e., it lacks the sagittal crest seen in most protrogomorphs), suggesting decreased emphasis on the temporal muscles and increased emphasis on the masseters. The incisors of *Protosciurus* are transversely compressed and have uniserial enamel, as in extant squirrels, suggesting that the mechanical advantage gained by shortening the rostrum created a need for stronger incisors, also necessary for sciuromorphy. It seems that *Protosciurus* was well on its way to becoming sciuromorphous, lacking only the forward shift of the origin of the masseter muscles.

In cranial characters other than those associated with the masseter muscles, *Protosciurus* is essentially like extant squirrels, though its cheek teeth retain some of the features seen in paramyid rodents; e.g., the cross lophs, particularly the metaloph, of the upper teeth have one

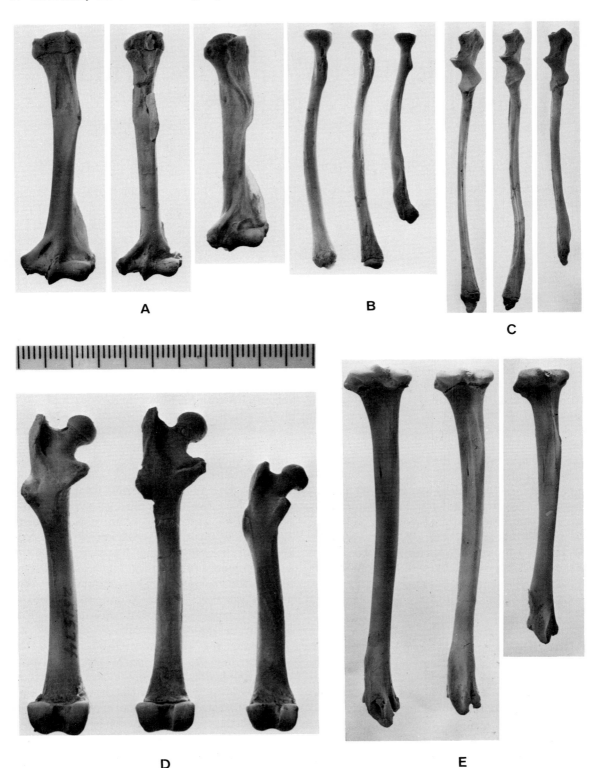

Fig. 1. Comparative limb elements of three squirrels to show the relative proportions. (A) humeri; (B) radii; (C) ulnae; (D) femora; (E) tibiae. In each instance the bone on the left is of the fox squirrel *Sciurus niger* (USNM 251574), that in the center is of the fossil *Protosciurus jeffersoni* (USNM 243981), and that on the right is of the ground squirrel *Spermophilus beecheyi* (USNM 484951). Scale in mm; all bones are to the same scale.

to several distinct conules, as in many genera of Paramyinae (though this characteristic is also reminiscent of some flying squirrels), and the lower teeth have hypolophids that are more distinct than in other squirrels.

It is in the postcranial skeleton that *Protosciurus* is most strikingly like *Sciurus*. *Protosciurus* was the size of a typical fox squirrel, *Sciurus niger*. In the vertebral column, we could detect no important differences; as in *Sciurus*, the anterior few caudal vertebrae of *Protosciurus* are very short, allowing extreme dorsiflexion of the tail. The limb elements of the two genera are strikingly similar (Fig. 1) in their morphology and proportions (e.g., in the relative lengths of upper to lower limb elements, the relative length of fore and hind limbs, and in their general slender, gracile construction). In these characteristics *Protosciurus* is like *Sciurus*, differing from the shorter, stouter limbs of ground squirrels. *Paramys*, and the other paramyids for which sufficient postcranial material is known, are more nearly comparable to ground squirrels in these features. The hands and feet of *Protosciurus* are narrow as in extant arboreal squirrels. We cannot be certain of the hand (neither fourth metacarpal is complete), but at least in the foot of *Protosciurus*, the fourth digit is longest. The longer fourth digit is characteristic of arboreal squirrels, whereas in ground squirrels and paramyids, the hand and foot are more nearly symmetrical, centered around a strong third digit that is longest.

In postcranial osteology, the few minor differences noted between *Sciurus* and *Protosciurus* are in details of joint construction. Perhaps the most significant of these is in the astragalar–navicular joint, which in *Sciurus* is modified into a concavo-convex "saddle joint." This is not seen in *Protosciurus*, but neither is it seen in many other squirrels, even in most arboreal squirrels. It appears to be a derived character only of the tribe Sciurini. It is obviously possible to be a successful arboreal squirrel, even to completely evert the foot when the leg is extended, as squirrels do while descending a tree head first, without the special tarsal joint of the Sciurini. *Protosciurus* is similar to all squirrels except Sciurini in this feature only because they all share the primitive character state. In the carpus, *Protosciurus* has separate scaphoid and lunar bones (the primitive rodent condi-

tion), whereas in all extant squirrels, and in all but a few rodents, the bones are fused. Except for a few such minor differences, *Protosciurus* is exceedingly similar to *Sciurus* in its postcranial osteology. *Sciurus* seems to have changed very little from what appears to be the primitive sciurid morphotype.

Time

The skeleton of *Protosciurus* (USNM 243981) is from the Chadronian (approximately Early Oligocene) White River Formation, Flagstaff Rim Area of Wyoming, at 13.5 m (45 ft) below ash B (Emry 1973:29). This volcanic ash has potassium–argon dates of 35.2 million years (biotite) and 33.3 million years (sanidine) (Evernden et al. 1964). The mammalian fauna occurring with it suggests a time slightly older than the Pipestone Springs fauna of Montana, which has the same, or a very closely related, species of *Protosciurus*. If the correlation is correct, USNM 243981 is probably the oldest recognized squirrel fossil. The radiometric dates on ash B are consistent with other ash dates in the same sequence, and with radiometric dates in other sequences that correlate paleontologically. *Protosciurus* and its extant relative *Sciurus* are therefore believed to be separated by about 35 million years.

Relationships

Because *Protosciurus* is not sciuromorphous, Wood (1980) excludes it from the squirrel family, Sciuridae. We would not exclude it on this basis, which is retention of protrogomorphy, the primitive character state; rather, we classify it in the Sciuridae because it has a number of important derived characters that it shares with other members of the family. The auditory bullae are enlarged and firmly fused to the periotics, and the stapedial artery is enclosed in a bony conduit through the middle-ear cavity. This combination of auditory characters occurs elsewhere in rodents only in Sciuridae. The skull of *Protosciurus* is broad interorbitally and postorbitally and the frontals have postorbital processes. The scapula has a subscapular spine.

The incisors are transversely compressed and have uniserial enamel. All of these characters occur in the Sciuridae, some occur only in Sciuridae, and, so far as we can determine, they all occur together only in Sciuridae.

At the same time, *Protosciurus* lacks some of the derived characters that seem to be present in all extant squirrels; it is not sciuromorphous, has separate scaphoid and lunar bones in the carpus, and its cheek teeth are not characteristic of any of the modern squirrel groups. We therefore consider *Protosciurus* the sister-group of all other Sciuridae, having many features of Sciuridae but lacking some of the derived characters seen in all others.

Among those conversant with fossil rodents, there is little dissension with the derivation of squirrels from the Paramyidae (Ischyromyidae of some authors), most likely from within the subfamily Paramyinae, though there is some disagreement as to the exact source within the family. In any case, the sister-group of Sciuridae is almost certainly the Paramyidae and its other derivatives.

Fossil History

The fossil record of the Sciuridae is too imperfectly known to allow generalizations about species level diversity through time. Despite the emphasis of the last two decades on collecting small mammal fossils, which has resulted in a manyfold increase in the knowledge of fossil rodents, fossil squirrels remain relatively uncommon. Very little can be added to Black's (1972) summary. The first known members of the family appeared in the Early Oligocene of North America (±35 million years ago). They may be present in the early Oligocene of Europe as well, and were certainly present there in Middle and Late Oligocene times (Black 1972). The earliest known members in North America are very similar to extant tree squirrels (Sciurini), and the earliest European records are most like the earliest North American material. By the end of the Oligocene, ground squirrels had appeared in both North America and Europe. In North America, Late Oligocene *Miospermophilus* is the earliest known member of the Spermophilini. Black (1972) recognized two lineages arising from *Miospermophilus*,

each undergoing a modest radiation in the Miocene, with one lineage eventually giving rise to the marmots and the other to ground squirrels. In Europe, the tribe Xerini (the extant African ground squirrels) first appeared in the Late Oligocene. According to Black (1972), several lineages of xerines can be traced through the Neogene of Spain and France, and one genus, *Getuloxerus*, occurred in Morocco as well as in Spain. This genus is very similar to the extant *Atlantoxerus*. Fossil xerines are known only in western Europe and Africa.

The squirrel fossil record is most diverse in the Miocene. In North America the Miocene record is dominated by ground squirrels (Black 1972). This increase in diversity of ground squirrels is probably related to the increase in grassland and savannah environments, and the better record is probably related to the greater likelihood of ground squirrel remains being preserved, compared with those of tree squirrels. In Europe, the Miocene record is dominated by fossil flying squirrels. Chipmunks are known from very sparse records of fragmentary material in the North American Neogene and in the Late Pliocene of Poland and China (Black 1972). The fossil chipmunks are all assigned to extant genera.

The Pleistocene record is predominantly of modern types, mainly modern species.

The extant genus *Sciurus* has a long fossil history. Fossils that cannot be distinguished from the living genus are recognized as far back as the Miocene in both North America and Europe. The Sciurini seems to have been the conservative lineage, which was probably never morphologically diverse and which was probably always restricted to forested habitat, probably contributing to the relatively sparse fossil record of tree squirrels throughout the history of the family. Among the living squirrels, the Sciurini seem to have a narrow range of adaptive types, body size, and so on. The ground squirrels were more diverse, with several different lineages, possibly independently derived, and with a broad range of adaptive types, body size, and so forth. The known fossil record suggests that the Sciurini have changed very little from the earliest known squirrels, and that the other squirrel groups (ground squirrels, marmots, chipmunks, flying squirrels, and the unusually adapted members of arboreal squirrel groups) have diverged from the primitive morphotype.

Diversity and Distribution of Recent Sciuridae

The Sciuridae is usually divided into two subfamilies, the Petauristinae, or flying squirrels, and the Sciurinae, or tree and ground squirrels (e.g., Simpson 1945, after Pocock 1923). McLaughlin (1967) estimated that there are 51 Recent genera with 261 species. The family is cosmopolitan except for the Australian region, Madagascar, and South America south of 35°S.

Some authors (e.g., Black 1963; Hight et al. 1974) have suggested that the Petauristinae (approximately 13 genera and 34 species) may not be monophyletic, perhaps being derived more than once from different tree squirrels. Mein (1970) even doubts that they and the Sciurini share a common ancestor within the Sciuridae. However, all taxa now classified in the Petauristinae share a suite of derived characters used for gliding, including long limbs and a gliding membrane (Thorington and Heaney 1981), and the similarity among all members of Petauristinae, even in unique details of these gliding adaptations (Thorington, in preparation) suggests that at least all living taxa assigned to the subfamily are monophyletic.

The Sciurinae (approximately 38 genera and 227 species) are generally spoken of as tree-squirrel and ground-squirrel groups, but this is an oversimplification. Following Moore's (1959) classification (although we place *Tamiasciurus* in the Sciurini and do not recognize Tamiasciurini), we consider five of the tribes to be radiations of tree squirrels and two tribes to be ground squirrels.

The best known tree squirrels are those of the tribe Sciurini, of the Holarctic and Neotropical realms. South America has three genera, *Sciurus, Microsciurus,* and *Sciurillus. Sciurillus* is a pygmy squirrel with derived morphology. *Sciurus* ranges as far south as Argentina. In North America and Central America there are four genera, *Sciurus, Syntheosciurus, Microsciurus,* and *Tamiasciurus. Sciurus* and *Tamiasciurus* are widespread, so that the tree squirrels occur throughout the continent almost ev-

erywhere there are trees. Only *Sciurus* occurs in Eurasia, but it ranges from England to Japan, making it by far the most widespread genus of squirrel. Moore (1959) included *Rheithrosciurus* in the tribe Sciurini, but we doubt that it belongs there; it is a large, terrestrial, cursorial squirrel of Borneo.

Africa has six genera of tree squirrels classified in two tribes. The Protoxerini includes the dissimilar genera *Protoxerus* and *Heliosciurus,* which occur in most forests from the Sahara south to Angola, Zimbabwe, and Mozambique. *Protoxerus,* the "giant" tree squirrel of Africa, is highly arboreal but is closely related to the third genus of its tribe, *Epixerus,* which is a cursorial, completely terrestrial squirrel with derived locomotor specializations (Emmons 1975, 1980). The Funambulini includes *Funisciurus* and *Paraxerus,* generally small scansorial to arboreal squirrels, ranging through most of sub-Saharan Africa. *Aethiosciurus* includes small to large squirrels, some of which are highly arboreal. *Myosciurus,* the African pygmy squirrel, is one of the smallest of all squirrels, weighing about 16 g. It is a bark gleaner with derived morphology in skull and limbs (Anthony and Tate 1935; Moore 1959; Emmons 1975, 1979, 1980). The last genus of the tribe is *Funambulus,* the striped squirrels of India.

The tribe Callosciurini of southeast Asia includes approximately 14 genera. Less than half these are true tree squirrels, like *Callosciurus* and *Sundasciurus,* which share many primitive traits with *Sciurus.* The tribe includes such highly derived forms as the terrestrial insectivorous squirrel *Rhinosciurus,* and the two genera of pygmy squirrels, *Nannosciurus* and *Exilisciurus.* Some members of the tribe reached the Celebes and underwent a small radiation (*Hyosciurus, Prosciurillus,* and *Rubrisciurus*).

The tribe Ratufini includes the single genus *Ratufa,* which occurs in tall forests from Ceylon to Nepal and Java.

The ground squirrels are classified in two tribes, the Marmotini and Xerini, representing two radiations. It is not clear whether they are independently derived from tree squirrels or whether their common ancestor was a ground squirrel. Moore (1959) notes that they share some cranial features but suggests that these may be independently derived. Compared to tree squirrels, both tribes of ground squirrels appear to be derived in their shorter, more robust limbs, probably associated with their burrowing habits, and many have derived features in their teeth, probably associated with a greater emphasis on leafy material in their diets.

The Xerini occur today in Africa (*Atlantoxerus* and *Xerus*) and in southwest Asia (*Spermophilopsis*). Fossils occur in Europe. The Marmotini includes the chipmunks (*Tamias* and *Eutamias*) of the Holarctic. *Tamias* nests in burrows, where it spends considerable time, but it is also known to forage extensively in trees (Elliott 1978). *Eutamias* is generally more scansorial and occasionally nests above ground. A second subtribe includes three genera, *Spermophilus*, *Ammospermophilus*, and *Cynomys*, the first with a Holarctic distribution, the other two North American. Although these are the "prototype" ground squirrels, some are known to be good climbers. Finally, the marmots (*Marmota*) comprise the third subtribe of Marmotini. These are the Holarctic giant ground squirrels that occur in meadows and a variety of montane habitats. Though generally terrestrial, *Marmota monax* can climb trees, descending head first, somewhat like a tree squirrel. Thus, it seems best to characterize ground squirrels generally by their burrowing proclivities, rather than by any reluctance to climb trees.

Species Ranges

Many tree squirrels have large to very large ranges. The eastern gray squirrel, *Sciurus carolinensis*, occurs throughout the eastern half of the United States. The eastern fox squirrel, *S. niger*, is almost as widespread. The North American red squirrel, *Tamiasciurus hudsonicus*, ranges throughout the coniferous forests of Canada, from the Atlantic to the Pacific, and through much of the northern United States (Hall 1981). The European red squirrel, *Sciurus vulgaris*, occurs from England to Japan (Ellerman and Morrison-Scott 1951). The South American squirrel, *Sciurus aestuans*, ranges from Venezuela and the Guianas to southern Brazil and Argentina (Cabrera 1961). The red-legged sun squirrel, *Heliosciurus rufobrachium*, is found from Senegal throughout the Congo basin to Angola and the Rift Valley of Kenya (Thorington, unpublished observations). The

giant squirrel, *Ratufa bicolor*, occurs from Nepal to Java (Lekagul and McNeely 1977).

In parts of the world where forests are discontinuous or where particular forest habitats are isolated from one another, as in mountainous areas, the ranges of tree squirrel species are much smaller. Central America provides a number of examples: *Sciurus alleni*, *S. yucatanensis*, *S. colliaei*, and *Syntheosciurus brochus* (Hall 1981).

At the generic level, *Sciurus* has by far the greatest range of any squirrel, from South America through North America and across Eurasia to England. This range is among the greatest for mammal genera, probably exceeded by that of *Canis* and possibly that of *Felis*.

Considering the probable distribution of forests during the Oligocene Epoch, it is likely that *Protosciurus* had a broad range, possibly throughout Holarctica. It is also probable that the species of *Protosciurus* had extensive ranges.

Intraspecific Variation

Phenotypic variation of tree-squirrel species can be dramatic but is often only skin deep. Coat color and pattern vary geographically and have been the basis for many named subspecies. It is particularly dramatic in *Callosciurus finlaysoni* from southeast Asia (Lekagul and McNeely 1977) and in *C. prevostii* (Medway 1969), but is also distinctive in *Sciurus aureogaster* (Musser 1968) and *S. niger*. Local color variations are likewise common; melanistic, erythristic, albino, and other coat color forms are found in varying frequencies in tree squirrel populations (Searle 1968).

Variation in size is also noteworthy for some species. *Sciurus carolinensis*, the eastern gray squirrel, is smallest in Florida and largest in Wisconsin and Minnesota. Our measurements of condylobasilar length of skull show that Wisconsin squirrels average 16% larger than Florida ones. Musser's (1968) data on cranial lengths show samples of *Sciurus colliaei* differing by 12%, and of *S. aureogaster* by 9%. Our data on condylobasilar length of *Tamiasciurus* skulls in the eastern United States show that the largest ones, which are from North Dakota, average 6% larger than the smallest ones, which are from New Hampshire. A larger survey of tree squirrels would undoubtedly show that some species are more uniform and others more variable than those cited here. Morphological variation within local populations is documented by Musser (1968) for *Sciurus aureogaster*. The coefficients of variation of cranial length vary from slightly less than 1.0 to more than 3.0 for 58 samples. We found similar coefficients of variation, averaging close to 2.0 for condylobasilar lengths of skull in 45 samples of *S. carolinensis* and *Tamiasciurus hudsonicus* from the eastern United States. Variability of postcranial bones of tree squirrels was documented by Thorington (1972) and Thorington and Heaney (1981).

Samples of fossil squirrels are too small to allow any meaningful interpretation of variation at the specific or population level.

Ecology

The ecology of *Protosciurus* is probably best represented today by tree squirrels such as *Sciurus*, *Callosciurus*, or *Heliosciurus*. These arboreal squirrels usually nest in trees, though some species forage and scatter-hoard food extensively on the ground. Nests are frequently in tree hollows but may also be placed on branches and constructed of small twigs and leaves (e.g., *Sciurus vulgaris*: Raspopov and Isakov 1980; *S. carolinensis*: Barkalow and Shorten 1973). The nest is generally lined with finely shredded bark, wood fiber, dry grass, or moss. Nests provide protection against cold and predators; thus the nest may vary seasonally and geographically. Tree hollows in which squirrels nest usually have tooth marks near the entrance; such marks could conceivably be identified in well-preserved fossil wood, but to our knowledge none have been.

The diet of some squirrels is very restricted. *Sciurus aberti* is almost completely dependent on ponderosa pine, feeding on its seeds, phloem, buds, and male cones (Keith 1965). *Sciurus granatensis* depends heavily on the seeds of two species of palm and one legume, *Dipteryx* (Heaney and Thorington 1978; Glanz et al., in press). Other tree squirrels are much more catholic in their tastes. *Sciurus carolinen-*

sis and *S. niger* rely heavily on hickory, beech, and oak (together about 10 species) in southeast Ohio (Nixon et al. 1968). Nuts of these species were found in 76% of stomachs examined and comprised 67% of the bulk, averaged over all seasons. Seasonal variation was extreme, however, with these nuts found in less than 25% of stomachs between March and July, when other foods became more important. Animal material, mostly insects, was found in 87% of stomachs in June, comprising only 5% of the bulk, but perhaps contributing a significant amount of protein.

Most African tree squirrels rely heavily on hard seeds. For example, Emmons (1975, 1980) noted that *Protoxerus stangeri* feeds extensively on the seeds of three species, *Panda oleosa*, *Entada gigas*, and *Pentaclethra eetveldeana*. *Heliosciurus rufobrachium* eats more insects than *Protoxerus stangeri* does, but it also uses hard nuts and becomes a pest in secondary forest because of its predeliction for palm nuts and cocoa seeds.

Among the squirrels of southeast Asia, there are species (e.g., *Ratufa bicolor*, *R. affinis*, and *Sundasciurus hippurus*) that feed extensively on hard nuts (MacKinnon 1978). However, other species (e.g., *Callosciurus notatus* and *C. prevostii*) seem to feed extensively on soft fruits and very little on hard nuts. The Callosciurini also includes *Rhinosciurus*, which has probably become the most insectivorous of all squirrels, as well as the pygmy squirrels *Exilisciurus* and *Nannosciurus*, which are probably highly insectivorous bark gleaners, like their African equivalent *Myosciurus*.

In general, then, the large tree squirrels are able to feed on the hardest nuts in the forest. In many forests they have relatively few competitors for this rich source of nutrients, which may have led to selection for the more powerful sciuromorphous musculature, uniserial incisor enamel, and other morphological features that enable them to exploit this resource. It is tempting to hypothesize that squirrels and hard nuts have coevolved, as argued by Smith (1970) for *Tamiasciurus* and the hard serotinous cones of lodgepole pine. However, the hard nuts of some extant genera of trees (e.g., *Carya*) probably predate the origin of squirrels; the plants protect their seeds with toxins as well as mechanically (e.g., *Quercus* and *Carya*); and insects such as weevils may also have played an important role in the evolution of nuts.

Since many tree squirrels have such similar ecologies, it is interesting that different species coexist in some forests. Smith and Folmer (1972) investigated gray and fox squirrels and found that the important differences are probably in foraging behavior and predator avoidance, both of which would not be distinguished in the fossil record. In Gabon and Malaya, the coexistence of several species involves altitudinal stratification and size differences (Emmons 1975, 1980; MacKinnon, 1978). Only the size difference would be recognized in paleoenvironments, although it may be noteworthy, and a clue to fossil interpretation, that in most cases of altitudinal stratification, morphologically distinct genera are involved (e.g., *Protoxerus, Helioiscurus,* and *Funisciurus; Ratufa, Sundasciurus,* and *Callosciurus*).

Summary

By about 35 million years ago, squirrels had evolved that seem to differ in no important ways from their living relative *Sciurus.* Since *Sciurus* is so similar to what is apparently the primitive squirrel morphotype, it seems to fit the concept of "living fossil." *Sciurus* presently has one of the largest geographic ranges of any mammal, and fossils that have not been distinguished from the modern genus go back as far as the Miocene in both Europe and North America. Arboreal squirrels seem to be adapted for utilizing the hardest nuts and seeds of the forest, for which there seems to be little competition.

Literature

Anthony, H. E., Tate, G. H. H. 1935. Notes on South American Mammalia. No. 1, *Sciurillus.* Amer. Mus. Novit. 780:1–13.

Barkalow, Jr., F. S., Shorten, M. 1973. The world of the gray squirrel. Philadelphia: L. B. Lippincott.

Black, C. C. 1963. A review of the North American Tertiary Sciuridae. Bull. Mus. Comp. Zool. 130:109–248.

Black, C. C. 1972. Holarctic evolution and dispersal of squirrels (Rodentia: Sciuridae), pp. 305–322. In: Dobzhansky, T., Hecht, M. K., Steere, W. C. (eds.), Evolutionary biology, Vol. 6, New York: Appleton-Century-Crofts.

Cabrera, A. 1961. Catalogo de los Mamiferos de America del Sur. II. Cien. Zool. IV:309–732.

Ellerman, J. R., Morrison-Scott, T. C. S. 1951. Checklist of Palearctic and Indian mammals, 1738–1946. Brit. Mus. (Nat. Hist.) London.

Elliott, L. 1978. Social behavior and foraging ecology of the eastern chipmunk (*Tamias striatus*) in the Adirondack Mountains. Smithsonian Contr. Zool. 265:1–107.

Emmons, L. H. 1975. Ecology and behavior of African rainforest squirrels. Diss. Cornell U., Ithaca, NY.

Emmons, L. H. 1979. A note on the forefoot of *Myosciurus pumilio.* J. Mam. 60:431–432.

Emmons, L. H. 1980. Ecology and resource partitioning among nine species of African rainforest squirrels. Ecol. Mon. 50:31–54.

Emry, R. J. 1973. Stratigraphy and preliminary biostratigraphy of the Flagstaff Rim Area, Natrona County, Wyoming. Smithsonian Contr. Paleobiol. 18:1–43.

Emry, R. J., Thorington, Jr., R. W. 1982. Descriptive and comparative osteology of the oldest fossil squirrel, *Protosciurus* (Rodentia: Sciuridae). Smithsonian Contr. Paleobiol. 47:1–35.

Evernden, J. F., Savage, D. E., Curtis, G. H., James, G. T. 1964. Potassium argon dates and the Cenozoic mammalian chronology of North America. Amer. J. Sci. 262:145–198.

Glanz, W. E., Thorington, Jr., R. W., Madden, J., Heaney, L. R. 1982. Seasonal food use and demographic trends in *Sciurus granatensis.* In: Leigh, Jr., E. G., Rand, A. S., Windsor, D. M. (eds.), Ecology of a tropical forest: seasonal rhythms and long-term changes. Washington, DC: Smithsonian Institution Press. pp. 239–252.

Hall, E. R. 1981. The mammals of North America, 2nd ed., Vol. 1. New York: Wiley.

Heaney, L. R., Thorington, Jr., R. W. 1978. Ecology of neotropical red-tailed squirrels, *Sciurus granatensis,* in the Panama Canal Zone. J. Mam. 59:846–851.

Hight, M. E., Goodman, M., Prychodko, W. 1974. Immunological studies of the Sciuridae. Syst. Zool. 23:12–25.

Keith, J. O. 1965. The Abert squirrel and its dependence on ponderosa pine. Ecology 46:150–163.

Lekagul, B., McNeely, J. A. 1977. Mammals of Thailand. Bangkok: Karusapha Ladprao.

MacKinnon, K. S. 1978. Stratification and feeding differences among Malayan squirrels. Malay. Nat. J. 30:593–608.

McLaughlin, C. A. 1967. Aplodontoid, Sciuroid, Geomyoid, Castoroid, and Anomaluroid rodents.

In: Anderson, S. A., Jones, Jr., J. K. (eds.), Recent mammals of the world. Ronald Press Co., N.Y. pp. 210–225.

Medway, Lord. 1969. The wild mammals of Malaya and offshore islands including Singapore. London: Oxford U. Press.

Mein, P. 1970. Les sciuropteres (Mammalia, Rodentia) Neogenes d'Europe Occidentale. Geobios 3:7–56.

Moore, J. C. 1959. Relationships among the living squirrels of the Sciurinae. Bull. Amer. Mus. Nat. Hist. 118:153–206.

Musser, G. G. 1968. A systematic study of the Mexican and Guatemalan gray squirrel, *Sciurus aureogaster* F. Cuvier (Rodentia: Sciuridae). Misc. Publ. Mus. Zool., U. Mich. 137:1–112.

Nixon, C. M., Worley, D. M., McClain, M. W. 1968. Food habits of squirrels in southeast Ohio. J. Wildl. Mgmt. 32:294–305.

Pocock, R. J. 1923. The classification of the Sciuridae. Proc. Zool. Soc. London 1923:209–246.

Raspopov, M. P., Isakov, Y. A. 1980. Biology of the squirrel. New Delhi: Amerind. (Translated from Russian).

Searle, A. G. 1968. Comparative genetics of coat colour in mammals. London: Academic.

Simpson, G. G. 1945. The principles of classification and a classification of mammals. Bull. Amer. Mus. Nat. Hist., 85:1–350.

Smith, C. C. 1970. The coevolution of pine squirrels (*Tamiasciurus*) and conifers. Ecol. Mon. 40:349–371.

Smith, C. C., Folmer, D. 1972. Food preferences of squirrels. Ecology 53:82–91.

Thorington, Jr., R. W. 1972. Proportions and allometry in the gray squirrel. Nemouria, Occas. Papers Delaware Mus. Nat. Hist., 8:1–17.

Thorington, Jr., R. W., Heaney, L. R. 1981. Body proportions and gliding adaptations of flying squirrels (Petauristinae). J. Mam. 62:101–114.

Wood, A. E. 1980. The Oligocene rodents of North America. Trans. Amer. Phila. Soc. 70:1–68.

3
The Tree-Shrew, *Tupaia:* A "Living Model" of the Ancestral Primate?

Ian Tattersall

Department of Anthropology, American Museum of Natural History, New York, NY 10024

Although the common tree-shrew, *Tupaia*, does not qualify as a "living fossil" in the sense that it shows a close identity with a known ancient fossil species (indeed, the tree-shrew fossil record is exceptionally poor), over the past several decades these mammals have regularly been held up as a "living model" of the "ancestral primate." The tree-shrews have thus been regarded widely as approximating (usually in some unspecified way or ways) the ancestral primate morphotype. Perhaps surprisingly, this viewpoint is not confined to the diminishing number of systematists and others who would admit the tree-shrews to membership in the order Primates. Hence it may be useful to evaluate the claims of the tree-shrews to living fossil status of this kind.

Tupaia itself, it should be noted, is a speciose genus that belongs to a group of some six genera, all allocated to the family Tupaiidae and distributed quite widely in forested regions of southeast Asia and its outlying islands. All tupaiids are small-bodied, scansorial, vaguely "squirrel-like" mammals, and most are poorly known morphologically, as well as behaviorally and ecologically. The term "tree-shrew" may, of course, be applied to any tupaiid, but most of what is known about tree-shrews has been learned from genus *Tupaia* (notably the species *Tupaia glis*), and unspecific references to tree-shrews may normally be taken to apply to species of that genus.

Historical Background

In the early years of this century, most authors followed Gregory (1910) in allocating the tree-shrews, together with the elephant-shrews (Macroscelidea) to a taxon Menotyphla. There was some disagreement, however, over the level at which it was appropriate to separate this group from the lipotyphlous (roughly, "true") insectivorans. Various authors, including Gregory, had pointed out at one time or another that tree-shrews exhibited some primate-like characters (and Gregory himself had gone so far as to suggest that the primates were derived from "insectivores resembling in many ways *Tupaia* and *Ptilocercus*"). However, it was not until 1922 that Carlsson first actually classified these animals as primates, largely on the basis of comparisons with the strepsirhine genus *Lemur*. This transfer was shortly thereafter strongly supported by the work of Clark (1924a, 1924b, 1925, 1926) on the musculoskeletal system, brain, and skull of *Tupaia* and its relative *Ptilocercus,* the pen-tailed tree-shrew. Clark's elegant and eloquent contributions were instrumental in entrenching, for a half-century, the idea of the tree-shrews as the most primitive of the primates.

Within the last decade or so, however, the tide of opinion has turned firmly against the inclusion of the tree-shrews in Primates. The most exhaustive recent exploration of the problem is the volume edited by Luckett (1980), in which over a dozen authors approach the question of tree-shrew affinities from a variety of standpoints, both morphological and molecular. The unanimous judgment arrived at in this volume is that the tree-shrews are *not* primates; but what they actually are remains unclear. Most contributors would prefer to allocate them to their own order, Scandentia; and there is some support for a clade Archonta (first proposed by Gregory) that would include colugos, primates, tree- and possibly elephant-shrews, and bats. But Archonta is shaky, at best, and for the present the tree-shrews must remain scandentians, with judgment suspended on what this means in the context of wider relationships.

Nonetheless, even though it is clear that the tree-shrews are not primates, the possibility remains that in some respects these small mammals do display certain attributes that also characterized the ancestral primate; in other words, that in some very limited sense they may indeed serve as a "model" for early primates. This possibility is briefly examined below.

Morphology

In 1940 Evans was able to list 40 characters of the skull and postcranial skeleton that had been alleged to ally *Tupaia* with the lemuroid primates and to separate it from the elephant-shrews. Evans concluded that the characters involved in fact suggested that if *Tupaia* were to be aligned with the lemuroids, then so too should the macroscelideans be, since most of the supposedly lemuroid features of the tree-shrews were also shared with the latter. Subsequent reanalyses have been less noncommittal, and although it is impossible to discuss all of the characters involved here, I will briefly touch on a few of the most important.

Those characteristics of the skull to which most attention has been paid are those of the auditory and orbital regions. Perhaps the most striking feature of the tupaiid cranium (and that which, together with the relatively rounded braincase and reduced snout length, most clearly imparts to it whatever primate-like gestalt it possesses) is the presence of a complete bony postorbital bar (Fig. 1). But whereas this

is a character that tupaiids certainly hold in common with all living primates (and a variety of other mammals too, for that matter), it is not a character shared with the plesiadapiform primates of the Paleocene and was clearly convergently acquired among tree-shrews and "primates of modern aspect." Similarly, while the sutural pattern on the medial orbital wall of *Tupaia* is closely similar in certain respects to that seen in *Lemur* (but not in many other lemuroids), both Cartmill and McPhee (1980) and Novacek (1980) have produced cogent arguments for rejecting this as evidence of affinity.

Attention has been called to certain dental resemblances between tree-shrews and primates. However, the only one of these that approximates an apomorphy of the ancestral primate is a general lowering of molar cusp relief seen in *Ptilocercus*. Gregory (1910) was sufficiently impressed by this "omnivorous" modification to hazard that primate molars "perhaps primitively resembled those of the modern *Ptilocercus* in many characters"; but it is clear that primitively tupaiid molars possessed high, pointed cusps, probably accompanied by substantial stylar elaboration such as is seen, for instance, in *Tupaia*. *Ptilocercus* remains, in molar as in many other characters, uniquely apomorphic within the group. One other dental point may be worth mentioning. Tupaiids, like the strepsirhine primates, possess "dental combs" composed of elongated, procumbent anterior lower teeth. However, the tree-shrew comb consists of four teeth (usually interpreted as I_{1-2} bilaterally), while the standard (and primitive) condition among strepsirhines is six (usually interpreted as $I_{1-2} + \overline{C}$ bilaterally). I_3 in tupaiids is normally highly reduced, if sometimes procumbent, while in the strepsirhines the third (lateral) tooth of the comb is largest of all. And in any event, the dental comb is an apomorphy in the living strepsirhines, clearly absent in the ancestral primate.

The two supposedly shared characters of the auditory region that have weighed most heavily in the allocation of tree-shrews to Primates are the possession of a petrosal-derived bulla, and the manner in which the complexity of the internal carotid arterial system is reduced. Doubt was cast on the validity of such comparisons by Van Valen (1965), and the question has since been investigated in detail by Cartmill and

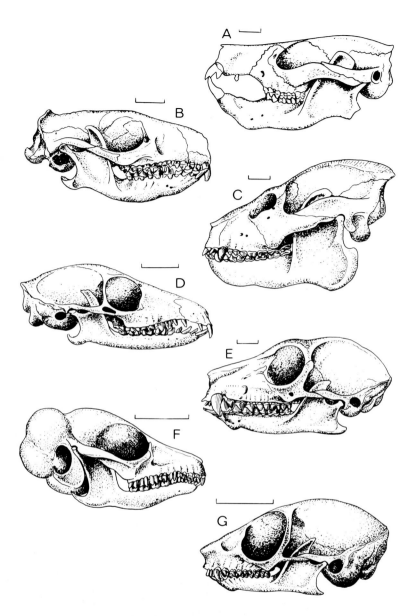

Fig. 1. Lateral view of the skull of *Tupaia glis* (D, Burma), compared with those of four primates: *Plesiadapis tricuspidens* (A, Paleocene of Europe); *Adapis parisiensis* (C, Eocene of Europe); *Lemur fulvus* (E, Madagascar); *Microcebus murinus* (G, Madagascar); and two insectivores *s.l.: Erinaceus europaeus* (B, Europe); and *Macroscelides proboscideus* (F, South Africa). Each scale represents 10 mm. Drawing by Nicholas Amorosi.

McPhee (1980). In the case of the bulla, Cartmill and McPhee have demonstrated that, whereas the bulla in lemuroid primates is formed entirely through the fusion of the rostral and caudal processes of the petrosal, the bulla in *Tupaia* derives from the coalescence of the caudal petrosal process with an independent entotympanic element; the two structures are thus nonhomologous. The same authors also provide a detailed description of carotid circulation in *Tupaia,* showing that it does not support primate affinities.

Clark (1924, 1934) leaned heavily on evidence from the brain of *Tupaia* in reinforcing his hypothesis of tree-shrew–primate relationships. Characters to which Clark drew attention included relatively large brain size; expansion and elaboration of the neopallium, with the development of a distinct temporal pole and the downward displacement of the rhinal sulcus; substantial elaboration of the nuclear elements of the thalamus; well-defined cellular lamination of the lateral geniculate nucleus, and, indeed, a pronounced elaboration of all the visual areas of the brain. More recent studies (Campbell 1980 and references therein), however, have shown that those ways in which the brain of *Tupaia* contrasts with that of insectivores are related virtually entirely to the possession by the former of a well-developed visual apparatus, and that such characters are to be found in all mammals that emphasize vision. Moreover, *Tupaia* shows none of the characters of the brain that so far have been found to distinguish the primate visual system from that of nonprimates.

The reproductive system has also been invoked in arguments for the primate status of the tree-shrews. However, neither the features of the external genitalia, nor details of the reproductive cycle, nor characters of the fetal membranes and placenta provide any argument for primate affinities (Martin 1968; Luckett 1974). Moreover, Martin (1966, 1968) showed that maternal care in *Tupaia* takes a unique form, one highly unlikely to represent anything that has ever occurred among primates. Infants are left in a nest constructed by the male parent, and are suckled by the mother only once per 48 hours.

Behavior and Ecology

The brief and highly incomplete notes above serve at least to indicate that, beyond being small-bodied eutherians, the tree-shrews show virtually nothing in their anatomy to suggest that they might usefully fill a role as a "model" for the ancestral primate condition. But it might legitimately be asked whether, the question of primate affinities aside, these small mammals might nonetheless fill an ecological role not dissimilar to that of the earliest primates. Here, of course, we enter a potential minefield, involving the vexed questions of the "insectivore–primate boundary" and the "ecological shift" at the origin of Primates (see discussion of these topics in Schwartz et al. 1978). But first, what do we know of the ecology of the tree-shrews?

Martin (1968) summarized what was then known of the ecology and habits of the tree-shrews, and in 1979 Kawamichi and Kawamichi reported on the first systematic field study of a tree-shrew (*Tupaia glis*). All tree-shrews except the crepuscular *Ptilocercus* are diurnal in habit. Some tupaiids are typically terrestrial; others are semiarboreal or arboreal. Kawamichi and

Kawamichi found *Tupaia glis* to be active mostly on the ground; 96% of sightings were on the ground or low in the trees, below 1.5 m. Individual home ranges were small, about 1 ha; the ranges of adult males and females overlapped completely, while those of adults of the same sex did not. This produced either "solitary-ranging [male–female] pairs" or single male–multiple female "harems," depending on the number of female ranges overlapped by a male; each adult attempted to exclude others of the same sex from its range. Activity was generally solitary, and scent marking was frequent among both sexes but more so among males.

Tupaiids seem generally to be omnivorous. The more arboreal species, however, appear to be quite highly insectivorous, while the most terrestrial forms may prey extensively on small terrestrial vertebrates and invertebrates, for which they dig and forage in the leaf litter on the ground. The semiarboreal forms, such as *Tupaia glis*, appear to feed largely on fruit, supplemented by insects and small vertebrates, for which they also forage. Martin (1968) reports that captive *Tupaia belangeri* exhibit hair and weight loss if deprived of adequate animal protein. He also cautions that the family Tupaiidae shows an extensive spectrum of adaptation both as to diet and habitat, but suggests that the more generalized semiarboreal forms probably most closely reflect the ecological niche primitive for the family.

Opinions differ widely as to the primitive primate ecology. Szalay (1968), for example, sees the origin of primates as lying in a shift from an insectivorous diet to one consisting of fruits and leaves. Certainly, the most characteristic feature of the earliest primates is the lowering and rounding of the cusps of their cheek teeth: a modification that indeed suggests a deemphasis of insectivory. Cartmill (1972), on the other hand, proposes that it was visually directed manual predation in the lower canopy and marginal growth of tropical forest that gave rise to the "primates of modern aspect." The main shortcoming of this scenario as it applies to the origin of primates in general is that it ignores the Paleocene plesiadapiforms, which were undoubtedly primates, but which lacked many of the characters of the primates of modern aspect that Cartmill's hypothesis was intended to explain.

In 1975, Cartmill provided a possible scenario to reconcile the two views, whereby early plesiadapiforms took to the trees to exploit fruit but retained a dietary interest in insects, in search of which they visited the forest floor; subsequently, some plesiadapiforms began to engage in visual arboreal predation, which in turn gave the impetus to the evolution of primates of modern aspect. R. W. Sussman (personal communication) proposes, more simply, that the early primates were essentially omnivores, among several groups of eutherians that from the beginning of the Tertiary began to exploit the wealth of new resources made available by the angiosperm (flowering) plants, and as yet largely untapped by mammals. This suggests a largely arboreal habitus, although possibly not an exclusive one.

Among living primates, the degree of insectivory exhibited is negatively correlated with body size (and even then, the tiny *Microcebus* eats under 50% insects); and it seems reasonable to conclude that the same applied early on in primate evolution. Just how large the ancestral primate was is a matter of conjecture, but if it ate significant quantities of fruit, as the dental evidence of its descendants suggests, it was probably relatively large by insectivore standards—plausibly in the tree-shrew range. As concerns social organization, it seems possible that the primitive pattern for mammals in general may not have been vastly different from that primitive for primates: male ranges overlapping those of females and excluding other males. If so, then the "solitary ranging pair" of *Tupaia glis* seems close to the pattern.

Amid all these speculations, how good an ecological approximation to the ancestral primate might the more generalized tree-shrews, such as *Tupaia*, be? To the extent (and no more) that these animals are opportunistic frugivores, and that they are clawed (like the plesiadapiforms), moderately small-bodied arboreal eutherians that live in solitary-ranging pairs, they may well provide a reasonable living model. But in many ways these putative resemblances represent retentions from a primitive eutherian condition, in which the ancestral primate also remained primitive. And in others, such as their diurnality (and hence possibly also in their elaborate visual apparatus), the tree-shrews depart from the presumed primitive primate pattern.

The Tree-Shrew Fossil Record

The tupaiid fossil record, as already noted, is poor. Since the early part of this century, several fossil genera have been touted as possible tupaiids, but virtually all pre-Miocene forms have by now been rejected from consideration as tree-shrews. Jacobs (1980), Chopra et al. (1979), and Chopra and Vasishat (1979) have described various cranial fragments and dentitions from the Miocene of the Pakistani and Indian Siwaliks, including an anterior cranium made the type of a new genus and species, *Palaeotupaia sivalicus,* by Chopra and Vasishat. The Indian material has been assigned to the subfamily Tupaiinae (to which all the extant genera but *Ptilocercus* belong), and all seems to be on the order of 10 million years old. Jacobs (1980) believes that a facial fragment he describes is primitive for Tupaiidae in having a robust snout and large teeth; and in noting certain resemblances to both the relatively primitive tupaiine *Dendrogale* and to *Ptilocercus,* he suggests that not only the modern tree-shrew genera, but also the two subfamilies, differentiated within the last 10 million years. More food for thought for those who would embrace the living model concept.

Literature

Campbell, C. B. G. 1980. The nervous system of the Tupaiidae: its bearing on phylogenetic relationships, pp. 219–242. In: Luckett, W. P. (ed.), Comparative biology and relationships of tree shrews. New York: Plenum.

Carlsson, A. 1922. Uber die Tupaiidae und ihre Beziehungen zu den Insectivora und den Prosimiae. Acta Zool. 3:227–270.

Cartmill, M. 1972. Arboreal adaptations and the origin of the order Primates, pp. 97–122. In: Tuttle, R. (ed.), The functional and evolutionary biology of primates. Chicago: Aldine-Atherton.

Cartmill, M. 1975. Primate origins, pp. 1–39. Minneapolis: Burgess.

Cartmill, M., McPhee, R. D. E. 1980. Tupaiid affinities: the evidence of the carotid arteries and cranial skeleton, pp. 95–132. In: Luckett, W. P. (ed.), Comparative biology and relationships of tree shrews. New York: Plenum.

Chopra, S. R. K., Vasishat, R. N. 1979. Sivalik fossil tree shrew from Haritalyangar, India. Nature 281:214–215.

Chopra, S. R. K., Kaul, S., Vasishat, R. N. 1979. Miocene tree shrews from the Indian Sivaliks. Nature 281:213–214.

Clark, W. E. LeGros. 1924a. The myology of the tree-shrew (*Tupaia minor*). Proc. Zool. Soc. London 1924:461–497.

Clark, W. E. LeGros. 1924b. On the brain of the tree-shrew, *Tupaia minor*. Proc. Zool. Soc. London 1924:1053–1074.

Clark, W. E. LeGros. 1925. On the skull of *Tupaia*. Proc. Zool. Soc. London 1925:559–567.

Clark, W. E. LeGros. 1926. On the anatomy of the pen-tailed tree-shrew (*Ptilocercus lowii*). Proc. Zool. Soc. Lond. 1926:1179–1309.

Clark, W. E. LeGros. 1934. The early forerunners of man, pp. 1–296. London: Bailliere, Tyndall, and Cox.

Evans, F. G. 1940. The osteology and relationships of the elephant shrews (Macroscelididae). Bull. Amer. Mus. Nat. Hist. 80:85–125.

Gregory, W. K. 1910. The orders of mammals. Bull. Amer. Mus. Nat. Hist. 27:1–524.

Jacobs, L. L. 1980. Siwalik fossil tree shrews, pp. 205–216. In: Luckett, W. P. (ed.), comparative biology and relationships of tree shrews. New York: Plenum.

Kawamichi, T., Kawamichi, M. 1979. Spatial organization and territory of tree shrews (*Tupaia glis*). Anim. Behav. 27:381–393.

Luckett, W. P. 1974. The comparative development and evolution of the placenta in primates. Contrib. Primat. 3:142–234.

Luckett, W. P. 1980. Comparative biology and relationships of tree shrews, pp. 1–314. Plenum: New York.

Martin, R. D. 1966. Tree shrews: unique reproductive mechanism of systematic importance. Science 152:1402–1404.

Martin, R. D. 1968. Reproduction and ontogeny in tree-shrews (*Tupaia belangeri*), with reference to their general behavior and taxonomic relationships. Zeitschr. Tierpsychol. 25(4):409–495; (5):505–532.

Novacek, M. J. 1980. Cranioskeletal features in tupaiids and selected Eutheria as phlogenetic evidence, pp. 35–93. In: Luckett, W. P. (ed.), Comparative biology and relationships of tree shrews. New York: Plenum.

Schwartz, J. H., Tattersall, I., Eldredge, N. 1978. Phylogeny and classification of the primates revisited. Yrbk. Phys. Anthrop. 21:95–133.

Szalay, F. S. 1968. The beginnings of primates. Evolution 22:19–36.

Van Valen, L. 1965. Treeshrews, primates, and fossils. Evolution 19:137–151.

4
What is a Tarsier?

Jeffrey H. Schwartz

Department of Anthropology, University of Pittsburgh, Pittsburgh, PA 15260

Introduction and Overview

The living tarsier is represented by three species—*Tarsius spectrum, T. bancanus,* and *T. syrichta*—distributed throughout the islands of southeast Asia. Historically, *Tarsius* has been considered a primate that is somehow intermediate between the lower lemurs and lorises and the higher anthropoids. There have been two major recent views of the broader relationships of *Tarsius* (and, by association, fossil taxa) within Primates: (1) following Pocock (1918), extant tarsiiforms and Anthropoidea are considered sister-taxa (the clade Haplorhini) primarily because the nostrils are aborally rounded, without a slit (Hofer 1980); and (2) tarsiiforms are most closely related to the generally Paleocene "archaic" plesiadapiform primates because of common possession of an enlarged anterior tooth (only in the upper jaw of *Tarsius*); this group is the sister of all other primates (Gingerich 1975; Schwartz 1978a).

With regard to Haplorhini, Luckett (1976) has argued that various similarities in amnion formation and hemochorial placentation are synapomorphies uniting *Tarsius* and Anthropoidea and that there would be virtual identity in placentation and fetal membrane development were it not for the fact that *Tarsius* possesses the common eutherian bicornuate uterus rather than the simplex uterus of anthropoids. Schwartz (1978a) has, however, suggested that

these data do not unite a haplorhine clade since blastocyst attachment and details of the developmental sequences of placentation and amniogenesis of *Tarsius* and anthropoids are dissimilar. Szalay (1975a) has emphasized a posteromedial position of the carotid foramen as a haplorhine synapomorphy; although this configuration does apparently characterize anthropoids, the internal carotid artery penetrates *Tarsius'* auditory bulla centrally (Fig. 1) (Schwartz et al. 1978). Hershkovitz (1974) and, more recently, Cartmill (1981) have concluded that postorbital closure in anthropoids (which is accomplished primarily from expansion of the malar in concert with the alisphenoid, maxilla, and frontal) could be derived from that seen in *Tarsius,* wherein there is some expansion of the malar anteriorly as well as of the alisphenoid posteriorly. Others, however, have found specimens that provide a different description: "The malar is laterally displaced from the side of the cranium and maintains a broad contact with the correspondingly laterally flared frontal [and] the frontal [also] grows downward to overlap the malar posteriorly" (Schwartz et al. 1978). Perhaps the variance of opinion here reflects variation within the taxon.

Although *Tarsius* is unique among extant primates in the "formation of the tubus olfactorius," Starck (1975:151) concluded that, because its highly specialized skull contrasts markedly with those of strepsirhines, *Tarsius* is more closely related to anthropoids; similarities, such as reduction of the nasal fossa, are due in *Tarsius* to the enlarged orbits impinging upon this area while, in anthropoids, they are due to an absolute reduction in size. Other characters that have been cited as uniting *Tarsius* and Anthropoidea include: (1) a perbullar course of the internal carotid artery, development of an anterior accessory cavity of the middle ear, and (perhaps?) the prenatal loss of a functional stapedial artery (Cartmill and Kay 1978; Cartmill et al. 1981); and (2) the possession of a retinal fovea (Cartmill 1981). To this we might add syncheilism, but it is not certain that anthropoid syncheilism and that of *Tarsius* are homologous (Hofer 1980). *Tarsius* ("the classical case for haplorhinism") is not, however, typically haplorhine, but, rather, strepsirhine and platyrrhine nasally. Thus, the notion of Strepsirhini versus Haplorhini is not, in the strict sense, valid (Hofer 1980).

The results of molecular and biochemical studies are contradictory with regard to tarsier affinities: Some support the haplorhine clade hypothesis (e.g., Baba et al. 1975), while others suggest that strepsirhines and anthropoids are closely related and that *Tarsius*, *Tupaia*, and *Cynocephalus* diverged from the last common ancestor of all (e.g., Sarich and Cronin 1976). Whatever its affinities, it is noteworthy that *Tarsius* has the highest chromosomal diploid number (2N = 80, with 14 submetacentrics and 66 acrocentrics) of any primate (Egozque 1974). The most recent contribution by Baba et al. (1982) admits that molecular data generate three alternative hypotheses of tarsier affinities: (1) *Tarsius* + Anthropoidea, (2) *Tarsius* + Strepsirhini, (3) *Tarsius* + (Strepsirhini + Anthropoidea). They also admit that the hypothesis that they ultimately favored (*Tarsius* + Anthropoidea) is dependent on morphological criteria. If a morphologically based phylogeny is necessary in order to select the "best" molecularly based phylogeny, it would seem that the molecular data are not at present sufficiently sensitive to generate robust hypotheses or capable of falsifying competing hypotheses.

In contrast to the more morphological studies in support of a *Tarsius*–Anthropoidea sister relationship, the older hypothesis of Prosimii (Strepsirhini + *Tarsius*) versus Anthropoidea was based on little more than a vague notion that *Tarsius*, although intermediate between lower and higher primates, was not much more advanced than lemurs and lorises. That *Tarsius* occupied an intermediate phylogenetic position leading to the evolution of higher primates gained support from seeming similarities: (1) a reduced snout (but see above), which, with the large eyes, creates a marmoset-like face—and marmosets were considered primitive anthropoids; (2) possession of a form of partial postorbital closure; and (3) a habitual posture on vertical supports that is in line with a scala naturae that saw a trend in increasing tendencies toward erect posture and habitual bipedalism. The idea that the order Primates cannot be defined by discrete characters but only by perceived trends of increasing morphological and behavioral complexity has received strong support (e.g., Napier and Napier 1970; Simpson 1945).

Yet, amid the competing hypotheses that relegate *Tarsius* to a relationship closer to or farther away from Anthropoidea, but which similarly seek anthropoid descent from something tarsier-like, there remain characters that are perceived as unique among Primates and others that pose problems of parallelism. For example, the hind foot and especially the calcaneus of *Tarsius* are extraordinarily elongate; the tibia and fibula are fused; the digits of the pes and manus bear large terminal pads; scale-like areolae reminiscent of the scales of edentates and reptiles are present in adult *Tarsius spectrum* and have been observed during the ontogeny of *T. bancanus;* the head can rotate 180°; and, as mentioned above, there are uniquenesses in the development of the tubus olfactorius, the orbit, and the lateral wall of the cranium (see Clark 1962; Niemitz 1979; Schwartz et al. 1978; Starck 1975 and references therein). There is also the possibility that *Tarsius* possesses a compound auditory bulla (of petrosal and entotympanic elements), whereas all other extant and most fossil primates are supposed to develop a totally petrosal bulla (McKenna 1966; Schwartz 1978b; Starck 1975). (Presley [1982] has, however, introduced serious doubt as to the viability of the hypothesis that primates are united by common possession of a totally petrosal auditory bulla.) And the dentition of *Tarsius* is notably different from that of fossil and extant strepsirhines and especially anthropoids, as is its diet.

Although the main part of a tarsier's diet consists of insects (beetles, grasshoppers, cockroaches, butterflies, moths, praying mantises, phasmids, cicadas, and, sometimes, ants—even though some of these prey may retaliate), tarsiers, given individual variation, also eat a variety of birds (sometimes even larger than the tarsier itself), bats, shrimp, fish, and, quite unexpectedly, snakes, including neurotoxic species (Niemitz 1979 and references therein). Indeed, in its diet as well as its "noiseless locomotion" and "ambush-type predation by moving about at nighttime above ground level," *Tarsius* seems to fill the econiche of a small owl (Niemitz 1979:641). *Tarsius* dispatches its prey with its eyes closed and kills it with powerful bites inflicted by its pointed antemolar teeth.

The dentition of *Tarsius* (Fig. 1) lacks anything that, in contrast to the upper and lower teeth of anthropoids and the uppers of most strepsirhines, can be described as incisiform. The upper anteriormost tooth of *Tarsius* is the tallest in the upper jaw and is ringed by a band of enamel at its base; behind this tooth sit three single-cusped teeth with small posterior heels, bearing cingula inferiorly, and then two more premolariform antemolar teeth. The three upper molars are somewhat transverse and morphologically simple, bearing compressed and U-shaped protocristae connecting the trigon cusps, and lingual cingula that are swollen a bit in the hypocone region, where a tiny cusp may also be present. A weak prehypocone crista adorns M^{1-2}. The five lower antemolar teeth mirror their upper counterparts in morphological complexity as well as in size: The last two antemolar teeth are the bulkiest and most premolariform and are preceded by a smaller, simpler tooth bearing a shallow, vertical talonid basin and a tiny heel, in front of which is a larger, but simple, tooth. The lower molars bear distinct, subcentrally positioned and somewhat anteriorly displaced paraconids that open the trigonid lingually. On M_{1-2}, the buccolingually broad talonid basin is ringed by a complete, compressed cresting system that incorporates the entoconid and hypoconid and, with the cristid obliqua, terminates at the base of the protoconid. M_3 tapers distally to an elongate, somewhat buccally emplaced heel that is incorporated into the talonid crests.

Although the dental formula commonly given for *Tarsius* is 2.1.3.3/1.1.3.3, the morphologies of this animal's teeth have led to another interpretation: *Tarsius* lacks at least the incisor tooth class. In an earlier series of papers (see references in Schwartz 1980), I suggested that correlations in tooth morphology and postbudding sequences of growth and eruption indicated that the trenchant upper anteriormost tooth of *Tarsius* was a canine and the rest of its dentition was composed of molar-class teeth. New embryological data (Luckett and Maier 1982) provide better evidence for the interpretation that all of *Tarsius'* teeth belong to the molar class (Schwartz, 1983), and this seems to accord better with the morphology.

The Fossil Record

Szalay (1976) provides the most recent review of the history of investigations on presumed fossil tarsioids, to which the reader is referred for more detail than space here allows. Suffice it to say that consensus can hardly characterize opinion on most of the taxa that have been studied. What may be one systematist's tarsioid is another's (or later the same systematist's) lemuroid or lorisoid (e.g., compare Gazin 1958; Gregory 1922; Robinson 1968; Simpson 1940, 1955; Simons 1961a, 1961b; Szalay 1976). An incredible array of supposed tarsioids was created when Simpson (1940) grouped them in the family Anaptomorphidae. Inclusion of the subfamilies Paromomyinae, Omomyinae, Anaptomorphinae, Necrolemurinae, and Pseudolorisinae was based on "a balance of dental resemblances" (Simpson 1940:197) and because "most of the genera . . . [were] . . . considered to be tarsioids principally because of dental resemblances or linking by apparent annectant types to the three genera *known* to be tarsioids" (Simpson 1940:205) (emphasis mine). These three "known" tarsioids—*Necrolemur, Pseudoloris,* and *Tetonius*—were and are, however, considered to be *Tarsius*-like (and thus *Tarsius*-related), primarily because of general similarities in the shape of the skull and auditory region (*Necrolemur*) and in their having V-shaped dental arcades (Gregory 1922; Simons 1961a; Simons and Russell 1960). But even after

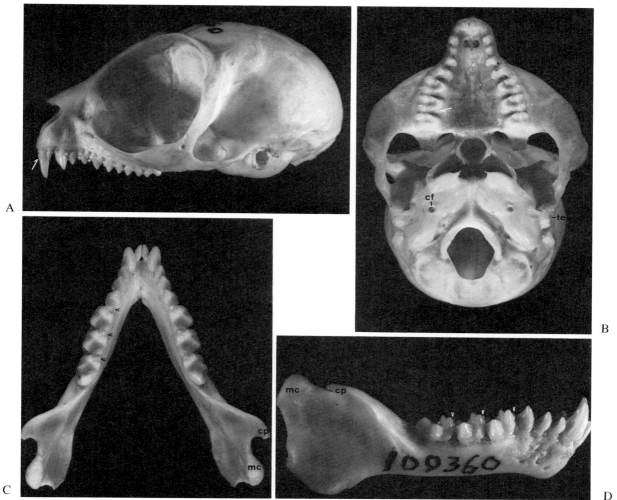

Fig. 1. Tarsius sp. (Amer. Mus. Nat. Hist. 109360).
(A) Cranium, lateral view (× 2.2); arrow points to
cingulum that is present at least buccally on all upper
teeth; note orbital flare and the extent to which the
enlarged orbit impinges upon the face and nasal cap-
sule. (B) Cranium, basal view (× 2.1); arrow points
to the prehypocone crista on the right M^2 (this structure is also faintly developed on the first molar); note
the central position of the carotid foramen (cf) and
the "tubular ectotympanic" (te). (C) Mandible, occlusal view (× 2.9); V's point to the distinct paraconids on M_{1-3}; note how laterally compressed the
mandibular condyle (mc) is and how the coronoid
process (cp) is flaring laterally. (D) Mandible, right
lateral view (× 3.4); V's point to the paraconids on
M_{1-3} and the arrow to cingulid found to some extent
on all lower teeth; note the angularity of the articular
components of the mandibular condyle (mc) and how
the coronoid process (cp) does not rise above the
level of the condyle.

stressing this "general" similarity, these authors (and Simpson 1940, as well) point to specific features (e.g., orbital size, postorbital closure, mastoid pneumatization, ectotympanic
enlargement, development of the hypotympanic
sinus, fewer teeth in the lower jaw [*Necrolemur, Pseudoloris*], large lower anterior tooth) in
which the fossils and *Tarsius* differ significantly

from each other. Each scholar has also felt compelled to comment on how the extinct taxa compare well, or even better, in various features with either "lemuroids" (Simpson's [1940] "lemuroids" is basically equivalent to Strepsirhini) in general or, more specifically, *Galago*. For example, Gregory (1922) pointed out that, although the then known skulls of fossil "tarsioids" (*Tetonius, Necrolemur*) looked generally similar to those of *Tarsius* and *Galago* in rostral elongation, the fossils were more similar to *Galago* in developing petromastoid inflation.

In this regard it is of special interest to note that all calcanei, astragali (tali), and navicular bones attributed to these taxa as well as to dentally presumed fossil tarsioids are most similar to those of *Galago* and/or *Microcebus* in specific details of elongation and articular facet configuration (S. M. Ford, personal communication; Savage and Waters 1978; Schlosser 1907; Schmid 1979; Simpson 1940; Szalay 1975b, 1976; Teilhard de Chardin 1927; Weigelt 1933). These postcranials also mirror in specific detail the known foot bones of Miocene lorisoids (Szalay 1976; Walker 1970). For example, the astragalus of presumed tarsioids displays a long, narrow trochlear facet for articulation with the tibia, a feature that is characteristic of *Galago* (S. M. Ford, personal communication); this facet in *Tarsius* is broad and shorter (Gingerich 1981) and thus represents a condition that appears to be primitive for mammals (Novacek 1980). Although fusion of tibia and fibula distally characterizes postcranials attributed to *Necrolemur* and *Nannopithex* as well as those of *Tarsius*—and this shared similarity has been important to the hypothesis of the existence of fossil tarsioids—a long-overlooked fact is that such fusion is also found in *Galago* (Howell 1944). With further regard to postcranial morphology, it is of interest to note that *Tarsius* does not display the cuboidocalcaneal facets and the "well-developed socket for the pivot" of the cuboidocalcaneal articulation characteristic of primates, including, for example, *Hemiacodon, Teilhardina* and *?Tetonius* (Szalay 1976:401; Szalay and Decker 1974). Effectively, Simpson's (1940:196) comment on the postcranials attributed to *Hemiacodon* pertains to all postcranials thought to represent fossil tarsioids: "If these bones were judged entirely

on their own merits . . . probably no one would refer them definitely to the Tarsioidea."

Although the known auditory bullae of fossil "tarsioids" differ from that of *Tarsius* in certain details (e.g., entry of the carotid artery, lack of hypertrophied hypotympanic sinus, degree of internalization of the tympanic ring [subtympanic extension of the tympanic cavity]), the one character that has been taken as "proof" of tarsioidness is the presence in these fossils and the extant taxon of a tubular ectotympanic that extends beyond the lateral margin of the bulla from an "intrabullar" tympanic ring (Gregory 1922; Simons 1961a; Szalay 1975a). There are, however, differences among taxa in how far laterally the tubular ectotympanic extends. The cheirogaleid *Allocebus* and lorisids (Cartmill 1975), as well as *Plesiadapis* and *Phenacolemur* (Russell 1964; Szalay 1972), also have, or had, tubular ectotympanics. And, in the degree of internalization of the tympanic ring by subtympanic extension of the tympanic cavity, *Plesiadapis* closely resembles *Necrolemur* (Gingerich 1975).

In spite of the fact that it possesses a large lower anterior tooth and one less antemolar tooth, *Pseudoloris* has long been considered dentally most similar to *Tarsius* (Simons 1961a; Simpson 1940; Teilhard de Chardin 1916–1921). Recently, however, Szalay (1975b) pointed out that *Pseudoloris* differs from *Tarsius* in its possession of a well-developed hypocone. In addition, *Pseudoloris* lacks the cingular thickening around the base of the protocone anterolingually that characterizes *Tarsius*. While *Tarsius* possesses a distinct paraconid M_{1-3} that remains separated from the metaconid on M_3, the lower molars of *Pseudoloris* bear stout and broad paracristids that subtend the metaconid and protoconid inferiorly. The mandible of *Pseudoloris* is not thinned out below the molar series as is that of *Tarsius* (beneath M_3 the mandibular corpus is barely thicker than the tooth is tall) (Fig. 1). The mandibular condyle of the fossil is not narrow and anteroposteriorly elongate, the coronoid process is not low and flared laterally, the ascending ramus is quite elevated, and the goneal region is hooked posteriorly, not squared up. M^{1-2} of *Pseudoloris* do, however, bear a faint prehypocone crista, and the bone of the upper jaw is distended downward at its anterior margin. These features are developed in

Tarsius, as well, and have also recently been suggested as among those characters that unite cheirogaleids, galagids, and lorisids (Schwartz and Tattersall, in preparation).

Anchomomys gaillardi has also been thought of as being similar dentally to *Tarsius* (Gregory 1922; Stehlin 1916). Both taxa are small and have relatively simple upper molars and elongate, posteriorly tapering M_3's, but the similarity ends there. The cusps of *Anchomomys* are lower, M^3 is smaller than M^{1-2}, the molar protocone regions are swollen and do not bear any cingulum, and the lower molars lack paraconids. In the configuration of its upper molars, but, more specifically, in the disposition of the paracristid (which courses down the face of the protoconid and then flexes sharply to proceed back to and then up the face of the metaconid), *A. gaillardi* can be argued to be united with the cheirogaleid–galagid–lorisid clade and is probably best interpreted as a primitive, cheirogaleid-like member of that group (Schwartz and Tattersall 1983).

Another candidate for matching *Tarsius* in dental morphology has been *Omomys* (M. C. McKenna, personal communication; Schwartz et al. 1978). In its upper molars, *O. carteri* does approach *Tarsius* in general shape and relative tooth size within the molar series, and both taxa do develop a faint prehypocone crista on M^{1-2}. However, the cusps of the fossil taxon are less acute, the paracone and metacone are further apart and not connected by a sharp, well-developed centrocrista, the protocone is lower and broader, the protocristae are more broadly divergent, and the pre- and postcingula are confluent around the base of the protocone. Furthermore, the lower premolars are more elongate and the posterior premolars are more distended anteriorly; and although M_1 bears a distinct (though sometimes ledgelike) paraconid, the anterior disposition of this cusp differs markedly from the configuration in *Tarsius*: In the series M_{1-3}, the paraconid becomes markedly reduced in size, shifts medially, and becomes increasingly incorporated in the paracristid. In general, however, lower molar cusps are more bulbous and crests are less crisp in the fossil form. *Omomys minutus* (known only from M_{1-3}) maintains a paraconid distinct from the metaconid on its molars, but on M_{2-3}, these two cusps are not as far apart as they are in *Tarsius*.

Although the fossil's M_3 bears a long, narrow heel, the talonid cusps, as they are on M_2 (M_1 is broken), are bulbous and fairly isolated, not compressed and incorporated into a continuous crest that contains the talonid basin. M_{1-2} bear a small but identifiable hypoconulid that, as preserved on M_2, forms a relatively deep notch with the entoconid; this latter character is, I think, potentially of significance. The tips of the protoconid and metaconid are rounded, project markedly upward, and are not compressed or incorporated into a high crest between the cusps. *Omomys lloydi* is equally distinct from *Tarsius* in its lower molars; upper molars are not at present known.

Surprisingly, and contrary to received wisdom, there are no fossils so far known that mirror *Tarsius* dentally, cranially, or postcranially in more than a few features. One could continue citing taxa that appear at first glance to be similar to *Tarsius*, but the truth is that the dissimilarities are overwhelming.

If Not a Fossil Tarsioid—What?

It is a curious historical fact that virtually all (with the notable exception of Szalay [1976]) who have worked on fossil tarsioids, especially anaptomorphids/omomyids, have commented on how morphologically disparate were many of the taxa included in this group. Thus precise, workable family-level diagnoses have not been forthcoming. For example, taxa may or may not have a protocone fold on their upper molars; they may or may not enlarge the last premolar; they may have a large, procumbent anterior tooth or two small, orthally implanted anterior teeth. Gazin (1958) and, more briefly, Robinson (1968) argued that some semblance of order within the assemblage would be achieved if a group represented by *Anaptomorphus, Tetonius, Absarokius*, and others was separated, leaving *Omomys, Washakius, Hemiacodon*, and the rest. The former group could be united by such characters as a protocone fold on at least M^1; reduced upper and lower third molars; enlarged last premolars; and increased melding of the paraconid with the metaconid from M_{1-3}. Gazin (1958) allocated these taxa to the family Anaptomorphidae. The other taxa, subsumed in Omomyidae, were by default left as a group and thus remained a much more disorderly morphologic assemblage. In apparent frustration, Robinson (1968) concluded that the reason all these taxa had been grouped together and considered tarsioids was because they were small and from Palearctic deposits!

With the exception of being typically small, the teeth of most fossil "tarsioids" are in general dissimilar to those of galagids for the same reasons they are similar to the teeth of *Tarsius*: i.e., because of primitive retentions (e.g., a paraconid on M_1). Galagids are united with the larger lorisoid clade by the lack of paraconids and the presence of a distinct, anterobuccally broad paracristid that turns severely at the base of the protoconid to proceed posteriorly to the metaconid (Schwartz and Tattersall, in preparation). Among presumed fossil tarsioids, *Pseudoloris* alone can also be characterized by these features. Indeed, *Pseudoloris* also displays a faint prehypocone crista on at least M^1; this is another synapomorphy of lorisoids. If the cranial and postcranial morphologies discussed above for *Pseudoloris* are taken into consideration along with the dental ones, there appears to be more than a superficial case for suggesting that this fossil taxon is in some way related to lorisoids. Since similarities in cranial shape have also been pointed to among *Pseudoloris*, galagids, and *Tarsius*, and *Tarsius* also develops a prehypocone crista on M^{1-2}, perhaps this is not an either/or situation, but rather a reflection of synapomorphy among these taxa. Thus, for example, orbital enlargement and frontation, reduced interorbital septum, and rostral elongation characterize *Tarsius*, *Pseudoloris* and many lorisoids because of inheritance from a common ancestor. Calcaneonavicular elongation (perhaps to the extent seen in *Galago* and *Microcebus*) would also unite these taxa, with *Tarsius* being autapomorphic in the extraordinary expression of this feature (lorisids and *G. crassicaudatus*, however, would have secondarily shortened these bones). In the shape of the astragalar trochlear facet, however, *Tarsius* remains primitive relative to lorisoids; even the shorter tarsaled *G. crassicaudatus* possesses the derived (i.e., narrower, longer) trochlear facet.

If these suggested synapomorphies reflect reality, then the skulls of *Tetonius* and *Necrolemur* and the postcranials of these two genera and *Nannopithex*, *Microchoerus*, *Teilhardina*, *Arapahovius*, and *Hemiacodon* can be accommodated by a hypothesis of relatedness to lorisoids. Interestingly, the lower molars attributed to these taxa as well as to *Pseudoloris* are characterized by tall protoconids and metaconids that are broadly melded at their bases and connected by a crest; these cusps form a steep wall that faces upon the talonid. The lower posterior premolars appear bulky, bearing rather truncated posterior heels (thus appearing compressed anteroposteriorly), and are thus somewhat ovoid in cross section inferiorly. These features describe *Tarsius* as well and have been suggested as being among the dental synapomorphies that unite lorisoids (Schwartz and Tattersall, in preparation). These dental features also typify most of those taxa commonly interpreted as fossil tarsioids. Fossil tarsioids are, however, more derived in lower molar morphology than is *Tarsius* in that *Tarsius* displays a distinct paraconid that remains apart from the metaconid on M_{1-3}, whereas fossil taxa shift the paraconid inward (e.g., *Nannopithex*), meld it with the metaconid on M_{2-3} (e.g., *Absarokius*), or do not develop it on M_{2-3} (e.g., *Necrolemur*) or even on M_{1-3} (e.g., *Pseudoloris*).

A Rough Phylogenetic Hypothesis: I

If *Tarsius*, (at least some) anaptomorphids/omomyids, microchoerines, and lorisoids constitute a monophyletic group, *Tarsius* (by virtue of, for example, the configuration of the astragalar trochlear facet, the possession of two pedal grooming claws, and the presence of distinct, well-developed, separate paraconids on M_{2-3}) is logically interpreted as the sister-taxon of the rest of the clade. In many of its features (e.g., mode of placentation, hypertrophied hypotympanic sinus, orbital enlargement, partial postorbital closure, constitution of its dentition, and extreme tarsal elongation), *Tarsius* is, however, quite uniquely derived. And, since it seems logical to unite as a monophyletic group all toothcombed primates, *Tarsius* is also autapomorphic in losing these lower anterior teeth—which, as independently and separately discussed above, is indicated by developmental data.

The possibility that *Tarsius* and its commonly presumed fossil allies are nested within a clade that is itself united by modifications of the lower anterior dentition (toothcomb) leads to an unexpected corroboration of a recent suggestion: The enlarged lower anterior tooth characteristic of microchoerines and many anaptomorphids/omomyids, which is morphologically comparable with the lateral tooth of a strepsirhine's toothcomb, is, as reflected in this morphological congruity, homologous with the lateral tooth of a strepsirhine toothcomb (Schwartz 1980)—whatever the identity of that lateral tooth may be! Thus, (some) anaptomorphids/omomyids and microchoerines would be united by lack of possession (development) of the slender, subparallel-sided teeth that are subtended by the larger, more robust, laterally flaring and margo-cristid-bearing lateralmost teeth of the strepsirhine toothcomb. There seem, therefore, to be three variants of a strepsirhine toothcomb, the differences lying in the number (four or two) or absence of the slender central teeth. Although use in feeding would be expected, these different toothcombs have also been found to be used similarly in grooming (Rose et al. 1981; Schmid 1983 and personal communication).

A review of all large (or apparently large) anterior-toothed anaptomorphids/omomyids and microchoerines reveals that some can be distinguished by the development on at least M_1 of a cristid obliqua that courses to meet the metaconid and, when upper molars are also known, the presence on at least M^1 of a distinct protocone fold. These taxa include *Utahia,* "*Uintanius vespertinus,*" *Anemorhysis, Chlororhysis* (referred), some specimens referred to *Uintanius ameghini, Altanius*(?), *Trogolemur, Arapahovius, Gazinius, Strigorhysis, Hemiacodon, Nannopithex, Necrolemur,* and *Microchoerus.* In some specimens referred to *Tetonius* and *Absarokius,* the cristid obliqua terminates just prior to meeting the metaconid. Since a centrally terminating cristid obliqua characterizes all other anaptomorphids/omomyids, the majority of lorisoids, as well as *Pseudoloris* and *Tarsius,* and the upper molars of these taxa are unadorned by a protocone fold, taxa listed above are clearly united by the possession of an M^1 protocone fold and a lingually directed M_1

cristid obliqua. (With regard to a protocone fold, it would seem logical to conclude that it is a more developed, and thus derived, expression of the prehypocone crista that otherwise unites the larger clade.) The monophyly of these primates is further suggested by the common possession of a distinctive depression on the last lower premolar that descends buccally from the centrally emplaced cristid obliqua, which, in the more prevalent condition, subtends only the lingually oriented talonid basin. A subclade within this group appears to be constituted by *Necrolemur, Microchoerus,* and *Hemiacodon* (cristid obliqua-to-metaconid on M_2, some enamel crenulation), within which *Necrolemur* and *Microchoerus* (e.g., quadrate M^{1-2}, more enamel crenulation) are most closely united. The relationships of the remaining anaptomorphids/omomyids (basically, *Omomys, Ourayia, Macrotarsius,* and *Mytonius*) are less decipherable, but their common possession of broadly parabolic protocristae and a more medially emplaced paraconid in the series M_{1-3} may reflect the unit of these taxa (Krishtalka and Schwartz 1978).

Problems: I

In the above, I have not mentioned *Loveina, Anaptomorphus, Shoshonius, Chumashius, Washakius,* or (type) *Chlororhysis,* taxa that have been regarded as "good" anaptomorphids/omomyids. Unlike the taxa I have discussed, these six do not possess a large lower anterior tooth (or, at least, preserve a large lower anterior alveolus that would presumably have borne such a tooth). Rather, these taxa preserve two small alveoli (or parts of roots) at the front of the jaw, behind which is a larger alveolus and then, when preserved, recognizably premolariform antemolar teeth.

Until the morphologies of the missing anterior teeth are better known, whatever is said about these taxa must be recognized as hyperhypothetical. But on the basis of the relative sizes of lower anterior teeth, *Loveina, Anaptomorphus, Shoshonius, Chumashius, Washakius* (teeth known), and (type) *Chlororhysis* are not cladistically anaptomorphid/omomyid. I do not wish to engage here in an attempt to unveil the identities of the large versus the small anterior teeth, but the present state-of-the-art in dental development and developmental theory (see reviews by Osborn 1978; Schwartz 1982, 1983) prevents me from lumping all of these tiny North American Eocene taxa together and then noting the resultant variability in anterior tooth size (and, when known, shape). Instead, it seems more logical to suggest that *Loveina*, *Anaptomorphus*, *Shoshonius*, *Chumashius*, *Washakius*, and (type) *Chlororhysis* possess (or retain) the condition characteristic of "adapids" and most anthropoids: i.e., two anterior teeth (identified as incisors) that are smaller than the next tooth back in the jaw (commonly identified in all as the canine).

I will be the first to admit that the removal of these taxa from Anaptomorphidae/Omomyidae is worrying, because *Loveina*, *Anaptomorphus*, *Shoshonius*, *Washakius*, and (type) *Chlororhysis* have a cristid obliqua that meets the metaconid on M_1 (in *Loveina* and *Shoshonius*, this crest bifurcates as it does in *Hemiacodon*); *Loveina*, *Anaptomorphus*, *Shoshonius*, and *Washakius* develop a protocone fold on M^1; and the last lower premolar of *Loveina* and *Anaptomorphus* bears a buccal depression. (The type of *Chumashius* is a lower jaw fragment lacking the molars.) On the other hand, *Pelycodus* (*Cantius*) possesses (or retains) two small lower anterior teeth that are followed by a much larger tooth, and this taxon has long been recognized by its M^1 protocone fold and an M_1 cristid obliqua that courses to meet the metaconid. *Loveina*, *Anaptomorphus*, and *Pelycodus* are even further similar in the transverseness of M^{1-2} and in the width buccolingually and the squareness lingually of M^2. Additionally, *Pelycodus* is characterized by lack of M_{1-2} talonid cusp distinctiveness and complete cristid enclosure of the talonid basin; these features are also present in *Loveina* and *Anaptomorphus*. These taxa appear to be sisters.

The affinities of *Washakius* and possibly *Shoshonius* may lie with yet another "adapid." *Smilodectes*, which also has two small anterior lower teeth followed by a much larger tooth, is distinguished in its lower molars by: (1) cristids obliquae on M_{1-2} that join the metaconid; (2) a medially directed cristid obliqua on M_3; (3) a "notch" on M_{1-2} formed between the entoconid and the inferiorly descending and thickening posthypocristid that terminates just behind the cusp; and (4) thickened paracristids that become lower in the series M_{1-3}. The squared upper molars of *Smilodectes* bear mesostyles and protocone folds that course to a cingular hypocone, and M^{1-2} are subequal in size, with M^3 not much smaller.

The lower molars of *Shoshonius* are accurately described by features 1 and 2, and those of *Washakius* by 1, 2, and 4; in some specimens of *Washakius*, there may be a small notch formed by a small "hypoconulid" and the entoconid. The upper molars of *Washakius* and *Shoshonius* are (primitively) more transverse, but M^{1-2} bear a protocone fold, M^2 is not greatly enlarged relative to M^1, and M^3 is not disproportionately small; the opposite of the latter two features characterizes most anaptomorphids/omomyids. In addition, M^{1-3} of *Shoshonius* and M^1 of *Washakius* bear mesostyles.

I have reserved comment on *Teilhardina* because of difficulties that exist in delineating the anterior lower teeth, at least their relative sizes. Depending on what one interprets as representing what remains of the anteriormost part of the lower jaw and the alveolar septa, one can conclude that (at least the relative sizes of) the anterior lower dentition of *Teilhardina* was like that of *Pelycodus*, or of a typical anaptomorphid/omomyid, or even of *Tarsius*. What complicates matters further is that the postcranials attributed to *Teilhardina* are similar to those of various anaptomorphids/omomyids, microchoerines, *Pseudoloris*, and galagids; and although M^{1-2} are extraordinarily transverse, narrow mesiodistally, and lack hypocones, M^1, at least, bears what would be identified as a prehypocone crista. This crest as well as the detail of tarsal elongation certainly suggests that *Teilhardina* may be tied to the lorisoid–fossil "tarsioid" clade at large, but further speculation seems at present unwarranted.

A Rougher Phylogenetic Hypothesis: II

If the majority of anaptomorphids/omomyids, the microchoerines, and *Pseudoloris* are united by developing a large lower anterior tooth, and if some of these taxa are further distinguished as a subclade by the possession of a protocone fold on at least M^1 and an M_1 cristid obliqua that courses to meet the metaconid, and if, further, these shared similarities do indeed reflect phylogeny, then one must entertain the hypothesis (Gingerich 1975; Schwartz 1980) that plesiadapiforms are phylogenetically related in some way to this group. The "some way" lies not just in details of the large lower anterior tooth (Schwartz 1980), but in the M_1^1 characters cited. It is thus interesting to note that Gazin (1958) suggested a possible association between *Uintanius ameghini* and carpolestids because of details of upper and lower premolars as well.

The plesiadapiforms in which the molar configurations I cited are unarguably present are Plesiadapidae, Carpolestidae, and Paromomyidae. Picrodontids develop what can be interpreted as a protocone fold in a very displaced position on all upper molariform teeth. The tooth that is traditionally identified as the picrodontid M_2, but which might in reality be the M_1 (Schwartz and Krishtalka 1977), has the cristid obliqua coursing to the metaconid; the boat-shaped "M_1," with its diminutive trigonid and elongate talonid, has its cristid obliqua buccally emplaced. I believe these observations contribute further corroboration of the hypothesis that the picrodontid "M_2" is homologous with the M_1 of other plesiadapiforms; a corollary homology is implied for the upper molars. Microsyopids do not develop an upper molar protocone fold or a lingually directed cristid obliqua, thereby making them at best the sister-taxon of all other plesiadapiforms.

Problems: II

A series of problems arise from the above. (1) Microsyopids are dentally primate (see Schwartz et al. 1978) but because, within Primates, they are either primitive morphologically or autapomorphic, the only seemingly viable hypothesis is that they are cladistically plesiadapiform. (2) The inclusion of plesiadapiforms within Strepsirhini, much less within Anaptomorphidae/Omomyidae, leads to the conclusion that such characters as tarsal elongation and the development of a postorbital bar, grasping hands and feet, and nails on some digits, evolved independently many times (if *Plesiadapis* and, to a less well-known extent,

Cynodontomys, are reflective of the otherwise dental plesiadapiforms), and that the development of intrabullar carotid tubes and (possibly) a petrosal or petrosal/entotympanic bulla occurred in parallel in even more instances. (3) Multiple and numerous dental, cranial, and postcranial parallelisms would also arise if a plesiadapiform–"tarsiiform" clade was posited as the sister-taxon of Strepsirhini or of a strepsirhine–anthropoid clade; for the most part, these would be the same parallelisms that result from a hypothesis of Prosimii or of Haplorhini.

Conclusion

Prior to this undertaking, I was a passive adherent of the commonly held belief that *Tarsius* was a living fossil. Since I cannot identify any so far known taxon that matches *Tarsius* in more than a few characters—and none resembles *Tarsius* in its autapomorphies—I must reject the living fossil hypothesis. And, given the preceding pages, I am forced to conclude that, at present, there are no identifiable fossil tarsioids. In short, the "living fossil" has no fossil record! And, rather than being overly primitive (as would be expected of a living fossil), *Tarsius*, in, for example, its diet, social behavior (pair-bonding), as well as in many of its morphologies, is outstandingly apomorphic.

What has become clearer to me is that choosing among alternative hypotheses because of a "principle of parsimony" that is based on a "minimum number of parallelisms" may be less productive, if not less accurate, than favoring a hypothesis that may generate a greater number of parallelisms, but which is also based on a greater and more robustly diverse number of synapomorphies. But, having taken more time and space than originally intended, I leave my suggestions and loose ends in their present states of development and/or resolution. Others can deal with *Rooneyia* and *Ekgmowechashala*, and all are invited to pick away at the parallelisms. I would, however, hope that detailed analyses and rigorous debate will not be compromised by the convenience of conventional wisdom.

Acknowledgments: I thank J. Eaton (*Washakius*) and R. Stucky (*Loveina* and *Shoshonius*)

48   J. H. Schwartz

for information on recently collected specimens, and M. C. McKenna and G. Musser (The American Museum of Natural History) and M. R. Dawson (Carnegie Museum of Natural History) for access to specimens in their charge, and S. Daley for photography. S. M. Ford kindly provided unpublished data on postcranial morphology, as did P. Schmid on "grooming" in microchoerines. I. Tattersall graciously took the time to calmly offer helpful criticism.

Literature

Baba, M., Goodman, M., Dene, H., Moore, G. W. 1975. Origins of the Ceboidea viewed from an immunological perspective. J. Hum. Evol. 4:89–102.

Baba, M., Weiss, M. L., Goodman, M., Czelusniak, J. 1982. The case of tarsier hemoglobin. Syst. Zool. 31:156–165.

Cartmill, M. 1975. Strepsirhine basicranial structures and the affinities of the Cheirogaleidae, pp. 313–354. In: Luckett, W. P., Szalay, F. S. (eds.), Phylogeny of the primates. New York: Plenum.

Cartmill, M. 1981. Morphology, function, and evolution of the anthropoid postorbital septum, pp. 243–274. In: Ciochon, R. L., Chiarelli, A. B. (eds.), Evolutionary biology of the New World monkeys and continental drift. New York: Plenum.

Cartmill, M., Kay, R. F. 1978. Cranio-dental morphology, tarsier affinities and primate suborders, pp. 205–214. In: (Chivers, D. J., Joysey, K. A. (eds.), Recent advances in primatology, Vol. 3. London: Academic.

Cartmill, M., MacPhee, R. D. E., Simons, E. L. 1981. Anatomy of the temporal bone in early anthropoids, with remarks on the problem of anthropoid origins. Amer. J. Phys. Anthrop. 56:3–21.

Clark, W. E. LeGros. 1962. The antecedents of man, 2nd edn. Edinburgh: Edinburgh U. Press.

Egozcue, J. 1974. Chromosomal evolution in primates, pp. 857–864. In: Martin, R. D., Doyle, G. A., Walker, A. C. (eds.), Pittsburgh: U. Pittsburgh Press.

Gazin, C. L. 1958. A review of the middle and upper Eocene primates of North America. Smithsonian Misc. Coll. 136:1–112.

Gingerich, P. D. 1975. Systematic position of Plesiadapis. Nature 253:111–113.

Gingerich, P. D. 1981. Early Cenozoic Omomyidae and the evolutionary history of tarsiiform primates. J. Hum. Evol. 10:345–374.

Gregory, W. K. 1922. The origin and evolution of the human dentition. Baltimore: Williams and Wilkins.

Hershkovitz, P. 1974. The ectotympanic bone and origin of higher primates. Folia Primatol. 22:237–242.

Hofer, H. O. 1980. The external anatomy of the oronasal region of primates. Z. Morph. Anthrop. 71:233–249.

Howell, A. B. 1944. Speed in animals. New York: Hafner.

Kay, R. F. 1981. Platyrrhine origins: a reappraisal of the dental evidence, pp. 159–188. In: Ciochon, R. L., Chiarelli, A. B. (eds.), Evolutionary biology of the New World monkeys and continental drift. New York: Plenum.

Krishtalka, L., Schwartz, J. H. 1978. Phylogenetic relationships of plesiadapiform–tarsiiform primates. Ann. Carnegie Mus. 47:515–540.

Luckett, W. P. 1976. Cladistic relationships among primate higher categories. Folia Primatol. 25:245–276.

Luckett, W. P., Maier, W. 1982. Development of deciduous and permanent dentition in Tarsius and its phylogenetic significance. Folia Primatol. 37:1–36.

McKenna, M. C. 1966. Paleontology and the origin of the Primates. Folia Primatol. 4:1–25.

Napier, J. R., Napier, P. H. 1970. A handbook of living primates, 3rd edn. London: Academic.

Niemitz, C. 1979. Outline of the behavior of Tarsius bancanus, pp. 631–660. In: Doyle, G. A., Martin, R. D. (eds.), The study of prosimian behavior. New York: Academic.

Novacek, M. J. 1980. Cranioskeletal features in tupaiids and selected Eutheria as phylogenetic evidence, pp. 35–93. In: Luckett, W. P. (ed.), Comparative biology and evolutionary relationships of tree shrews. New York: Plenum.

Osborn, J. W. 1978. Morphogenetic gradients: fields versus clones, pp. 171–201. In: Butler, P. M., Joysey, K. A. (eds.), Development, function, and evolution of teeth. New York: Academic.

Pocock, R. I. 1918. On the external characters of the lemurs and of Tarsius. Proc. Zool. Soc. London 1918:19–53.

Presley, R. 1982. Review of auditory regions of primates and eutherian insectivores (by R. D. E. MacPhee). Internatl. J. Primatol. 3:509–513.

Robinson, P. 1968. The paleontology and geology of the Badwater Creek area, central Wyoming. Part 4. Late Eocene primates from Badwater, Wyoming, with a discussion of material from Utah. Ann. Carnegie Mus. 39:307–326.

Rose, K. D., Walker, A., Jacobs, L. L. 1981. Function of the mandibular tooth comb in living and extinct mammals. Nature 289:583–585.

Russell, D. E. 1964. Les mammifères Paleocènes d'Europe. Mem. Mus. Natl. Hist. Nat., n.s. 13:1–324.

Sarich, V. M., Cronin, J. E. 1976. Molecular systematics of the Primates, pp. 141–170. In: Goodman, M., Tashian, R. E. (eds.), Molecular anthropology. New York: Plenum.

Savage, D. E., and Waters, B. T. 1978. A new omomyid primate from the Wasatch Formation of southern Wyoming. Folia Primatol. 30:1–29.

Schlosser, M. 1907. Beitrag zur Osteologie und systematischen Stellung der Gattung *Necrolemur*, sowie zur Stammesgeschichte der Primaten überhaupt. Neues Jb. Miner. Geol. Palänt. Festband 1907:197–226.

Schmid, P. 1979. Evidence of microchoerine evolution from Dielsdorf (Zürich Region, Switzerland)—a preliminary report. Folia Primatol. 31:301–311.

Schmid, P. 1982. The front dentition of the Omomyiformes (Primates). Folia Primatol. 40:1–10.

Schwartz, J. H. 1978a. If *Tarsius* is not a prosimian, is it a haplorhine?, pp. 195–202. In: Chivers, D. J., Joysey, K. A. (eds.), Recent advances in primatology, Vol. 3. London: Academic.

Schwartz, J. H. 1978b. Dental development, homologies and primate phylogeny. Evol. Theory 4:1–32.

Schwartz, J. H. 1980. A discussion of homology with reference to primates. Amer. J. Phys. Anthrop. 52:463–480.

Schwartz, J. H. 1982. Morphological approach to heterodonty and homology, pp. 123–144. In: Kurtén, B. (ed.), Teeth: form, function, and evolution. New York: Columbia U. Press.

Schwartz, J. H. 1983. Premaxillary-maxillary suture asymmetry in a juvenile *Gorilla:* implications for understanding dentofacial growth and development. Folia Primatol. 40:69–82.

Schwartz, J. H., Krishtalka, L. 1977. Revision of Picrodontidae (Primates, Plesiadapiformes): dental homologies and relationships. Ann. Carnegie Mus. 46:55–70.

Schwartz, J. H., Tattersall, I. 1983. A review of the European genus *Anchomomys* and some allied forms. Anthrop. Pap. Amer. Mus. Nat. Hist. 57:344–352.

Schwartz, J. H., Tattersall, I., Eldredge, N. 1978. Phylogeny and classification of the primates revisited. Yrbk. Phys. Anthrop. 21:95–133.

Simons, E. L. 1961a. Notes on Eocene tarsioids and a revision of some Necrolemurinae. Bull. Brit. Mus. (Nat. Hist.) Geol. 5:45–69.

Simons, E. L. 1961b. The dentition of *Ourayia*: its bearing on relationships of omomyid prosimians. Postilla 54:1–29.

Simons, E. L., Russell, D. E. 1960. Notes on the cranial anatomy of *Necrolemur*. Breviora (Mus. Comp. Zool.) no. 127:1–14.

Simpson, G. G. 1940. Studies on the earliest primates. Bull. Amer. Mus. Nat. Hist. 77:185–212.

Simpson, G. G. 1945. The principles of classification and a classification of the mammals. Bull. Amer. Mus. Nat. Hist. 85:1–350.

Simpson, G. G. 1955. The Phenacolemuridae, new family of early primates. Bull. Amer. Mus. Nat. Hist. 105:411–442.

Starck, D. 1975. The development of the chondrocranium in primates, pp. 127–155. In: Luckett, W. P., Szalay, F. S. (eds.), Phylogeny of the primates. New York: Plenum.

Stehlin, H. G. 1916. Die Säugetiere des schweizerischen Eocaens. Kritischer Katalog der Materialien, 7. Theil, Zweite Hälfte, *Caenopithecus-Necrolemur*. Abh. Schweiz. Pal. Ges. 41:1297–1552.

Szalay, F. S. 1972. Cranial morphology of the early Tertiary *Phenacolemur* and its bearing on primate phylogeny. Amer. J. Phys. Anthrop. 36:59–76.

Szalay, F. S. 1975a. Phylogeny of primate higher taxa: the basicranial evidence, pp. 91–125. In: Luckett, W. P., Szalay, F. S. (eds.), Phylogeny of the primates. New York: Plenum.

Szalay, F. S. 1975b. Phylogeny, adaptations, and dispersal of the tarsiiform primates, pp. 357–404. In: Luckett, W. P., Szalay, F. S. (eds.), Phylogeny of the primates. New York: Plenum.

Szalay, F. S. 1976. Systematics of the Omomyidae (Tarsiiformes, Primates): taxonomy, phylogeny, and adaptations. Bull. Amer. Mus. Nat. Hist. 156:157–450.

Szalay, F. S., Decker, R. L. 1974. Origins, evolution, and function of the pes in the Eocene Adapidae (Lemuriformes, Primates), pp. 239–259. In: Jenkins, Jr., F. A. (ed.), Primate locomotion. New York: Academic.

Teilhard de Chardin, P. 1916–1921. Sur quelques primates des Phosphorites du Quercy. Ann. Paleont. 10:1–20.

Teilhard de Chardin, P. 1927. Les mammifères de l'Eocene inferieur de la Belgique. Mem. Mus. Roy. Hist. Nat. Belg. 36:1–33.

Walker, A. C. 1970. Post-cranial remains of the Miocene Lorisidae of East Africa. Amer. J. Phys. Anthrop. 33:249–262.

Weigelt, J. 1933. Neue Primaten aus der mitteleozänen (oberlutetischen) Braunkohle des Geiseltals. Nova Acta Leop. Carol. Halle n.f. 1:97–156, 321–323.

5
Are There Any Anthropoid Primate Living Fossils?

Eric Delson and Alfred L. Rosenberger

Department of Anthropology, Lehman College, CUNY, Bronx, NY 10468, and Department of Vertebrate Paleontology, American Museum of Natural History, New York, NY 10024
Department of Anthropology, University of Illinois, Chicago, IL 60630

Introduction

The concept of "living fossil" as employed by Simpson (1953) and others has been somewhat altered for the purposes of this volume. As we understand it, the implication is of a living taxon that differs only slightly if at all in known morphology from an early fossil member of its clade, at whatever taxonomic rank. In this spirit, we will examine the "higher" or anthropoid primates to determine if any taxa, including some previously suggested, qualify for this status. Following Szalay and Delson (1979), the Order Primates is divided into three suborders, the extinct Plesiadapiformes, the Strepsirhini (lower primates), and the Haplorhini, including the infraorders Tarsiiformes (tarsiers and fossil relatives—see Schwartz, this volume), Platyrrhini (New World anthropoids), and Catarrhini (Old World anthropoids). The formal taxon Anthropoidea was not recognized in order to reduce the number of ranks allowed, but it may be considered a hyporder (between suborder and infraorder; see Delson 1977) including Platyrrhini and Catarrhini. The nomen Simiiformes Hoffstetter 1974, may be substituted for Anthropoidea Mivart 1864, if desired, to avoid confusion with earlier contrasts between Anthropoidea and "Prosimii", a grade term including all non-anthropoids. We shall briefly review the evolutionary history of each anthropoid infraorder here, searching for taxa that may fall under the expanded living fossils rubric and commenting on any that have been implied as such previously; the contrasting patterns of evolution in these two groups will then be analyzed. Unless otherwise indicated, background material and references for this chapter may be found in Szalay and Delson (1979).

Evolution and Living Fossils Among the Catarrhini

The evolutionary record of the Old World anthropoids reveals a pattern of temporal replacement of one successful, radiating group by a distant "cousin." Briefly, the Oligocene parapithecids were more numerous and more diverse taxonomically than their relatively derived pliopithecid contemporaries (*Propliopithecus*, including *Aegyptopithecus*, species), but the former do not appear to have left any descendants or close relatives. During the Early and Middle Miocene, pliopithecids occurred rarely alongside a third radiation, the early Hominidae (Dryopithecinae or Proconsulinae) especially in Africa. By the Middle Miocene, two new groups arose that, for the first time, represent close relatives of living taxa: the modern-ape-like *Sivapithecus* group and the Cercopithecidae (Old World monkeys). One cercopithecid tooth is known from the Early

50

Miocene, and there are suggestions of links to some Oligocene taxa, but no morphology amenable to analysis is known until after 16 million years ago.

The *Sivapithecus* group is currently the subject of much debate (see papers in Ciochon and Corruccini 1983), and both its position in ape phylogeny and a meaningful family-group nomen within Hominidae are still uncertain. This group of species appeared in the later Middle Miocene, diversified in the earlier Late Miocene, and then essentially disappeared; it is clearly related to the Asian orangutan (Ponginae) but perhaps also to the African ape-human lineage (Homininae) of the Pliocene and Pleistocene. The later Late Miocene saw the diversification of the Cercopithecidae (and the brief flowering of the cercopithecoid *Oreopithecus*), with a peak in generic diversity probably occurring in the Late Pliocene. The sequence of replacement therefore might be seen as: Parapithecidae, Pliopithecidae, Dryopithecinae, *Sivapithecus* group, and Cercopithecidae, with Homininae coexisting with the latter and eventually dominating all surviving forms ecologically. The living Hylobatidae (gibbons) has essentially no fossil record, other than rare Pleistocene teeth, although it has often been incorrectly linked to the pliopithecids. This pattern of taxic succession (linked to ecological adaptations by Andrews 1981 and Ripley 1979) leads to a picture of mainly short-lived modern groups mostly separated from their closest fossil relatives. There are three taxa, however, that merit closer examination in terms of this paper: *Hylobates*, *Pongo*, and *Macaca*.

The Gibbons

Hylobates has been linked to Cenozoic fossils as old as the Oligocene by Simons (1965), among others. The genus *Aeolopithecus* is now seen to be merely a species of the common Fayum taxon *Propliopithecus* (Szalay and Delson 1979; Kay et al. 1981), and its putative gibbonlike features have been refuted as better material appeared. The Miocene *Pliopithecus* (and *Dendropithecus*) have been linked to *Hylobates* on the bases of shared conservative features of the dentition and face, along with postcranial characters that are now seen to depend

mainly on the relative gracility of long bones that do not show the extreme elongation typical of living gibbons (Simons and Fleagle 1973). Thus, there are no significant fossils known that are cladistically (as opposed to merely phenetically) linked to the hylobatids. In light of this review, then, *Hylobates* (only known genus of the Hylobatidae) cannot be readily termed a "living fossil," even under the broad definition of this volume.

On the other hand, can one argue that the gibbons are conservative enough of ancestral catarrhine morphologies in general to be treated as a "living morphotype"? Their facial architecture has long been known as retentive of various features deduced for the catarrhine morphotype (see Vogel 1966; Delson and Andrews 1975), but it also presents such derived characters as the protruding circumorbital rims and shallow mandibular corpus. Dentally, *Hylobates* species are also relatively conservative compared to the inferred common ancestor of eucatarrhines (nonparapithecids), with many similarities to pliopithecids. Their derived features, however, include some lengthening of upper cheek teeth and cingulum reduction, loss of protoconule, reduction of M_3 length, and especially near loss of canine sexual dimorphism. The cerebral contours and sulcal pattern of gibbons are probably the most conservative of living catarrhines, although the relative brain size is larger than in cercopithecids. The diploid chromosome number and presence of ischial callosities also conform to reconstructions of the ancestral eucatarrhine conditions. On the other hand, below the neck, so to speak, *Hylobates* is one of the most derived catarrhines. It shares the hominoid synapomorphies of shoulder, elbow, and thorax morphology, presents a slightly different wrist articulation (with a lunula between ulna and carpus), and is highly derived in its strongly elongated limbs (especially the antebrachium) and other adaptations to ricochetal brachiation (such as relative elongation of metacarpals and nonhallucal phalanges). Thus, although *Hylobates* does retain a number of specific features nearly unchanged from the Oligocene eucatarrhine ancestor, it shows derived features even within the same systems and a highly derived locomotor–behavioral complex that clearly removes it from consideration as a living morphotype.

The Orangutan

Until recently, it was widely assumed that the ancestry of the orangutan, *Pongo pygmaeus,* lay among the Asian "dryopithecines" of 13–8 million years ago, but without any known fossil showing especially close affinities. Numerous isolated teeth known mainly from Pleistocene deposits in China and Java were the only "real" orangutan fossils. In the last few years, however, more detailed studies by Pilbeam, Andrews, Wu, and others on older and newly found specimens of what is here called the *Sivapithecus* group showed that these materials shared detailed similarities with the derived orangutan facial architecture (Pilbeam 1982; Andrews and Cronin 1982; Wu et al. 1983). In addition, the Chinese specimens display a pattern of enamel wrinkling that may be similar to that of living orangutans. It is important to note, therefore, that the orangutan lineage extends back to at least 8 million years ago based on known derived facial morphology and to around 14 million years ago if the apparent identity of 9–14 million-year-old gnathic remains with those in 8 million year old crania imply identity of facial structure as well. No author has yet claimed or even implied that *Pongo* is a living fossil, nor do we do so at this stage in the discovery and study of the fossils, but the possibility should be seriously considered.

The Macaques

The genus *Macaca* is represented today by about 15 species ranging across southern and southeast Asia and one in North Africa. The Asian forms occur in a wide variety of local environments, both arboreally and terrestrially, with sympatry among members of several species groups, although the underlying morphology of the taxa is fairly uniform (Fooden 1982). In a wide variety of characters, macaques retain what Delson (1975a and later) inferred to be the ancestral condition for the subfamily Cercopithecinae, if not the family as a whole. These features include all dental structures except for the reduction of lingual lower incisor enamel that characterizes the tribe Papionini and perhaps the cercopithecine enlarged incisors overall. Cranially, *Macaca* preserves the facial morphology of a hypothesized ancestral cercopithecine, an inference supported in part by its similarity to the Pliocene *Parapapio* (Fig. 1), the potential ancestor of the living African papionins *Papio, Theropithecus,* and *Cercocebus,* which have divergently derived faces. The brain, postcranium, chromosomes, and other soft tissues of macaques are also essentially unchanged from the cercopithecine morphotype condition.

To be a living fossil, however, a modern taxon must match early fossil relatives. Macaque fossils in Asia are relatively few in number and fragmentary, although a Middle Pleistocene form may indicate a link between two modern subgroups (Delson 1980). The circum-Mediterranean record is more continuous, with a sampling of mandibular remains that cannot be specifically distinguished from the living North African *M. sylvanus* stretching back to nearly 5 million years ago (Delson 1980). In addition, the Late Miocene site of Marceau, Algeria, has yielded a collection of isolated teeth

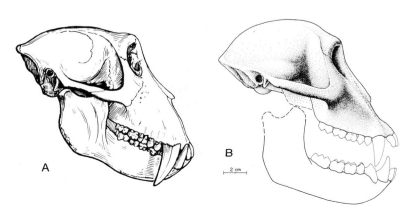

Fig. 1. Right lateral views of male skulls in Frankfurt orientation, at approximately same scale. (A) *Macaca nemestrina leonina,* living today in Indochina, after Gregory (1951, Fig. 23.44b, cut D3, courtesy of the American Museum of Natural History; (B) *Parapapio broomi,* middle to late Pliocene of southern Africa, from Szalay and Delson (1979; Fig. 169A, courtesy Academic Press). The facial profiles are nearly identical.

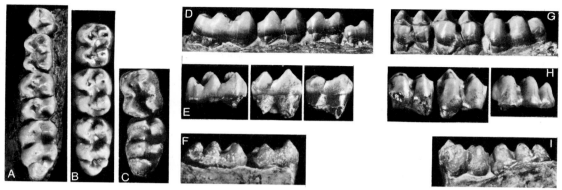

Fig. 2. Morphology of left lower molars (and P₄) of macaques and early cercopithecids. (A–C) occlusal views; (D–F) lingual; (G–I) buccal. (A, D, G) *Macaca sylvanus florentina,* early Pleistocene of Italy, P₄–M₃ from partial corpus (P₄ removed in D). (B, E, H) *?Macaca* sp., late Miocene of Algeria, isolated teeth aligned as M₁₋₃. (C, F, I) *"Victoriapithecus" leakeyi,* middle Miocene of Kenya, M₂₋₃ of partial corpus (scaled up to molar length equal to two younger fossils). Some retouching of the prints has covered portions of corpus and supporting clay, but not affected morphology, which is quite constant across this 13 million year span.

(Fig. 2) that are morphologically inseparable from those of modern macaques of comparable size. It is not clear that these specimens confirm the presence of *Macaca,* because they might also be assigned to the phenetically identical *Parapapio* (of the sub-Saharan Pliocene), but biogeographic indicators suggest that the Saharan region was already a barrier to migration at this date, thus enhancing a referral to *Macaca* (Delson 1975b; Thomas 1979; Thomas et al. 1982; but compare Geraads, 1982).

Finally, the earliest evidence for modern-type cercopithecids comes from the Middle Miocene (ca. 15 million years ago) site of Maboko, Kenya. Here are found two morphs that have been termed *Victoriapithecus macinnesi* and *"V." leakeyi.* The former appears to show some derived dental features in common with colobines but the latter is more "generalized" and thus cercopithecine-like. This second form is represented by only a few teeth and uncertainly referred limb bones, but a partial mandible (Fig. 2) and elbow fragments are hard to separate from those of modern macaque species. The overall indication is that in terms both of morphotype conservatism and phenetic similarity to ancient cercopithecines, *Macaca* may be termed a living fossil under our working definition. The earliest fossils of *Papio* are about 2.5 million years old and so close to modern forms that they are placed in the same species.

Nonetheless, the time scale involved is so much shorter than we would hardly consider the baboon a living fossil.

Evolution and Living Fossils Among the Platyrrhini

The fossil record of the platyrrhine primates is exceedingly meager compared with that of the catarrhines. Nonetheless, it hints at a broadly different pattern of diversification that, in some ways, seems confirmed by the taxonomic and morphologic composition of the surviving forms. The record opens in the early Oligocene with *Branisella,* whose affinities are not demonstrably near any of the other ceboid monkeys. Morphologically highly primitive (e.g., Hoffstetter 1980), *Branisella* possibly represents an early branch antedating the last common ancestor of all other fossil and living New World monkeys (Rosenberger 1981a). From the Late Oligocene onwards, however, the record reveals an intriguing number of examples that are surprisingly modern in appearance and assignable to extant clades. Thus platyrrhine evolution may not have unfolded in waves of successive adaptive radiations but rather as long-stemmed branches of persistent lineages (Rosenberger 1980). The impressive anatomical variety of the living genera may be a manifestation of this historical pattern.

The Marmosets

The living platyrrhines comprise some 15 or 16 genera (Szalay and Delson 1979; Rosenberger 1981b) and may be grouped into two families (Cebidae and Atelidae) each containing a pair of subfamilies. In contrast to many other taxonomic schemes (e.g., Napier and Napier 1967; Napier 1976; Hershkovitz 1977), our divisions are designed to conform to a cladistic hypothesis of the affinities of both living and extinct forms, and we think it very likely that all of the higher taxa that we recognize are monophyletic. Of these, the Callitrichinae (marmosets) have long been regarded as the most conservative (Hershkovitz 1977), hence the best candidates for preserving living fossils. However, many authorities have also argued that marmosets are a highly modified lineage. Independent analyses of numerous anatomical systems (reviewed in Rosenberger, in press), including the skull, dentition, aspects of the postcranium and reproductive system, which together define what may be termed the Marmoset Anatomical Complex, indicate that callitrichines are indeed a very derived assemblage. While the fossil record, given its scarcity and incompleteness, cannot be called upon to "prove" or "disprove" this notion, none of the craniodental characters of the complex appear among the Paleogene fossil platyrrhines, and none are elements of the euprimate morphotype (Szalay and Delson 1979). The evidence thus does not uphold Hershkovitz's theory that marmosets are a plesion, and we would predict that none of the living callitrichines will be found to be living fossils with any significant temporal dimension.

The Squirrel Monkey

The subfamily Cebinae, sister-group of the callitrichines, is represented in the fossil record by two genera, *Dolichocebus* of the Late Oligocene and *Neosaimiri* of the Middle Miocene. (Another species, *"Saimiri" bernensis* of the latest Pleistocene, probably is a third.) Of the two living genera, *Cebus* and *Saimiri,* the latter

seems to conserve some of the dental traits characterizing the platyrrhine morphotype, such as a relatively low hypocone, highly cuspate occlusal relief, and elevated trigonids. On the other hand, as with all other ceboid genera, *Saimiri* presents a suite of unique characters or mosaics that distinguishes it from all its living relatives.

The Middle Miocene *Neosaimiri* is known only by a nearly complete mandible lacking the rami and a few teeth. Although slightly larger than *Saimiri,* it differs from the living form solely in a few minor occlusal details, such as cuspal acuity, basin constriction, and cingular development (Fig. 3). Given the limited material, and by analogy with the diversity and taxonomy of the catarrhines, it is reasonable to suggest that *Neosaimiri fieldsi* be ranked as a subgenus of *Saimiri.* We do not take this step formally here, as full revision of the Miocene ceboids is under way by Rosenberger and Setoguchi, but we do wish to emphasize the continuity of morphology within this lineage.

In fact, this continuity may extend back even farther into the Middle Cenozoic. There are important similarities in the cranial anatomy of the Late Oligocene *Dolichocebus gaimanensis* and *Saimiri* (Fig. 4). At least two derived diagnostic features of the modern genus, a dolichocephalic neurocranium and a pattern of circumorbital traits including a very narrow interorbital pillar (Fleagle and Rosenberger, 1983), narrow and elongate nasals, a prolonged frontal process, reduction of the interorbital sinus, and a probably fenestrated interorbital septum (Rosenberger 1979) are present in both (but see Hershkovitz 1982). Other details of the neurocranium suggest additional *Saimiri*-like aspects in the masticatory apparatus and the soon to be described (Rosenberger and Mills, in preparation) middle-ear region. The natural endocast of *Dolichocebus* also suggests a frontal lobe that is relatively enlarged and positioned much as it is in *Saimiri* as well as a markedly creased Sylvian sulcus, perhaps a uniquely *Saimiri*-like feature. The evident differences in the configuration of the facial skull and in what can be discerned of the auditory bulla that set the fossil apart from the living *Saimiri* are easily transformable into a fully modern pattern. Thus we regard *Dolichocebus* not only as a close relative of the living

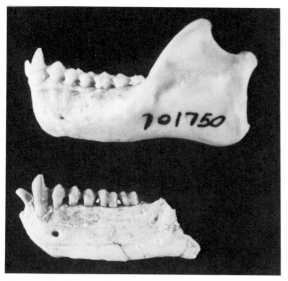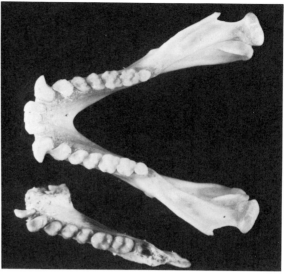

Fig. 3. Left lateral and occlusal views of *Saimiri sciureus* (complete jaw of living widespread Neotropical species) and *Neosaimiri* (=*Saimiri*?) *fieldsi,* Middle Miocene of Colombia.

Saimiri, but possibly its direct Oligocene ancestor; no autapomorphies are yet known that would preclude this hypothesis. In either case, *Saimiri* is an excellent example of an anthropoid living fossil and perhaps the sequence *Dolichocebus–Neosaimiri–Saimiri* represents one of the longest generic lineages among all primates.

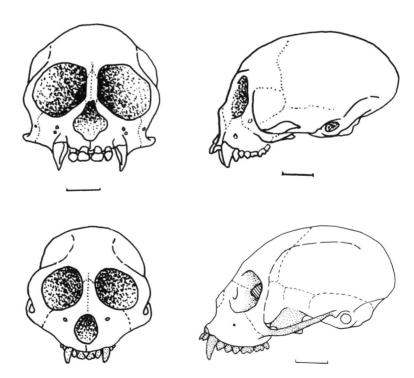

Fig. 4. Frontal (to left) and right lateral views of male *Saimiri sciureus* (above, living taxon) and ?female *Dolichocebus gaimanensis* (below, restored, Late Oligocene of Argentina). Scale bars represent 1 mm; reconstructed areas are uniformly stippled; the interorbital fenestra is hatched. Lower right figure after Rosenberger (1979).

The Howler Monkey

Among the atelids, both pitheciine and ateline subfamilies are represented in the Tertiary record. The atelines are known from the Miocene form *Stirtonia,* which very closely resembles the living *Alouatta* in dental anatomy (but see Setoguchi et al. 1982), although not in mandibular form. Significantly, it is the shape of the mandible and presumably correlated modifications of the skull that set the living genus apart from all other platyrrhines most strikingly—a complex of characters that many have argued are related to the elaboration of the vocalization mechanism. We take this to mean that the howler lineage may also be as ancient as the Middle Miocene, but the evidence is still too spotty to discern if its most obvious autapomorphies were then existent; certainly its predilection for a folivorous diet was, as judged by the dentition of *Stirtonia.*

The Saki-Uakaris

The pitheciines, which present some of the most unusual dental and gnathic specializations of all platyrrhines, are also represented in the record by several genera. One of these, *Cebu-*

pithecia, comes from the same Middle Miocene fauna as does *Stirtonia.* It has been likened to *Pithecia* (e.g., Stirton and Savage 1951), and one can draw the erroneous conclusion that *Pithecia* is thus a living fossil. However, a recent reconsideration of the holotype (Rosenberger and Mills, in preparation) indicates that the genus shows no positive derived features that are exclusively shared with *Pithecia;* it may be a sister-taxon of the entire saki-uakari (Pitheciini: Rosenberger 1981b) radiation.

The Owl Monkey

Another pitheciine comes from an earlier period. *Tremacebus harringtoni* occurs in Late Oligocene beds and bears a remarkable resemblance to the living owl monkey, *Aotus.* Two of the more startling apomorphies of *Aotus* are its enlarged orbits, reflecting its shift to a nocturnal/crepuscular activity cycle, and its greatly enlarged incisors (of uncertain adaptive explanation). The latter are exceedingly broad teeth, especially the upper central, and require a much broader premaxilla and anterior maxilla. *Tremacebus* is known by a fossil cranium that displays both of these osseous characteristics

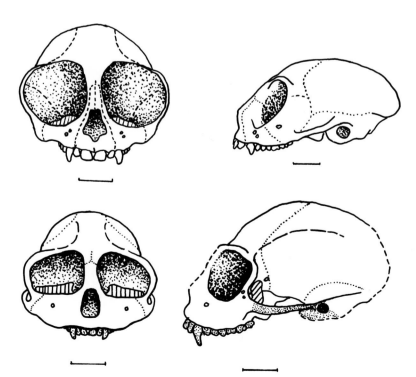

Fig. 5. Frontal (to left) and right lateral views of *Aotus trivirgatus* (above, modern, South America) and *Tremacebus harringtoni* (below, restored, late Oligocene of Argentina). Scale bars represent 1 mm; reconstructed areas are uniformly stippled or indicated by broken line; the orbitotemporal fenestra is hatched.

(Kraglievich 1951; Rosenberger 1980; see Fig. 5). Although the orbits are not quantitatively as enlarged as are those of some populations of *Aotus* (Fleagle and Rosenberger, 1983), both forms show the flaring orbital margin, enlarged secondary postorbital fenestra, capacious postorbital plate, and depressed orbital floor that are elements of the hypertrophic eyeball complex. Additionally, although anterior teeth are missing in the fossil, the anterior aspect of the palate is notably squared off, as in *Aotus,* despite the fact that the preserved canine roots are relatively small, implying large incisors. Thus it appears that at least two of the most important functional adaptations of *Aotus* are evident in *Tremacebus*. The few features visible on the badly damaged molar teeth of the fossil indicate that molar proportions would have been as predicted for the pitheciine morphotype and that crown morphology was much more primitive in *Tremacebus*—in having an offset hypocone, for example. Whatever else can be studied of the skull shows no important differences, however.

Therefore, *Aotus* becomes the second platyrrhine living fossil, paralleling *Saimiri* in its close morphological and phylogenetic linkage to a Late Oligocene fossil. The time depth of these two apparent lineages is truly astounding, especially by comparison to the more familiar catarrhine pattern of successive replacement. On the other hand, lest we be accused of making all platyrrhine fossils direct ancestors of living forms, it may be noted that the well-known Miocene *Homunculus* is a "primitive" pitheciine not readily linked more closely to any modern genus.

Adaptation, Ecology, and Time: A Comparison of the Two Patterns

Knowing full well that we are prone to grand error because the fossil record forever surprises us with new information, and especially because so little is available for Tertiary platyrrhines, we take this opportunity to explore some of the possible implications of our analysis and of the patterns of diversification evident among New and Old World anthropoids.

While none of the modern anthropoids can be unequivocally viewed as epitomizing arrested evolution, there seem to be indications that platyrrhines are overall more retentive of ancient Cenozoic morphologies than are the catarrhines. A combination of cladistic analysis and paleontology indicates that there have been two large-scale adaptive radiations among the catarrhines since the Late Oligocene, the hominids and then the cercopithecids. Modern subfamilies are not definitively represented until the Late Miocene. Neither line of evidence, in contrast, shows this pattern in the Neotropics: The only side lineage to the mainstream would have antedated the Late Oligocene, when two of the four modern subfamilies (if not generic lineages) first appear; a third dates at least to the Middle Miocene. Significantly, the cercopithecid and hominid radiations occurred in broadly distinct adaptive zones. Old World monkeys are probably a terrestrial diversion of the primitive arboreal way of life that apes retained, with only a few exceptions. No such ecological division occurred in the New World (Rosenberger 1980). Thus the earlier initiation of the monophyletic platyrrhine radiations would make it likely that should living fossils occur, they would be expected to be of more ancient origin in the New World than in the Old. Alternatively, or perhaps predictably, if Vrba's (1980, 1983) "effect hypothesis" of macroevolutionary trends has validity, the cercopithecids and perhaps other catarrhine groups may have been genetically more "disposed" toward producing numerous lineages continuously, while the platyrrhines radiated early and persisted.

A second factor devolves from these considerations. The highly successful cercopithecid radiation became ever more numerous in surviving taxa and also apparently reinvaded the arboreal milieu. This says as much for the severity of selective pressures in the changing Old World biosphere as it does of the competition between rather closely related primates. In both arboreal and terrestrial habitats, the cercopithecids tend to be more abundant than hominids in species and genera. This also implies that they may have outcompeted at least some of their ape contemporaries, leaving fewer possible living fossil survivors. While South America was also certainly subject to large climatic changes and faunal turnovers, the primates were apparently less affected, or were affected in other ways. No terrestrial sublineages appeared (so far as we know), despite the proliferation of savannah-like grasslands across the continent (Hershkovitz 1972). It is conceivable that some of the living marmosets may have diversified as a result of the opening of this new habitat, but on no level higher than the species or subgenus. Two points can be made: Competition between taxa occupying the same habitat would have been more intense, leading to more character divergence and finer niche partitioning; but no innovative higher morphological complexes emerged under novel selective pressures to enter into competition with established genetic potentials. Thus generic differences among the platyrrhines could have become marked over time without wholesale extinction eliminating large portions of the fauna. This makes for a greater opportunity to preserve living fossils.

A similar phenomenon would have resulted from the contrasting continental circumstances in the New and Old World. Essentially isolated throughout much of the Tertiary (Marshall et al. 1982; Patterson and Pascual 1972; Hershkovitz 1972), the South American primate fauna was self-contained and free from invasion by closely related forms. On the other hand, Europe, Asia, and Africa experienced intermittent contacts at various times since the Late Oligocene (Bernor, in press; Savage and Russell 1983), enabling faunas to mix and competitors to pressure taxa to transform or become extinct. The waves of extracontinental migrants may have severely affected the survivorships of early lineages especially. A possible test of both these hypotheses is offered by the several Latest Pleistocene Caribbean ceboids, all of which appear to have diverged quite strongly from their closest relatives despite rather short time spans involved (Rosenberger 1978; MacPhee and Woods 1982). This implies that once a novel ecozone became available, divergence occurred rapidly.

Another aspect may have influenced the diversification of cercopithecids in a restrictive sense. Although they have occupied much more continental land than the platyrrhines and invaded such contrasting ecological situations as the arboreal and terrestrial zones, they still exhibit less anatomical variety than do the platyrrhines. It may be suggested that this is because platyrrhines appeared earlier than cercopithecids. However, we think it is also significant

that Old World monkeys are possibly more canalized anatomically than the platyrrhines. For example, their bilophodont molar dentition manifests a surprising homogeneity in form, suggesting an all-purpose design irrespective of diet. Platyrrhines, in contrast, are highly diverse dentally (Rosenberger and Kinzey 1976; Hershkovitz 1977). Postcranially, cercopithecids are relatively uniform (e.g., Schultz 1970), whereas the platyrrhines display nearly all variations, except terrestriality, that the order Primates has produced (e.g., Erikson 1963). If true, this canalization might mean that cercopithecids are evolutionarily "interchangeable," producing short-lived taxa that may succumb to extinction if a competitor gains a relatively small adaptive advantage. This would again support the "effect hypothesis" interpretation noted above. On the other hand, wider adaptive differences separate platyrrhine generic lineages in which the potential for anagensis (slow phyletic evolution) is dominant so long as the essential ecological balance is not destroyed.

If we assume that extinction has more or less randomly influenced the survivorship of adaptive types in both the New and Old World, than what can explain the survivorship of living fossils in each group? *Macaca* represents an archetypal eurytope, or ecological generalist (see Eldredge 1979), whose species differ in minor ways (Fooden 1982) from a norm unchanged over millions of years. The African *Papio* has a much shorter known duration, although fossils nearly 3 million years old can be placed in the living species, and its degree of eurytopy is even greater as evidenced by its monotypy (Vrba 1980). *Saimiri* may have achieved success for much the same reasons. *Aotus,* however, has taken itself out of competition with close relatives by moving into an entirely different ecological realm, that of the night.

Summary

The anthropoid primates are not usually considered as candidates for the position of living fossils, which often implies great antiquity as well as a lineage that has shown morphological conservatism throughout its existence. Under the broadened definition of this volume, however, several taxa appear to qualify handily, being phenetically quite similar to relatively ancient cladistic relatives. Among the catarrhines, or Old World anthropoids, the hylobatids have often been suggested as tracing ancestry back to the Miocene or even Oligocene pliopithecids, but this concept has now been widely rejected; the many postcranial, behavior, facial, and dental apomorphies of gibbons far outweigh their several dental and cranial eucatarrhine symplesiomorphies, so that they cannot be considered as living morphotypes either. The orangutan lineage, on the other hand, can now be traced back at least 8 (if not nearly 15) million years to the *Sivapithecus* group of hominids; until the phenetic similarities have been analyzed in detail, we refrain from too readily considering the orangutan as a living fossil. Of all the catarrhines, only the cercopithecid genus *Macaca* appears to qualify for this role. It corresponds closely in dental, cranial, and postcranial details to the inferred morphotype of the cercopithecine or even cercopithecid ancestor of 10–15 million years ago. Moreover, the species *M. sylvanus* can be extended back to the beginning of the Pliocene on the basis of circum-Mediterranean gnathic (and partial postcranial) evidence, while the genus as a whole may be traced through Late Miocene North African teeth to approach the 15 million-year-old *"Victoriapithecus" leakeyi* of East Africa both dentally and in elbow morphology.

Among the New World platyrrhines, generic lineages are much more readily traced into the middle Cenozoic. Although the callitrichine cebids (marmosets) have been suggested by Hershkovitz as persistently primitive (essentially living fossils), they are in fact a highly autapomorphic group. The cebine *Saimiri,* on the other hand, is both relatively conservative dentally and so close to the Middle Miocene *Neosaimiri* as to bring their generic distinction into serious question. In addition, the Late Oligocene *Dolichocebus* presents a large number of specifically *Saimiri*-like features (several autapomorphic) in the skull, as well as the lesser known brain, suggesting a true continuation of the generic lineage over some 25 million years. A second clear case of a platyrrhine living fossil is the pitheciine atelid *Aotus.* The mosaic of cranial features related to this form's nocturnal adaptation are foreshadowed in *Tremacebus,* a contemporary of *Dolichocebus.*

The persistence of these two rather "specialized" lineages indicates unexpectedly early differentiation of the ceboids at fairly low taxonomic levels. The resultant implication of numerous other such lineages in the (now meager) fossil record is supported by the presence in the Middle Miocene of taxa rather similar to the living ateline *Alouatta* and the common pitheciin stock. Because these similarities seem less close than that seen between *Pongo* and *Sivapithecus,* although La Venta is comparable in age to the oldest sivapiths, the forms involved are not granted living fossil status.

Why was the pattern of differentiation so dissimilar in the New and Old World anthropoids? In the Old World, a set of sequentially replacing sister-taxa or collateral relatives characterized not only the family-group but also the generic history of the catarrhines. In South America, family-group and even generic lineages with unique specializations appeared early and persisted; apart from *Branisella,* only one early fossil genus, Early Miocene *Homunculus,* cannot be placed more closely than in a modern subfamily. The isolation of South America, as opposed to the freer intercontinental passage and competition in the Old World, more than anything else, appears to be at the root of the differences. In the New World, platyrrhines began to diverge earlier and were less directly affected by Miocene climatic shifts (no terrestriality) or intercontinental migration. Instead, they emphasized anagenesis except when offered wholly new geographic zones, as in the Caribbean, where two novel generic lineages are known from Latest Pleistocene (Mid-Holocene) fossils.

In contrast, the catarrhines (especially the cercopithecids) may have been more canalized toward producing numerous short-lived lineages that responded to competition mainly by speciating or becoming extinct, rather than through niche separation and character displacement. This follows Vrba's "effect hypothesis" model of macroevolution. The competition provided by intercontinental faunal exchange combined with internal replacement to reduce the chances for ancient catarrhine lineages to survive as living fossils. Moreover, the relative morphological homogeneity of the highly successful cercopithecids further suggests they were likely to replace each other as rather small adaptive novelties became selectively advantageous. Nonetheless, the several extreme eurytopes among the Cercopithecidae, such as *Macaca* and *Papio*, did manage to persist for reasonably long intervals with little change, once their underlying adaptations were fixed. Only additional fossils, as always, will tell if these interpretations are defensible.

Acknowledgments: We thank Niles Eldredge for requesting this paper and thus leading us to think about these patterns in a different light. We further thank Mr. Chester Tarka for help with preparation of the prints and composition of Figs. 1 and 2 and advice with Figs. 3–5; his unstinting demand for an approach to perfection keeps us honest. The photographs of Figs. 1 and 3 were taken by the authors; Ms. Biruta Akerbergs drew Fig. 2A, which is reproduced courtesy of Academic Press; Fig. 2B originally appeared in Gregory (1951) and is reproduced courtesy of the American Museum of Natural History; Ms. Lisa Calvert drew Figs. 4 and 5. We are deeply indebted to them all. The research reported here was financially supported, in part, by grants from the National Science Foundation (BNS 81-13628 to E.D. and BNS 80-16634 to A.L.R.) and the PSC-CUNY research award program (12988 and 13453 to E.D.).

Literature

Andrews, P. J. 1981. Species diversity and diet in monkeys and apes during the Miocene, pp. 25–61. In: Stringer, C. B. (ed.), Aspects of human evolution. London: Taylor and Francis.

Andrews, P. J., Cronin, J. E. 1982. The relationships of *Sivapithecus and Ramapithecus* and the evolution of the orangutan. Nature 297:541–546.

Bernor, R. L. In press. A zoogeographic theater and biochronologic play: the time/biofacies phenomena of Eurasian and African Miocene mammal provinces. In: Mein, P. (ed.), Proc. RCMNS Colloq. 1983.

Ciochon, R. L., Corruccini, R. (eds.) 1983. New interpretations of ape and human ancestry. New York: Plenum.

Delson, E. 1975a. Evolutionary history of the Cercopithecidae. Contrib. Primatol. 5:167–217.

Delson, E. 1975b. Paleoecology and zoogeography of the Old World monkeys, pp. 37–64. In: Tuttle,

R. (ed.), Primate functional morphology and evolution. The Hague: Mouton.

Delson, E. 1977. Catarrhine phylogeny and classification: principles, methods and comments. J. Human Evol. 6:433–459.

Delson, E. 1980. Fossil macaques, phyletic relationships and a scenario of deployment, pp. 10–30. In: Lindburg, D. E. (ed.), The Macaques: studies in ecology, behavior and evolution. New York: Van Nostrand.

Delson, E., Andrews, P. 1975. Evolution and interrelationships of the catarrhine primates, pp. 405–446. In: Luckett, W. P. and Szalay, F. S. (eds.), Phylogeny of the primates: a multidisciplinary approach. New York: Plenum.

Eldredge, N. 1979. Alternative approaches to evolutionary theory. In: Schwartz, J. H., Rollins, H. B. (eds.), Models and methodologies in evolutionary theory. Bull. Carnegie Mus. Nat. Hist. 13:7–19.

Erikson, G. E. 1963. Brachiation in New World monkeys and in anthropoid apes. Sympos. Zool. Soc. London 10:135–164.

Fleagle, J. G., Rosenberger, A. L. 1983. Cranial morphology of the earliest anthropoids, pp. 141–153. In: Sakka, M. (ed.), Morphologie evolutive, morphogenèse du crâne et origine de l'homme. Paris: C.N.R.S. (Centre National de la Recherche Scientifique).

Fooden, J. 1982. Ecogeographic segregation of macaque species. Primates 23:574–579.

Geraads, D. 1982. Paléobiogéographie de l'Afrique de Nord depuis le Miocène terminal, d'après les grands mammifères. Geobios Mem. Spec. 6:473–481.

Gregory, W. K. 1951. Evolution Emerging. New York: MacMillan.

Hershkovitz, P. 1972. The Recent mammals of the Neotropical Region: a zoogeographic and ecological review, pp. 311–431. In: Keast, A., Erk, F. C., Glass, B. (eds.), Evolution, mammals and southern continents. Albany: State U. of New York Press.

Hershkovitz, P. 1977. Living New World monkeys (Platyrrhini), Vol. I. Chicago: U. Chicago Press.

Hershkovitz, P. 1982. Supposed squirrel monkey affinities of the late Oligocene *Dolichocebus gaimanensis*. Nature 298:201–202.

Hoffstetter, R. 1974. *Apidium* et l'origine des Simiiformes (=Anthropoidea). C. R. Acad. Sci., Paris 278D:1715–1717.

Hoffstetter, R. 1980. Origin and deployment of the New World monkeys emphasizing the southern continents route, pp. 103–122. In: Ciochon, R. L., Chiarelli, A. B. (eds.), Evolutionary biology of the New World monkeys and continental drift. New York: Plenum.

Kay, R. F., Fleagle, J. G., Simons, E. L. 1981. A revision of the Oligocene apes of the Fayum Province, Egypt. Amer. J. Phys. Anthropol. 55:293–322.

Kraglievich, J. L. 1951. Contribuciones al conocimiento de los primates fosiles de la Patagonia. I. Diagnosis previa de un nuevo primate fosil del Oligoceno superior (Colhuehuapiano) de Gaiman, Chubut. Commun. Inst. Nac. Invest. Cien. Nat. Cien. Zool., 2:57–82.

MacPhee, R. D. E., Woods, C. A. 1982. A new fossil cebine from Hispaniola. Amer. J. Phys. Anthropol. 58:419–436.

Marshall, L. G., Webb, S. D., Sepkoski, Jr., J. J., Raup, D. M. 1982. Mammalian evolution and the great American interchange. Science 215:1351–1357.

Mivart, St. G. 1864. Notes on the crania and dentition of the Lemuridae. Proc. Zool. Soc. London 1864:611–648.

Napier, P. H. 1976. Catalogue of Primates in the British Museum (Natural History) and elsewhere in the British Isles. Part I. Families Callitrichidae and Cebidae. Brit. Mus. (Nat. Hist.), London.

Napier. J. R., Napier, P. H. 1967. A handbook of living primates. London: Academic.

Patterson, B., Pascual, R. 1972. The fossil mammal fauna of South America, pp. 247–309. In: Keast, A., Erk, F. C., Glass, B. (eds.), Evolution, mammals and southern continents. Albany: State U. New York Press.

Pilbeam, D. R. 1982. New hominoid skull material from the Miocene of Pakistan. Nature 295:232–234.

Ripley, S. 1979. Environmental grain, niche diversification, and positional behavior in Neogene primates: an evolutionary hypothesis. pp. 37–74. In: Morbeck, M. E., Preuschoft, H., Gomberg, N. (eds.), Environment, behavior and morphology: dynamic interactions in primates. New York: Gustav Fischer.

Rosenberger, A. L. 1978. New species of Hispaniolan monkey: a comment. An. Cien. U. Cent. Este Republica Dominicana 3(3):249–251.

Rosenberger, A. L. 1979. Cranial anatomy and implications of Dolichocebus, a late Oligocene ceboid primate. Nature 279:416–418.

Rosenberger, A. L. 1980. Gradistic views and adaptive radiation of platyrrhine primates. Zeitschr. Morphol. Anthropol. 71:157–163.

Rosenberger, A. L. 1981a. A mandible of Branisella boliviana (Platyrrhini, Primates) from the Oligocene of South America. Intl. J. Primatol. 2:1–7.

Rosenberger, A. L. 1981b. Systematics: the higher taxa, pp. 9–27. In: Coimbra-Filho, A. F., Mittermeier, R. A. (eds.), Ecology and behavior of neo-

tropical primates, Vol. 1. Acad. Brasil. Cien., Rio de Janeiro.

Rosenberger, A. L. In press. Aspects of the systematics and evolution of the marmosets. In: de Mello, M. T. (ed.), A primatologia no Brasil. Belo Horizonte: U. Federal de Minas Gerais.

Rosenberger, A. L., Kinzey, W. G. 1976. Functional patterns of molar occlusion in platyrrhine primates. Amer. J. Phys. Anthropol. 45:281–298.

Savage, D. E., Russell, D. E. 1983. Mammalian paleofaunas of the World. Reading: Addison-Wesley.

Schultz, A. H. 1970. The comparative uniformity of the Cercopithecoidea, pp. 39–51. In: Napier, J. R., Napier, P. H. (eds.), Old World monkeys. London: Academic.

Setoguchi, T., Watanabe, T., Mouri, T. 1982. The upper dentition of Stirtonia (Ceboidea, Primates) from the Miocene of Colombia, South America and the origin of the postero-internal cusp of upper molars of howler monkeys (Alouatta). Kyoto U. Overseas Research Reports of New World Monkeys, II:51–60.

Simons, E. L. 1965. New fossil apes from Egypt and the initial differentiation of Hominoidea. Nature 205:135–139.

Simons, E. L., Fleagle, J. G. 1973. The history of extinct gibbon-like primates. Gibbon Siamang 2:121–148.

Simpson, G. G. 1953. Major features of evolution. New York: Columbia U. Press.

Stirton, R. A., Savage, D. E. 1951. A new monkey from the La Venta Miocene of Colombia. Compilacion de los Estudios Geol. Oficiales en Colombia, Serv. Geol. Nac. Bogota 7:345–356.

Szalay, F., Delson, E. 1979. Evolutionary history of the primates. New York: Academic.

Thomas, H. 1979. Le rôle de barrière ecologique de la ceinture saharo-arabique au Miocène: arguments paleontologiques. Bull. Mus. Nat. Hist. Nat. Paris, Ser. 4, Sec. C, 2:127–135.

Thomas, H., Bernor, R., Jaeger, J-J. 1982. Origines du peuplement mammalien en Afrique du nord durant le Miocène terminal. Geobios 15:283–297.

Vogel, C. 1966. Morphologische Studien an Gesichtsschädeln Catarrhiner Primaten. Biblio. Primatol. 4:1–226.

Vrba, E. S. 1980. Evolution, species and fossils: how does life evolve? S. Afr. J. Sci. 76:61–84.

Vrba, E. S. 1983. Macroevolutionary trends: new perspectives on the roles of adaptation and incidental effect. Science 221:387–389.

Wu, R., Xu, Q., Lu, Q. 1983. Morphological features of Ramapithecus and Sivapithecus and their phylogenetic relationships. Acta Anthropol. Sinica 2:1–10.

6
Evolutionary Pattern and Process in the Sister-Group Alcelaphini–Aepycerotini (Mammalia: Bovidae)

Elisabeth S. Vrba

Transvaal Museum, Pretoria, 0001, South Africa

Sisters of a sister-group are taxa that share a more recent common ancestry with each other than either does with any other taxon. One of the most interesting aspects of evolution concerns the very different histories in terms of morphological diversification that such sister-taxa often have. Elsewhere I have suggested that the causes of different kinds of evolution may be especially well studied in low-ranking sister-groups that include a fossil record plus extant survivors, and that are still in a phase of evolutionary radiation (Vrba 1980a). The bovid tribes Alcelaphini (blesbuck–hartebeest–wildebeest group) and Aepycerotini (impalas) provide such a case.

The Extant Species

There appear to be seven extant alcelaphine species (see Table 1, 1–7, for Latin and common names). Irrespective of intraspecific variation, each species is distinguishable from any other by a number of phenotypic characters. For example, diagnosis is possible not only on whole skulls, but also parts thereof, such as half maxillae or mandibles and horn-core frontlets. Species pairs, generally regarded as closely related, have been recorded as occurring sympatrically today and/or historically (species 4 and 5, 6 and 7, e.g., Ansell 1971). I am not aware of

anyone who argues that there are less than seven extant species. Some have held that there are more, particularly that what are here taken as northern and southern subspecies (see Distribution below) of *Alcelaphus buselaphus* and *Damaliscus lunatus* may be specifically distinct.

In sexually reproducing, cross-fertilizing organisms the individuals in any one species share a unique specific-mate recognition system (Paterson 1978, 1980). I have argued (Vrba 1980a) that in cases where mating communication is primarily visual, a breakdown of communication, i.e., speciation, should involve a shift in morphology (for example, structure and color, that are not only visible to conspecifics and members of the parent species, but may also be visible to us). Many bovids, certainly those considered here, include a strong visual component in their communication systems (Walther 1974). In particular I want to suggest that certain aspects of horn morphology are important in specific-mate recognition and may be expected to vary between bovid species but not within. Thus some components of horn-core orientation show a variation across the subspecies of *A. buselaphus,* but others do not (Ruxton and Schwarz 1929; Wells 1959). The combination of a very elongated frontal boss supporting horns with strong clockwise torsion (in the right horn from the base up) and double curvature (Fig. 2:4) seems to be present in all

Table 1. Recognized extant and extinct species.

Species	Status and fossil site record
Alcelaphini	
1. *Damaliscus dorcas*[a] (Pallas 1766) bles- and bontebok	Extant; various Mid-Pleistocene to recent southern African sites
2. *Damaliscus lunatus* (Burchell 1832) topi, korrigum, tsessebe	Extant
3. *Damaliscus hunteri* (P. L. Sclater 1889); hirola	Extant
4. *Alcelaphus buselaphus*[a] (Pallas 1766); bubal, tora, red, etc., hartebeest	Extant; various Late Pleistocene north African sites
5. *Alcelaphus lichtensteini* (Peters 1849); Lichenstein's hartebeest	Extant; Broken Hill (now Kabwe, in Zambia)
6. *Connochaetes gnou*[a] (Zimmermann 1780); black wildebeest	Extant; Elandsfontein, Cornelia, Florisbad (all South African)
7. *Connochaetes taurinus* (Burchell 1823); blue wildebeest	Extant; Olduvai Middle Bed II to Bed IV, Peninj (Tanzania); several southern African Pleistocene to Recent sites; Temara (Morocco)
8. *Parmularius angusticornis* (Schwarz 1937)	Extinct; Olduvai Middle and Upper Bed II; Peninj; Isimila (Tanzania); Kanjera (Kenya); Swartkrans Member I (Transvaal, South Africa)
9. *Parmularius rugosus* Leakey 1965	Extinct; Olduvai Lower Bed II to Bed IV
10. *Parmularius altidens*[a] Hopwood 1934	Extinct; Olduvai Bed I; Member H of Shungura Formation, Omo (Ethiopia)
11. *Parmularius braini* Vrba 1977	Extinct; Makapansgat Limeworks Members 3 and 4 (Transvaal, South Africa); Member C of Shungura Formation Omo; Ileret and Koobi Fora Foundations, East of Lake Turkana (Kenya); Ain Boucherit (Algeria)
12. nov.sp. based on Laetolil (Lit.) 1959:277 (?*Parmularius* sp. in Gentry and Gentry 1978)	Extinct; Laetolil Beds *sensu stricto* (Tanzania)
13. *Damaliscus niro* (Hopwood 1936)	Extinct; Olduvai Middle Bed II to Bed IV; Peninj; several South African sites including Swartkrans Member 2, Cornelia and Florisbad
14. *Damaliscus gentryi* Vrba 1977	Extinct; Makapansgat Limeworks Member 4
15. *Beatragus antiquus* Leakey 1965	Extinct; Olduvai Beds I and II; Member G of Shungura Formation, Omo
16. nov.sp. based on Laetolil (Lit.) 1959:233.	Extinct; Laetolil
17. nov.sp. based on Hadar (AL 208-7) (?*Damalops* sp. in Gentry 1981)	Extinct; Hadar Formation, Afar (Ethiopia)
18. *Rabaticeras arambourgi*[a] Ennouchi 1953	Extinct; Rabat (Morocco); Olduvai Beds III and IV; Elandsfontein
19. *Megalotragus kattwinkeli* (Schwarz 1932)	Extinct; Olduvai Beds I to IV; Peninj, Chesowanja (Kenya); Member G and post G of Shungura Formation, Omo; Later Chiwondo, Malawi; probably present in Makapansgat Member 3, Sterkfontein Member 4
20. *Megalotragus priscus*[a] (Broom 1909)	Extinct; various Mid-Pleistocene to terminal Pleistocene South African sites
21. nov.sp. based on Olduvai 1970 Geologic Locality 208 (labeled S.208)	Extinct; from Lemuta Member in Bed II, Olduvai
22. *Connochaetes africanus* (Hopwood 1934)	Extinct; Olduvai Bed II (level unknown)
23. nov.sp. based on Olduvai FLKN I 1961: 7154; or on KNM-ER287 (from East of Lake Turkana)	Extinct; Olduvai Bed I and Middle Bed II; Koobi Fora Formation; probably Swartkrans Member 1

continued

Table 1. Recognized extant and extinct species. (*Continued*)

Species	Status and fossil site record
24. *Damalacra neanica*[a] Gentry 1980	Extinct; Langebaanweg (Cape Province, South Africa)
25. *Damalacra acalla* Gentry 1980	Extinct; Langebaanweg
26. *Damaliscus agelaius* Gentry and Gentry 1978	Extinct; Olduvai Lower Bed II to Beds III–IV (area where these Beds not divisible)
27. *Rabaticeras porrocornutus* Vrba 1971	Extinct; Swartkrans Member 1
28. *Damalops palaeindicus*[a] Falconer 1859	Extinct; Pinjor stage, Siwalik Hills, India; Tadzhikistan in the USSR
29. *Parmularius parvus* Vrba 1978	Extinct; Kromdraai A (Transvaal South Africa); Olduvai Bed IV (Tanzania)
30. *Parmularius/Damaliscus* sp. nov.	Makapansgat Member 3, Sterkfontein Member 4, Swartkrans Member 1 or 2 Elandsfontein, Cornelia (all South African)
31. *Damaliscus/Beatragus* sp. nov.	Elandsfontein, Swartkrans Member 2 (South Africa)
32. *Connochaetes tournoueri* Thomas 1884	Ain Jourdel (Algeria); possibly also several Late Pliocene occurrences in east Africa
33. *Aepyceros melampus* (Lichtenstein 1812); impala	Extant; Olduvai Beds I and II; Omo Shungura Formation from Member H onward; Peninj; some very late southern African sites
34, 35. *Aepyceros* sp.	Omo, Usno, and Mursi Formations; Omo Shungura Formation Member B to H; Hadar Formation, Afar (Ethiopia); Early and Late Kaiso Formation (Uganda); Early and Later Chiwondo (Malawi)

[a] Type species.

individuals, almost literally "from Cape to Cairo." It separates *A. buselaphus* from all other known alcelaphine morphologies, past and present. The nature of variation in horn morphology in the extant sample of alcelaphines (see also Fig. 2) suggests a possible criterion for recognizing species of Alcelaphini–Aepycerotini and other groups in the fossil record: Particular horn-core morphologies may characterize species across space and time, irrespective of geographic variation at any one time or any phyletic change through time, in other features such as size.

Most authorities recognize only one extant aepycerotine species, *Aepyceros melampus*. There is considerable size variation among modern subspecies of *A. melampus*.

The Fossil Record

In terms of how cranial variation distributes within and between species in the extant sample

(Laubscher et al. 1972), some 25 extinct species of Alcelaphini may be recognized to date (Table 1, 8–32). A high degree of sympatry is evident in many fossil assemblages (insofar as discovery of different morphologies in one stratum indicates sympatry). Thus, for example, in particular strata such as the subdivisions of the Olduvai Beds in east Africa (Gentry and Gentry 1978) and the individual cave breccia members of South Africa (Vrba 1975), there are typically four to seven decidedly distinct alcelaphine morphologies occurring together, often each represented by large numbers of individuals and without intermediates. Such morphologies have been taken to represent species.

At any one time level only one *Aepyceros* species seems to be present. Gentry and Gentry (1978) regard occurrences previous to near 2 million years ago as specifically distinct from *A. melampus*. In the light of distribution and abundance data, the magnitude and kind of variation in space and time (see below), and in terms of concepts of the species through time (Simpson

1951; Wiley 1978; Eldredge and Cracraft 1980), I am inclined to recognize only a single species of *Aepyceros* that still survives today. There may be unrecognized ancestor–descendant sequences of unbroken and unbranching reproductive continuity among the alcelaphine morphospecies as well. Thus, in the interests of comparability, I record two aepycerotine "morphospecies" in the cladogram of Fig. 1.

Phylogeny and Diversification

Both the monophyly (in the precise sense, Nelson 1971) of the seven extant alcelaphine species and the sister-group status of *Aepyceros melampus* with these, may be regarded as corroborated. Apart from cranial morphology, radioimmunoassy analysis of serum–antiserum reactions supports both hypotheses (Lowenstein and Vrba, in preparation). Fossil-plus-recent species 1–28 (Table 1) have been subjected to a detailed cladistic analysis using 58 cranial and mandibular characters (Vrba 1979). A phylogenetic hypothesis (cladogram) suggested by that analysis is represented by the dotted lines in Fig. 1. *Aepyceros* and additional alcelaphine species 29–32 have been added, although the latter have not yet been incorporated in the cladogram. Pronounced apomorph basioccipital and premaxillary morphologies (Vrba 1979; Table 2, 32, 46) are shared only by Alcelaphini and Aepycerotini among all Bovidae examined. Characters of tooth morphology, supraorbital foramina, and the cornual sinuses also lend strong support to alcelaphine–aepycerotine monophyly (see Vrba 1979 for further details). The African fossil record between 10 and 4 million years ago is poor. Thus, I can suggest only tentatively that the divergence may have occurred during the Late Miocene.

Without referring for the moment to what may or may not be termed a species, it is of interest to know whether any particular morphology has resulted from branching or unbranching evolution. A statistical test (Vrba 1980a) for transformational versus splitting evolution was applied to the alcelaphine cladogram. The outcome suggests that changes by transformation, or unbranching evolution, were in the minority. In terms of character state distribution in the cladogram of Fig. 1, some unbranching ancestor–descendant relationships

are possible, and indeed the temporal record would not contradict such a notion (19 → 20; 21 → 5; 23 → 7; 12 → 11). One can consider additional unbranching ancestor–descendant pairs, such as the couple suggested by Gentry and Gentry (1978), which in terms of the present analysis would necessitate minor reversals and/or reinterpretation of plesiomorph–apomorph character states. Even after such additions, by my count no less than 18 splitting events must have occurred to produce species 1–27, compared to none in the impala sister-group.

I can add that the discrepancy is most unlikely to be a taphonomic artifact; i.e., that a comparable range of impala morphologies was produced but not recovered. Impala fossils are plentiful over a wide area and long time but never suggest the presence of more than one species at a time.

Since the Alcelaphini and Aepycerotini diverged somewhere near the latter part of the Miocene, they have had remarkably different

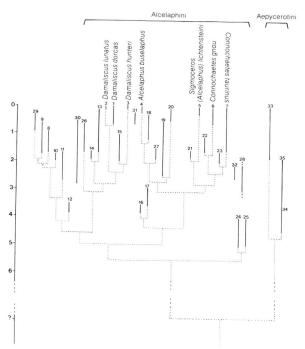

Fig. 1. Durations and cladogram of species. Solid lines give approximate times of occurrence of species (species numbered as in Table 1; extant species are named). Dotted lines indicate a cladistic relationship for most of the species (source for duration dates, as well as cladistic arguments, given in Vrba 1979). The diagram is not a phylogenetic tree (*sensu* Eldredge and Cracraft 1980).

evolutionary histories. One can argue that such discrepant evolutionary rates may well result from random causes. The data presented below should help to decide what kind of deterministic factor, if any, may have been operating.

Distribution

The entire group seems to have been wholly African throughout its evolution, apart from one alcelaphine species from the Pinjor stage of the Indian and Pakistan Siwaliks and from Tadzhikistan in the USSR (Table 1, 28), and the occurrence in the recent past of a hartebeest in Palestine, Jordan, and more doubtfully Lebanon (Garrod and Bate 1937; Ducos 1968; Clutton-Brock 1970; Hooijer 1961; Vrba 1979). Modern distributions are shown in maps in Dorst and Dandelot (1970) and are commented on in detail in Ansell (1971). *Connochaetes gnou* and *Damaliscus dorcas* seem to have been restricted in historic times, and earlier as well as far as we know, to a southern temperate distribution. Three alcelaphine species are and were found mainly in the tropics but also beyond: *Connochaetes taurinus* today and historically from north of the Orange River in the Cape Province and the Orange Free State, also Transvaal and northern Natal, South Africa, with a west to east distribution north of the Tropic of Capricorn, then along southeastern Africa towards the equator, and previously in north Africa (Table 1); *Alcelaphus buselaphus* is distributed through three main disjunct areas, (1) the temperate-to-tropic southwest of the continent, (2) a tropical distribution from east Africa west to Senegal, and (3) north to the Sahara from Morocco to Egypt, plus the Middle East records already mentioned; *Damaliscus lunatus* with a discontinuous historical and recent distribution, (1) from the temperate southwest, north of the Orange River, northwards to Zambia falling within the drier parts of the southern Savanna zone, and (2) from Tanzania east to Somalia, northwards and across to west Africa, largely in the Somali Arid and Sudanese Arid zones. Only two alcelaphine species are essentially tropical in present and former known distributions: *Damaliscus hunteri*, restricted to a small patch on the equatorial east coast; and *Alcelaphus lichtensteini* in the south-

ern Savanna on the east of the continent, and northwards without reaching the equator. The distribution of *Aepyceros melampus* during recent and historic times extends from south of the Tropic of Capricorn to just north of the equator, mainly in the southern Savanna, but also marginally in the southwest arid zone. The subspecies *A. m. petersi* is allopatrically restricted to northwestern Namibia and southwestern Angola.

It is curious that to date the southernmost record of *Aepyceros*, previous to the Middle Pleistocene, is Chiwondo in Malawi. Several extinct alcelaphine species occurred up to the southern or northern extremes of Africa, at least one in both areas (Table 1).

Morphological Change Along Lineages

There are no fossils known that could be those of the common ancestor of the impala and the alcelaphines. Nor are there any fossils known that might be closely related to that ancestry. The best one can do to gain some idea of the magnitude of evolutionary change along lineages is to compare the five species known earlier than 3.5 million years with the 18 lineage endpoints present later than 1.0 million years. Figure 2 gives an impression of diversification in horn-core and frontlet morphology. There are significant increases in size toward several alcelaphine lineage endpoints (e.g., species 8, 13, 4, 5, 6, and 7) and a truly spectacular one toward species 20. There is also evidence of size reduction (e.g., toward species 29 and 30). Alcelaphine horn-core twist, shape, orientation, and frontal support have undergone remarkable diversification, only partially shown in Fig. 2. Compared to the dentitions of species 24 and 25, dated 4–5 million years ago, many later forms show considerable reduction in the premolar–molar ratio. This is particularly marked in the *Parmularius, Megalotragus,* and *Connochaetes* lineages (such that in mandibles P_2 disappears and P_3 may be reduced to a peg). Face length has increased dramatically especially toward species like *Alcelaphus buselaphus* and *Connochaetes taurinus.* Dorsal braincase/face angles have become consider-

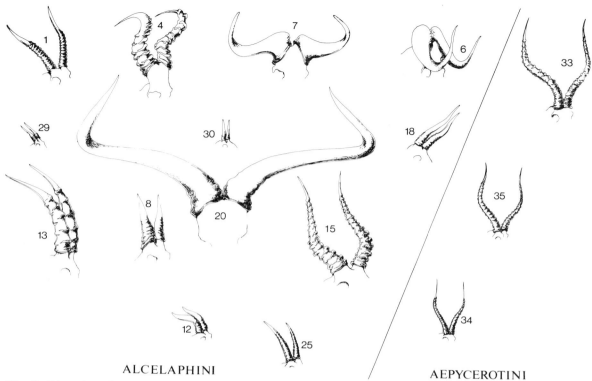

ALCELAPHINI   AEPYCEROTINI

Fig. 2. Diversity of horn and frontlet morphologies and relative sizes (drawn to scale). All are in antero-lateral view; numbers as in Table 1; earlier forms near the base of the page, extant species at the top; horn sheaths of fossils have been reconstructed according to horn-core morphology.

ably more acute in several lineages. Some of this evolution is very likely purely size related. That it cannot all be entirely allometric is shown by the diversity among species of similar size, and by many counter examples. Thus relative premolar length is shorter in the smallest *Parmularius* sp. than in larger ones; *Damaliscus niro*, which is at least during part of its temporal range one of the two largest species of the *Damaliscus* clade, has the least complex horn shape (Fig. 2).

The impala lineage also increased in average size. The horn cores evolved toward somewhat greater divergence and increased lyration. The premolars became reduced, without loss of P_2. Face length increased impressively and size-independently (Gentry 1976: Fig. 8). But none of these changes match in magnitude those seen towards diverse lineage endpoints of the alcelaphine radiation. I should add that the situation in *Aepyceros* is ambiguous. Some early horn cores are large, although not as lyrate as later ones, while some later ones are small. Gentry

(1976) records such size oscillation through time. He notes: "It is unfortunate that impala horn cores in the Shungura Formation [from near 3 million years until Lower Pleistocene] show no smooth transition from smaller and less lyrated ones to larger and more strongly lyrated ones. Indeed it is noteworthy that at present adult male impala from the Shaba (formerly Katanga) region of Zaire and adjacent parts of Zambia are small-boned with less lyrated horns than other living populations." (Gentry pointed out in a personal communication that "small-boned" is a printing error and should read "small-horned.") He shows in his Fig. 8 that several skull measurements of *A. melampus katangae* are very close to those taken on a skull aged nearly 3 million years. The modern skull however has a significantly longer face and somewhat more widely spaced supra-orbital pits. One should mention here that a trace of a preorbital fossa, present on the same fossil, is absent in all extant subspecies.

In sum, in spite of not knowing the morphology of the common ancestor, we can reasonably reach a conclusion. Much more change, at least in skull features, occurred along most individual ancestor–descendant lineages of Alcelaphini than it did from early to late *Aepyceros.* This, quite apart from the difference in splitting rate, is remarkable when one considers that both taxa started not so long ago in geological terms from the same species and appear to have evolved together in the African savanna, and in the case of some alcelaphine lineages probably in the same broad geographic area.

It may be noted that the much discussed "sudden appearances" of morphologies in the fossil record are certainly present in the present data on alcelaphine evolution. There are particularly many close to 2 million years ago. (I have shown elsewhere that before 2 million years ago there occurred a peak, unprecedented in African bovid evolution, in both extinction and origination of species in all African bovid groups; Vrba, in press). The fact that numerous African fossil assemblages span the 3–1 million year period, notably from the Omo, East Turkana, and the Transvaal cave breccia successions, gives special significance to such sudden appearances. There is certainly also evidence of gradual evolution, i.e., gradual average modification through time. But I wonder with what frequency in the present sample it occurred with smooth rate distribution and net unidirectionality (i.e., with comparable net rate and direction in the parent species, during and after speciation, and in the daughter species) from one morphology to another, here specified as alcelaphine species. Rather, the gradual average divergence occurs within species and is slow enough to be what I call "virtual equilibrium" (Vrba 1980a). I would suggest this description for the entire *Aepyceros* record in Fig. 1, as well as for several of the alcelaphine species of longer duration. There is also evidence of oscillation, such as in *Damaliscus niro,* which varies in size through time from smaller to larger to smaller, and in the impala.

Habitat Association

Recently I wanted to know whether extant African Bovidae, at the generic and tribal level, may be significantly associated with each other

and with particular habitats. To answer this question, I analyzed modern census data from 16 wildlife areas in sub-Saharan Africa. The nine tribes of Bovidae analyzed all originated either near the end of the Miocene or previously. A multidimensional graphic technique called correspondence analysis was performed on these data (Vrba 1980b; Greenacre and Vrba 1983). It can be called an objective method of statistical analysis in the sense that it does not *a priori* presume any structure (or causative factor) underlying the data. Instead it reveals any nonrandom structure *a posteriori.* In this particular case, it showed associations between different bovid taxa, and between bovid taxa and ecological variables such as vegetation cover, rainfall, altitude, soil nutrient status, and so on. A remarkable and consistent association of Alcelaphini and Antilopini (the gazelle–springbuck group) with open grassland was demonstrated. In areas with a combination of low altitude and low rainfall (0–400 mm mean annual) as well as those of high altitude with medium rainfall (400–800 mm mean annual) the resultant vegetational physiognomy is a low ratio of wood to grass cover. In such areas, alcelaphines plus antilopines never account, in this data set, for less than 65% of the total antelope frequency; while they never amount to more than 30% in predominantly bush-covered areas. In this analysis *Alcelaphus lichtensteini* emerges as the alcelaphine species most prevalent in areas of somewhat higher bush cover. *Connochaetes taurinus* appears to be the most versatile with respect to vegetational habitat. It occurs in both kinds of physiognomies, although with much higher frequencies in the open areas.

Was there such an association between alcelaphine occurrence and grasslands early in the history of the group, or is it only a recent phenomenon? To attempt an answer I scored the first and last appearances, over half-million-year intervals from Late Miocene to Recent, of all bovid species recovered from the fossil record of sub-Saharan Africa (Vrba, in press). When numbers of originations and extinctions are plotted against time, the resulting curve corresponds well with a graph of global temperature fluctuations during the Cenozoic compiled by Brain (1981). It has been suggested that over much of the African savanna low temperatures may have been overall associated with a reduction of bush and tree cover. In fact, around the end of the Miocene when temperatures plunged to a low point, unprecedented during the entire preceding Tertiary, a number of bovid tribes including the Alcelaphini and Aepycerotini are recorded for the first time. Around temperature maxima more species of alcelaphines and antilopines go extinct than of all other antelope tribes combined, while the latter are recorded in larger numbers. Around temperature minima the converse happens; i.e., more species of alcelaphines and antilopines are recorded than of all other tribes combined, while more species of the latter go extinct. The correlation appears to hold in both eastern and southern Africa. It appears that the association of alcelaphines as a whole with vegetationally open habitats may be one of long standing.

In the correspondence analysis, the impala emerged as strongly associated with areas of higher wood cover, associated with a number of other bovid tribes but decisively separated from Alcelaphini. Like the blue wildebeest, *C. taurinus*, it definitely occurs in both kinds of habitats, but in much higher abundance in open-to-moderately-closed woodland and thicket. Coe (1980) considers the impala over much of its range an ecotonal species because it often occupies the interface between patches of open grassland and scrub or woodland.

Resource Utilization

All extant alcelaphine species are specialist bulk and roughage eaters and purely grazers. They include species with the most advanced stomach forms among herbivores (Hofmann 1973). Some are mainly fresh-grass grazers, such as *Connochaetes taurinus*, while others are roughage grazers, such as *Alcelaphus buselaphus* and the species of *Damaliscus* (Hofmann 1973). All are able to go without drinking water for varying periods of time, perhaps *Connochaetes taurinus* less so than the others. Dorst and Dandelot (1970) note that during the dry season the latter may move over 30 miles a day to drink.

Impala belong in the ruminant class that Hofmann calls "intermediate feeders (seasonally and regionally adapted). . . . Their remarkable adaptability to a great variety of . . . plants according to availability, is reflected in apparently reversible adjustments in the structural components of the stomach" (1973:235). Impalas are dependent on permanent surface water. Where they occur in arid regions, for instance in Kenya and Namibia/Angola, they are restricted to the vicinity of river courses and springs (Stewart and Stewart 1963; Joubert 1971). It is interesting that impala skull characters concerned with feeding are but little modified from the original presumed browsing ancestry, while those of alcelapine species are changed considerably (Caithness, personal communication).

Abundance of Individuals

Populations in particular species, of both the alcelaphine "sister" and the impala one, are known to dominate the antelope frequencies numerically in certain ecosystems in which they occur. According to some recent census data (Vrba 1980b), impalas today comprise 72% and 82% of all bovid individuals in the Kruger National Park and the Mkuzi Game Reserve, both South African, respectively. A similar situation may be found in fossil assemblages: *Aepyceros* is the most common antelope at sites such as Later Chiwondo, Malawi (about 2.5–3.0 million years ago), and throughout Members B and G (spanning the period about 2.9–1.9 million years ago) of the Shungura Formation, Omo, Ethiopia. Today the topi, *Damaliscus lunatus*, is marginally the most abundant antelope east of Lake Turkana, while the blue wildebeest, *Connochaetes taurinus*, overwhelmingly dominates bovid numbers in areas like the Serengeti and the Ngorongoro Crater, Tanzania.

In the correspondence analysis, antelope census data were obtained from 16 game reserves of varying areas, spread over the greater portion of sub-Saharan Africa, from the south to beyond the equator, and from the east to the west (Vrba 1980b; see map in Greenacre and Vrba 1983). Within the limits of these census data (i.e., they cannot be claimed to constitute a random sample of African savanna ecosystems; the surface areas and census methods used differ from one reserve to the next; see Greenacre and Vrba 1983), some statistics are of interest. The blue wildebeest is by far the most abundant of all alcelaphines across most ecosystems here sampled. The total surface area of predominantly grass-covered reserves, heavily preferred by alcelaphines as a whole, is near 90,000 km². This is more than twice the total area of reserves censused with a higher bush cover (near 42,000 km²). Compare this with the fact that the total of impalas over all censuses is just less than half the number of all alcelaphine individuals together. The total of impalas is nearly four times as large as that of all alcelaphine species, excepting the blue wildebeest, combined. Over all reserves in which the impala occurs, its average density is 3.6 individuals/km². The corresponding figure for the blue wildebeest is 3.4; and for all other alcelaphine species combined, 1.9. Consider also the well-known fact that impala often flourish, in resilient and weed-like fashion, even in areas perturbed by human agricultural activity, as well as other large portions of Africa not censused here. This they do in spite of being, among bovid species, particularly vulnerable to overhunting (Ansell 1971). It seems possible that the single *Aepyceros* species leaves a number of genes to future generations comparable to all the diverse alcelaphine species put together.

Behavior and Mobility

Estes (1974) analyzed levels of social organization in Bovidae. He groups the impala together with *Damaliscus hunteri* and *Alcelaphus lichtensteini* as having attained an intermediate level of advancement (medium-sized herds, sedentary, i.e., nonmigratory). *Damaliscus lunatus* and *Alcelaphus buselaphus* are classed as being advanced in their level of social organization (large aggregations, migratory); and *Damaliscus dorcas* and the two species of *Connochaetes* among the most advanced of all Bovidae (large, dense aggregations, migratory). He and others further suggest that all Alcelaphini are unusual in having precocial, or "follower" young, except *Alcelaphus* (but there is some disagreement on this latter point; see Ansell 1970; Mitchell 1965). The literature suggests that the impala in some areas has "hider" young (the hiding phase lasting several days after birth; e.g., Schenkel 1966) and in other areas "follower" young (e.g., Jarman 1976; Fairall, personal communication).

Some general comments are germane to the present account. Several authors have observed that not only is the impala nonmigratory but even its daily movements are very much restricted. As a consequence contact between population units living in separate areas is restricted (Schenkel 1966). Many have remarked that social flexibility is a notable feature of the impala (Jarman and Jarman 1974). The fact that all alcelaphine species are less sexually dimorphic (e.g., having relatively strongly horned females) than the impala (with hornless females) is also relevant to these brief comments on behavior.

Capacity for Increase

The average longevity of alcelaphine species, as recorded in some animals in captivity, varies from about 15–18 years, according to Mentis (1972), who places *C. taurinus* as having the highest longevity. Age at sexual maturity varies from 15.7–28 months and age at first parturition from 24–37 months. Mentis records that the lowest among these values include those of *C. taurinus* individuals; but Estes (personal communication) suggests that most *C. taurinus* conceive at around 28 months. According to several authors (e.g., Mentis 1972), calving occurs once a year and is strictly seasonal in all species, although Estes has found two reproductive peaks in the topi, *Damaliscus lunatus*, in Kenya and Zaire (personal communication). Putting together these data with records of calving percentages, it appears that the blue wildebeest, the largest species, may have a higher rate of increase (at least in some areas) than the smaller, extant alcelaphine species.

A comparison of *Aepyceros melampus* with the comparably sized *Damaliscus dorcas* (recorded body weight ranges of 36–60 kg and 32–81 kg, respectively; Brain 1974) yields the following respective average figures: longevity in captivity 13 as against 14.8 years; age at sexual maturity 11–13 as against 28 months; age at first parturition 16–24 as against 36 months. Gestation in the impala is shorter than in all alcelaphines: 5–7 as against 7.7–8.5 months (Mentis 1972). It seems from these data that the impala has a higher rate of increase than its similarly sized alcelaphine relatives. How does it compare with the alcelaphine species recorded by

Mentis (1972) to have the highest r, the blue wildebeest? The latter has a very restricted breeding season, irrespective of whether it occurs near the equator or far away, and irrespective of rainfall distribution through the year (Ansell 1971). Estes (1976) suggests this as an adaptation to migratory behavior and large aggregations. He points out that in small herds, with a sedentary dispersed pattern, the lack of breeding synchrony has been observed to result in higher calf mortality. In contrast, breeding seasonality in the impala varies from place to place. For example, in the Transvaal it is strictly seasonal (Fairall 1968; Skinner et al. 1974). In east Africa, impalas have two annual breeding peaks during two rainy seasons, and territoriality throughout the year (Kayanja 1969; Leuthold 1977). The same phenomenon has been reported from Zululand, South Africa, by Anderson (1975) although Fairall questions the latter (personal communication). Similarly Jarman and Jarman (1974), on the basis of 6 years of fieldwork with extensive sampling of impala social organization in about 20 study areas from Kenya to Natal, conclude: "In eastern Africa, where rains fall over a greater part of the year or occur in more than one wet season, calving occurs throughout the year." Kayanja (1969) records an early post partum estrus (may be sooner than 21–26 days post partum) in Kenyan populations in which breeding is not rigidly seasonally restricted. It seems from all this that *Aepyceros melampus* may have a higher innate capacity for increase, i.e., r_m, than any extant alcelaphine species, including *C. taurinus*. The data of Grobler and Jones (1980) provide an opportunity to see this discrepancy in action (provided one assumes that the study area offered habitat of comparable quality of the two species): In the Whovi Wild Area, Rhodes Matopos National Park, Zimbabwe, 4 impalas and 12 blue wildebeest were introduced in 1960. By 1973 they numbered 200 and 145, respectively.

Discussion

There is some evidence of the past spread of open grassland across much of the African savanna around temperature minima, the converse during maxima. What we know about exact correlations with rainfall is conflicting (Brain 1981). The effect of different combinations of rainfall and temperature is commented on in Greenacre and Vrba (1981: Fig. 5). Today African areas with, for instance, a mean annual rainfall of 400–800 mm at lower altitudes have a higher wood/grass-cover ratio, while some reserves within the same mean annual rainfall range at higher altitudes are open grasslands. Brain (1981) has reviewed a curve of major global temperature fluctuations, with particular reference to Africa, against the Cenozoic time scale. Thus we have some estimate of the chronology of vegetational oscillations (at least of that component resulting from temperature and rainfall) that must have swept back and forth across the continent in a manner analogous to marine transgressions and regressions. Such data are very exciting to a student of herbivore paleontology because, when they are put together with a fossil record of sufficiently high quality (in terms of how much is recovered and the accuracy of dating), contrasting models of evolution can be tested. I do not regard the alcelaphine-aepycerotine data as adequate in this regard. But I want to use it nonetheless to illustrate the nature of possible tests.

Let us look at three combinations of hypotheses (being well aware that many others are possible) and at their predictions.

Model 1

Phenotypic change generally can occur equally well in large or small populations driven largely by directional natural selection—more slowly during periods of physical environmental stability and faster on average during times of environmental change.

Predictions: A. During times of slow environmental change, less contraction of organismal distribution ranges is expected than in terms of Models 2 and 3 below. Instead, most populations faced with slow habitat change will gradually adapt.

B. Thus the phenotypes in large populations may undergo considerable morphological change if the environmental change persists unidirectionally. If this happens often in large populations, without or across splitting events, the chance of finding evidence of the gradual change in the fossil record is good.

C. When the environment oscillates slowly from one extreme to another and back, phenotypic evolution in single unbroken lineages might be expected to oscillate too as it tracks the change.

D. Both generally faster and slower evolution potentially may occur at different times in the same lineage. But any one large temporal segment of that lineage should more or less reflect the rate of morphological change over any other such segment (including the whole lineage investigated), providing environmental oscillations of comparable magnitude and frequency have occurred in the segments compared.

Model 2

In large populations in relatively stable environments, phenotypes are under strong stabilizing selection. Extreme phenotypes are either not produced for some reason (e.g., developmental stability resulting from past stabilizing selection), or are produced but removed by selection. That is, stabilizing selection (as characterized by Charlesworth et al. 1982) is regarded as much more powerful than in Model 1; but gene flow, as a cohesive factor in producing equilibrium, is not accorded much importance. As habitats change, so distribution ranges of species become fragmented. Populations facing new environments are subject to directional selection. Small population size allows a greater frequency of fixations by random drift. Thus evolution occurs faster, on average, in these small populations (but still gradually from generation to generation—no saltation is here implied). Allopatric speciation may result. When the habitat of the new species spreads, its numbers and distribution range will increase. Large populations and renewed stabilizing selection will result in lower rates of directional evolution.

Predictions: A. When the environment changes, large-scale distribution range contraction will occur. Vicariance should result in some larger refuges, but also in small populations in continental habitat "islands."

B. The probability of seeing evidence of the faster portions of gradual change is low, because they mainly occurred in small populations.

C. Environmental oscillations will not be tracked to the same extent as under Model 1.

Rather, species ranges will contract and expand. The same species should reappear in any particular stratigraphic column as successive similar environments return and disappear in between.

D. Evolution over any time period may be considerably discordant with the overall rate of the lineage.

E. Small populations on their own, i.e., without directional selection from new environments, should not be found to be especially important in producing speciation.

Model 3

Model 3 is essentially similar to Model 2, except that gene flow is regarded as the primary stabilizing factor that overpowers selection in different directions in adjacent local populations.

Predictions: More or less as under Model 2, except that small population size, on its own without a new environment and strong directional selection, may be found to be important in bringing about speciation (small population sizes together with environmental change even more so).

The ideas of many workers are combined in these three models (such literature is reviewed in Vrba 1980a). They should not simply be equated with the models of phyletic gradualism and punctuated equilibria. The latter are hypotheses of phenotypic pattern, which patterns may have resulted from different causes. Thus, while Model 1 may produce a pattern of gradualism and Models 2 and 3 punctuated ones, they are not the only models that might be expected to result in such patterns.

The Predictions of the Models and Present Data

Let us see how the predictions of the various models accord with the present data. There is marked evidence of the contraction and expansion of ranges of antelope species. For example, *Aepyceros* is not known further southwards than Malawi (Kaufulu et al. 1981) around the time of the temperature minimum approximately 2.5 million years ago (see Brain 1981), as contrasted with the extent of its South African distribution in historical warmer times. Many examples of apparent range contraction in predictable synchrony with temperature changes, and others of extension, in the present alcelaphine sample may be cited (see distributions in Table 1). Temperature-correlated gaps in a species distribution, in the record of a single relatively complete stratigraphic succession, can be documented in many bovid groups, other African mammals like primates (Vrba, in press) as well as invertebrates (Coope 1979).

The early *Parmularius* lineage is of interest in this respect (onward from species 12, which is probably ancestral to 11; Fig. 1). Between 3.5–4.0 million years ago, around a temperature maximum (Brain 1981), the ancestor is found at Laetoli, Tanzania, and Lower Hadar, Ethiopia. Between about 3.5 and 2.6 million years, temperatures deteriorate to a minimum. Into this time gap fall later Hadar levels and a good assemblage from Omo Member B (Ethiopia). In neither of these is there sign of an unchanging or evolving *Parmularius* sp. 12. At around 2.5 million years, when temperatures are minimal, the new form *Parmularius braini* (11) is suddenly present across Africa, in Omo Member C, Makapansgat (Transvaal) and Ain Boucherit (Algeria). It appears to persist for the next million years with little change (according to fossils found post-KBS tuff in the East Turkana assemblages).

There is some evidence of evolutionary rate distribution in the *Connochaetes–Megalotragus–Alcelaphus* clade. Species 17 may be close to the ancestry of this clade. From Lower to Upper Hadar, from high to low temperatures as grasslands spread, this species does change gradually (e.g., the horns acquire a stronger clockwise torsion towards the morphologies of later members of the clade). *Connochaetes* and *Megalotragus* are very large and distinct from the time of their origins and undergo some gradual evolution throughout their duration (Fig. 1). The morphological gap between the Upper Hadar morphology and either of the other lineages is relatively large. The time gap of less than a million years (maybe only ±100,000 years in the case of *Megalotragus*) necessitates evolution that is substantially faster than that recorded within species *Megalotragus kattwinkeli, Connochaetes taurinus,* or the Hadar species. Either averagely faster change occurred in large populations, perhaps adapting to temperature/bush cover increase, of which, strangely, no trace is found in the several relevant assemblages; or contraction of ranges with speciation in small populations accounts for the gap in the record.

The much-cited power of gene flow as the preserver of species integrity (Mayr 1963; Stanley 1979) does not accord well with these data. The more mobile migratory Alcelaphini, which today potentially cover large distances annually, have split more rapidly. The impala lineage that must surely have been fragmented during past dry periods (if really as water-dependent as claimed), today with clumped distribution and apparently little contact between populations, has not. Thus the present study supports others which suggest that not gene flow, but another factor such as stabilizing selection, is the primary cohesion force (e.g., Ehrlich and Raven 1969; Vrba 1980a; Charlesworth et al. 1982). For the same reasons, the notion that small effective population sizes (Wilson et al. 1975), inbreeding, and random events on their own may have caused different splitting rates, is not satisfactory.

In sum, the predictions of Model 2 appear to be more nearly upheld than those of Models 1 and 3, by what we know so far of the morphological patterns in this group. I repeat that the present data are inadequate to provide an authoritative test. But this preliminary hypothesis of what happens during the evolution of large herbivore mammals is testable in numerous ways by future findings.

The positive correlation in other diverse groups of organisms, of splitting and extinction rates with each other, and with a high rate of phenotypic change along lineages, have been reviewed by Stanley (1975, 1979). Thus the proposition that such a correlation (consistently present within the alcelaphine–aepycerotine, as well as many other monophyletic groups) has a deterministic basis is entirely reasonable. I have discussed elsewhere that any observed correlation between magnitude of morphological diversification (splitting of lineages) on the one hand, and overall morphological change through time along lineages on the other, need not necessarily imply punctuated equilibria (Vrba 1980a:68). It "could mean one of three things: 1) The apparent difference in allopatric speciation rate may not be real but an artifact of different kinds of speciation (i.e., sibling versus morphospeciation). 2) There is a real difference in allopatric speciation rate, but the correlation between faster speciation and rates of morphological change is indirect. That is, the same environmental pressure results in high splitting rates as well as in faster phyletic evolution within species [e.g., under Model 1 above]. 3) There is a real difference in allopatric speciation rates. The correlation between faster speciation and morphological rates is direct." That is, more phenotypic change occurs during speciation and afterwards until establishment of the new species in large numbers than during the rest of that species' duration. In cases where alternative (3) pertains, a punctuated pattern might result in the fossil record—not rigidly "rectangular" or "catastrophic," merely "punctuated" in the sense of a differential rate distribution consistently associated with differential evolutionary mode (e.g., stabilizing selection during unbranching evolution on the one hand, versus directional selection to result in speciation on the other). Alternative (3) seems the most appropriate interpretation of the pattern available in the present analysis.

Somehow temperature/rainfall/vegetation shifts, and concomitant selection pressures, resulted in very different evolutionary rates in alcelaphines and *Aepyceros*. Let us look at the data on the intrinsic characteristics of the animals and their ecology, which must surely provide a key to the deterministic causes of this discrepancy.

Eldredge (1979) gave examples of resource specialists that have a higher species diversity than generalist relatives (see also Stebbins 1950; Simpson 1953; Rensch 1959; Vrba 1980a). Thus, the striking correlation in the present example hardly comes as a surprise. What model might link evolutionary rate with niche breadth? I investigated rate of morphological evolution in some African monophyletic mammalian genera and contrasted it with feeding behavior of extant relatives (Vrba, in press). What emerged cautions against a facile concept linking generalists with low rates, specialists with high ones. Instead the degree of *independence of the feeding niche from such alternative environments* as a lineage encounters through time appears to be of paramount importance. The data I analyzed accord well with the hypothesis that climate/vegetation oscillations provided recurrent alternative environments during Miocene–Recent African large mammal evolution. Herbivores were all predictably affected. Pure grazers or browsers evolved more rapidly than related clades that both graze and browse. In fact, other bovid sister-groups, showing precisely similar correlations to those in the Alcelaphini–Aepycerotini, emerged (e.g., the eland clade *Taurotragus* species that both graze and browse today, is reproductively flexible, occurs from arid grassland to dense bush cover habitats, and shows a low rate of species diversification; in contrast to its "sister" *Tragelaphus*, kudus and relatives, which displays more rapid diversification and is more narrow-niched in every respect). The analysis (Vrba, in press) suggests that *distribution of resource patches* in alternative environments is of primary importance. Thus the exclusively myrmecophagous specialist lineage of the extant aardvark, *Orycteropus*, today finds its feeding patches in a large range of vegetational physiognomies, has a trans-African distribution, and a fossil record indicating very slow evolution. Several authors (e.g., Coe 1980) have noted that the impala requires ecotonal "patches," at the interfaces between open grassland and scrub or woodland, with available nearby surface water. Such "patches" are incredibly widespread in Africa and must have persisted, though perhaps shrunk and expanded, for millions of years.

Whether one takes a gradual or punctuated view of evolution is not directly at issue in the present proposition. It may, for instance, be equally applied under Models 1 and 2 above, both of which stress the power of natural selection: Directional selection pressure acts on populations whose resource base has been removed, or severely altered, by environmental change. They respond by directional phenotypic change and speciation, or by extinction. Stabilizing selection prevails in populations, like those of the impala, whose broad niche encompasses alternative habitat parameters.

Only relative reproductive and feeding flexibility were concentrated on here. I can note in passing that the literature abounds in hints that modern alcelaphines are also narrowly tied to open grassland, the impala more broadly to both bush and grassland in ecotones, in adaptations of vision, hearing, locomotion, mating and predator avoidance behavior, thermoregulation, and disease resistance. When allopatric speciation occurs, the change that directly causes speciation must be that occurring in characters of the fertilization system (explored for example by Paterson 1978, 1980). Directional natural selection may act directly on crucial species-specific and habitat-specialized characters of the mating system, to result in speciation. Alternatively it may act to produce directional change in other resource-adaptive, phenotypic complexes, which, via genetic linkage, pleiotropy, or selective interaction with the loci of the fertilization system, results in speciation. The lineages that were infrequently subject to either of such selective episodes, like that of the impala, eland, aardvark, and others, provide the ''living fossils'' of African mammal evolution. This model of the influence on evolutionary rate of niche breadth and resource patch distribution (the varying breadth of ''habitat taboos'' explored by Rosenzweig 1975 may be included here) clearly predicts strongly correlated speciation and extinction rates.

A point concerning the terms *generalized* and *specialized* may be mentioned. They are terms pertaining to ecology and are often confused with the terms primitive or generalized, and advanced or specialized as used in systematics. It is clear that ecological generalization, which allows independence of environmental changes, is often conferred by an advanced, i.e., phylogenetically specialized, phenotypic character (*apomorphy*, as cladists would term it). The reversibility of stomach structure and the flexible breeding season of the impala, for instance, are almost certainly advanced characters in Bovidae; while the relative independence of water is a phylogenetic specialization that confers on Alcelaphini some independence of environmental change.

How may the potential rate of increase in numbers, higher in impalas than in any extant alcelaphine species, be correlated with evolutionary rate, if at all? I have tentatively implied that Model 2 above may accord reasonably well with the present pattern. This model lays stress on small populations *together* with new environments and concomitant directional selection. It does not so much stress purely random events as being of especial importance to speciation, but rather the role of natural selection in small populations in changed environments. Charlesworth et al. (1982) have argued convincingly that there is no conclusive evidence that small population size per se is especially important to speciation. One can note a similar lack of conclusive evidence that large populations should be. It seems that we are still largely ignorant of whether and how speciation and population size relate. So one might as well suggest models that invoke small and/or large population sizes. I tentatively return to a notion of the importance of small populations in many splitting events, without being able to support it at all in terms of population genetics, for the following reasons: I am impressed by (1) the spectacular contractions of distribution ranges of putative ancestral species, coincident with temperature maxima and minima, in African mammal evolution; and (2) the appearance of morphologically very different descendant species, often across the continent at first recording, which closely postdate ancestral range contraction. If there is any validity to a model invoking small population size, as linked to a higher probability of speciation, then the high potential increase rate of impalas becomes relevant to a discussion of its observed evolutionary rate. Paterson (1978) has argued that a small population present in or invading an environment similar to the parental one is unlikely to speciate, partly because there would be little selective pressure on the loci of the specific-mate recognition system and other adaptive characters. Also, he suggests, the rapid increase in numbers under such circumstances may prevent fixation of alternative alleles. Do lineages with r life-history parameters generally evolve more slowly than related ones with K characteristics? There is some evidence that this may be so (Vrba 1980a). Resource utilization and r versus K parameters are often associated. To test properly for the direct influence of increase rate on evolution in the impala or other groups, one needs to be able to separate these factors and to appreciate the implications at the population genetic level. This is one more case where the

76   E. S. Vrba

paleontologist perceives a pattern and a possible correlation with extant biological characters, but needs the experiments and theory of the population geneticist, so that a synthesis may be jointly approached. (Here is thus one more paleontologist, investigating whether a pattern of punctuated equilibria and species selection may or may not have occurred, who emphatically does not "negate the importance of [the study of] population level phenomena [as they may apply to] long term evolution" [Levinton and Simon 1980:131].)

The present data are not comprehensive enough to test the various models of how evolutionary trends come about. But a few remarks may be of interest. One could argue that there is a "minitrend" in these data, away from aepycerotine phenotypes and towards alcelaphine ones. After all, later in time there are more species characterized by the latter. It is possible that some kind of species selection was operating. Eldredge and Gould (1972) first suggested that species selection may favor superiorly adapted species. Eldredge (1979) and Eldredge and Cracraft (1980) explored the notion that eurytopic (generalist) clades may be expected to have low species diversity, stenotopic ones (specialists) the converse. They suggested a comparable rate of speciation for eurytopes and stenotopes. A high initial extinction rate of newly arisen eurytopic species, in their model, results from large niche overlap, and therefore competition, with other eurytopic species. On the other hand, incipient stenotopic species respond to competition by accommodation and resource subdivision, which permits sympatric existence without competitive exclusion. Stanley does not invoke resource utilization as an intrinsic control of evolutionary rate. He summarizes his concept of species selection as follows (1979:181): "The nonrandom (directive) components of species selection, analogous to the components of natural selection among individuals, are (1) differential rates of speciation among lineages and (2) differential rates of extinction (differential longevities) of lineages. The agents of species selection are the familiar limiting factors of ecology: competition, predation, habitat alteration, and random fluctuations in population size. These agents function within *both components of species selection,* causing differential extinction among established spe-

cies and selectively suppressing the multiplication of species" (italics mine). He makes it quite clear (e.g., p. 198) that low speciation rate (as recovered in the fossil record) results because adversity even more readily extirpates small isolates, i.e., incipient species, than fully established species. Like Eldredge (1979) and Eldredge and Cracraft (1980), Stanley visualizes the *removal* of young species, or "pre-species" to result in low diversity. All these proposals contain the notions of "species fitness," competitive superiority, success, and progress as being represented at the pinnacle of a trend. I have argued that species selection entails the presence of species level characters that are not merely simple sums of characters of organisms.

My own related suggestion, the "effect hypothesis of macroevolutionary trends," is different from that of species selection (Vrba 1980a, 1983). It requires no emergent properties of species, but suggests direct upward causation from characters and dynamics at the level of organisms to differential birth and/or death among species: Intrinsic attributes of individuals in species (such as characters originally selected for some individual adaptation like resource utilization, or indeed perhaps characters arising through random events) may have the *fortuitous* consequence of conferring on the lineage characteristic speciation and extinction rates. Thus trends may arise, in one sense and relative to selection within species, almost "by accident." There is no necessary connotation of among-species competition, superiority, or progress. The more numerous species at the pinnacle of the trend may not be "winners" in any sense of long-term genetic representation.

This latter principle may be well illustrated by alcelaphine–impala evolution. Were past impala species and incipient species outcompeted or removed by factors such as predation and vulnerability to habitat alteration? Under my suggestion of how ecology may influence evolutionary rate this seems unlikely. Lewontin (1980 Macroevolution Conference, Chicago) again put the question that has often been asked: We see gaps in phenotype space. Were the occupants of such gaps present but removed by selection, or were "they" never there at all? He was referring (see also diverse writings of Gould, e.g., 1980) to the evolution of individuals in species, where particular variation might be precluded from arising by factors such as developmental constraints. Analogous alternatives may be explored when we see gaps in species representation. The original statements of species selection (cited above), simply put, see such gaps as resulting from removal in one way or another of past variation. By contrast, if effect evolution occurred the gaps we see may have always been empty, the interspecific variation never produced at all. So perhaps there was no range of past impala species that were outcompeted, extinguished by predation, or which in incipient form were selectively suppressed before they could multiply (see Stanley, in the quote above). They may not be there in the record because the intrinsic and extrinsic conditions that lead to speciation, were not realized.

Notions of evolutionary "success" or "clade fitness" must surely include some concept of representation of replicators (i.e., genes; see Hull 1980) and organisms later in time. It seems that in this respect (see ABUNDANCE above), after much coming and going of species, the Alcelaphini may have achieved no notable "more-making" or greater "success," but merely "fortuitous subdivision" of gene pools, in comparison with the single impala living fossil species.

Acknowledgments: N. Eldredge, R. D. Estes, N. Fairall, A. W. Gentry, R. Lande, R. C. Lewontin, and G. C. Williams have given me useful comments.

Literature

Anderson, J. L. 1975. The occurrence of a secondary breeding peak in the southern impala. E. Afr. Wildl. J. 13(2):149–151.

Ansell, P. D. H. 1970. More light on the problem of Hartebeest calves. Afr. Wildl. 24(3):209–212.

Ansell, W. F. H. 1971. Part 15: Order Artiodactyla, pp. 1–93. In: Meester, J., Setzer, H. W. (eds.), The mammals of Africa: an identification manual. Washington, DC: Smithsonian Institution Press.

Brain, C. K. 1974. Some suggested procedures in the analysis of bone accumulations from southern African Quaternary sites. Ann. Transv. Mus. 29:1–8.

Brain, C. K. 1981. The evolution of man in Africa: was it a consequence of Cainozoic cooling? 17th Alex. L. du Toit Memorial Lecture. Annex. Transv. Geol. Soc. S. Afr. 84:1–19.

Charlesworth, B., Lande, R., Slatkin, M. 1982. A neo-Darwinian commentary on Macroevolution. Evolution 36:474–498.

Clutton-Brock, J. 1970. The fossil fauna from an Upper Pleistocene site in Jordan. J. Zool. 162:19–29.

Coe, M. 1980. African mammals and savannah habitats. Proc. Intl. Sym. Hab. Infl. Wildl. July 1980:83–109. Pretoria: The Endangered Wildlife Trust, U. Pretoria.

Coope, C. R. 1979. Late Cenozoic fossil Coleoptera: evolution, biogeography, ecology. Ann. Rev. Ecol. Syst. 10:247–267.

Dorst, J., Dandelot, P. 1970. A field guide to the larger mammals of Africa. London: Collins.

Ducos, P. 1968. L'origine des animaux domestiques en Palestine. Mém. Inst. Prehistor. U. Bordeaux 6:1–191.

Ehrlich, P. R., Raven, P. H. 1969. Differentiation of populations. Science 165:1228–1232.

Eldredge, N. 1979. Alternative approaches to evolutionary theory. Bull. Carnegie Mus. Nat. Hist. 13:7–19.

Eldredge, N., Cracraft, J. 1980. Phylogenetic patterns and the evolutionary process. New York: Columbia U. Press.

Eldredge, N., Gould, S. J. 1972. Punctuated equilibria: an alternative to phyletic gradualism, pp. 82–115. In: Schopf, T. J. M. (ed.), Models in paleobiology, Chap. 5. San Francisco: Freeman.

Estes, R. D. 1974. Social organization in African Bovidae. In: Geist, V., Walther, F. (eds.), The behavior of ungulates and its relation to management, Vol. 1. Morges, Switzerland: Intl. Un. Cons. Nat. Res.

Estes, R. D. 1976. The significance of breeding synchrony in the wildebeest. E. Afr. Wildl. J. 14:135–152.

Fairall, N. 1968. The reproductive seasons of some mammals in the Kruger National Park. Zool. Afr. 3:189–210.

Garrod, D. A. E., Bate, D. M. A. 1937. The Stone Age of Mount Carmel, I. Oxford: Clarendon Press.

Gentry, A. W. 1976. Bovidae of the Omo Group deposits, pp. 275–292. In: Coppens, Y., Howell, F. C., Isaac, G. L., Leakey, R. E. (eds.), Earliest man and environment in the Lake Rudolf Basin, Chap. 26. Chicago: U. of Chicago Press.

Gentry, A. W., Gentry, A. 1978. The Bovidae (Mammalia) of Olduvai Gorge, Tanzania. Part 1 and 2. Bull. Brit. Mus. (Nat. Hist) 29(4):289–446; 30(1):1–83.

Gould, S. J. 1980. The evolutionary biology of constraint. Daedalus 109:39–52.

Greenacre, M. J., Vrba, E. S. In press. A correspondence analysis of biological census data. Ecology.

Grobler, H. J., Jones, M. A. 1980. Population statistics and carrying capacity of large ungulates in the Whovi Wild Area, Rhodes Matopos National Park, Zimbabwe Rhodesia. S. Afr. J. Wildl. Res. 10(1):38–42.

Hofmann, R. R. 1973. The ruminant stomach. Nairobi: East African Literature Bureau.

Hooijer, D. A. 1961. The fossil vertebrates of Ksar'akil, a Palaeolithic rock shelter in the Lebanon. Zoöl. Verh. 49:1–67.

Hull, D. S. 1980. Individuality and selection. Ann. Rev. Ecol. Syst. 11:311–332.

Jarman, M. V. 1976. Impala social behaviour: birth behaviour. E. Afr. Wild. J. 14:153–167.

Jarman, P. J., Jarman, M. V. 1974. Impala behaviour and its relevance to management, pp. 871–881. In: Geist, V., Walther, F. (eds.), The behaviour of ungulates and its relation to management, Vol. 2. Morges, Switzerland: Intl. Un. Cons. Nat. Res.

Joubert, E. 1971. Observations on the habitat preferences and population dynamics of the black-faced impala. *Aepyceros petersi* Bocage, 1875, in South West Africa. Madoqua Ser. 1(3):55–65.

Kayanja, F. I. B. 1969. The ovary of the impala *Aepyceros malampus* (Lichtenstein, 1812). J. Reprod. Fer., Supp. 6:311–317.

Kaufulu, Z., Vrba, E. S., White, T. 1981. Age of the Chiwondo Beds, Northern Malawi. Ann. Transv. Mus. 33(1):1–8.

Laubscher, N. F., Steffens, F. E., Vrba, E. S. 1972. Statistical evaluation of the taxonomic status of a fossil member of the Bovidae (Mammalia: Artiodactyla). Ann. Transv. Mus. 28:17–26.

Leuthold, W. 1977. African ungulates. Berlin: Springer.

Levinton, J. S., Simon, C. M. 1980. A critique of the punctuated equilibria model and implications for the detection of speciation in the fossil record. Syst. Zool. 29(2):130–142.

Mayr, E. 1963. Animal species and evolution. Cambridge: Harvard U. Press.

Mentis, M. T. 1972. A review of some life history features of the large herbivores of Africa. Lammergeyer 16:1–89.

Mitchell, B. L. 1965. Breeding, growth and aging criteria of Lichtenstein's hartebeest. Puku 3:97–104.

Nelson, G. J. 1971. Paraphyly and polyphyly: redefinitions. Syst. Zool. 20:471–472.

Paterson, H. E. H. 1978. More evidence against speciation by reinforcement. S. Afr. J. Sci. 74:369–371.

Paterson, H. E. H. 1980. A comment on "mate recognition systems." Evolution 34(2):330–331.

Rensch, B. 1959. Evolution above the species level. London: Methuen.

Rosenzweig, M. L. 1975. On continental steady states of species diversity, pp. 121–140. In: Cody, M. L., Diamond, J. M. (eds.), Ecology and evolution of communities, Chap. 5. Cambridge: Belknap.

Ruxton, A. E., Schwarz, E. 1929. On hybrid hartebeests and on the distribution of the *Alcelaphus buselaphus* group. Proc. Zool. Soc. 2:567–583.

Schenkel, R. 1966. On sociology and behaviour in impala (*Aepyceros melampus* Lichtenstein). E. Afr. Wildl. J. 4:99–114.

Simpson, G. G. 1951. The species concept. Evolution 5:285–298.

Simpson, G. G. 1953. The major features of evolution. New York: Columbia U. Press.

Skinner, J. D., Van Zyl, J. H. M., Oates, L. G. 1974. The effect of season on the breeding cycle of plains antelope of the western Transvaal Highveld. J. S. Afr. Wild. Mgmt. Assn. 4(1):15–23.

Stanley, S. M. 1975. A theory of evolution above the species level. Proc. Natl. Acad. Sci. USA 72:646–650.

Stanley, S. M. 1979. Macroevolution: pattern and process. San Francisco: Freeman.

Stebbins, L. 1950. Variation and evolution in plants. New York: Columbia U. Press.

Stewart, D. R. M., Stewart, J. 1963. The distribution of some large mammals in Kenya. J. E. Afr. Nat. Hist. Soc. Coryndon. Mus. 24(3):1–52.

Vrba, E. S. 1975. Some evidence of chronology and palaeoecology of Sterkfontein, Swartkrans and Kromdraai from the fossil Bovidae. Nature (London) 254(5498):301–304.

Vrba, E. S. 1979. Phylogenetic analysis and classification of fossil and recent Alcelaphini (Family Bovidae, Mammalia). J. Linn. Soc. (Zool.) 11(3):207–228.

Vrba, E. S. 1980a. Evolution, species and fossils: how does life evolve? S. Afr. J. Sci. 76(2):61–84.

Vrba, E. S. 1980b. The significance of bovid remains as indicators of environment and predation patterns, pp. 247–271. In: Behrensmeyer, A. K., Hill, A. P. (eds.), Fossils in the making: vertebrate taphonomy and paleoecology, Chap. 14. Chicago, London: U. Chicago Press.

Vrba, E. S. In press. Palaeoecology of early Hominidae, with special reference to Sterkfontein, Swartkrans and Kromdraai. In: Coppens, Y. (ed.), L'environment des hominidés. Proc. Colloque Intl., Fondation Singer-Polignac, Paris, June, 1981.

Vrba, E. S. 1983. Macroevolutionary trends: new perspectives on the roles of adaptation and incidental effect. Science 221:387–389.

Walther, F. R. 1974. Some reflections on expressive behaviour in combats and courtships of certain horned ungulates, pp. 56–106. In: Geist, V., Walther, F. (eds.), The behaviour of ungulates and its relation to management, Vol. 1. Morges, Switzerland: Intl. Un. Cons. Nat. Res.

Wells, L. H. 1959. The Quaternary giant hartebeests of South Africa. S. Afr. J. Sci. 55:123–128.

Wiley, E. O. 1978. The evolutionary species concept reconsidered. Syst. Zool. 27:17–26.

Wilson, A. C., Bush, G. L., Case, S. M., King, M. C. 1975. Social structuring of mammalian populations and rate of chromosomal evolution. Proc. Natl. Acad. Sci. U.S.A. 72:5061–5065.

7
Tapirs as Living Fossils

Christine Janis

Division of Biology and Medicine, Brown University, Providence, RI 02912

Tapirs belong to the family Tapiridae of the order Perissodactyla (odd-toed ungulates). This order also includes the families Equidae (horses) and Rhinocerotidae (rhinos) among its extant members. Living tapirs are found primarily in forested areas in Central and South America and in southeast Asia. They comprise a single genus, *Tapirus,* with four species, three of which are American and one Asian. They are medium-sized perissodactyls, with a body weight of around 300 kg, and are apparently "primitive" among ungulates in various aspects of their morphology and behavior. Superficially, they resemble suoid artiodactyls in the degree of modification of their skeleton from the basic therian mammalian condition, with the general ungulate trend for adaptation to increasing body size and cursorial specialization. They are specialized among ungulates in their possession of a short, mobile proboscis, which they use to bring food to the mouth during feeding. However, despite their lack of derived ungulate specializations, in their general appearance and behavior they are much more reminiscent of plump, short-legged equids than suoids (Grzimek 1972), belying their perissodactyl affiliations (see Fig. 1).

Of the three New World species of *Tapirus,* the largest is the Central American tapir, *Tapirus bairdi,* which has a body weight of slightly over 300 kg. The most common species is the lowland tapir, *Tapirus terrestris,* and the smallest and most gracile species is the mountain tapir, *Tapirus pinchaque,* which has a body weight of approximately 240 kg. Unlike the other species, which are lowland forest dwellers, the mountain tapir lives at an altitude of 2000 to 4000 m; its ecology has not as yet been studied. All these New World species have a brownish colored coat, with a distinct neck ridge and a small mane (the mane is absent in *T. pinchaque*). In contrast, the Malayan tapir, *Tapirus indicus,* has a black coat with a large white patch on the rump and has no distinct mane or neck ridge. The proboscis is also longer and stronger than in the New World species, and the feet are more robust. The coat is short and slick in all the species except for the mountain tapir, where it is wooly (Grzimek 1972). As far as I am aware, the genetic differences between the different species of tapirs have not been studied, and little is known about the supposed times of divergence of the various species. The New World species appear to be more similar to each other in general appearance than to the Asian tapir. The genus *Tapirus* was found in the middle to late Tertiary in both North America and Eurasia, and it has been assumed (e.g., Thenius, in Grzimek 1972) that the geographic range of the living species represents a relict distribution of the genus on these continents (see Fig. 2), although I know of no systematic study that proves this to be the case. It may be that the Malayan tapir represents a

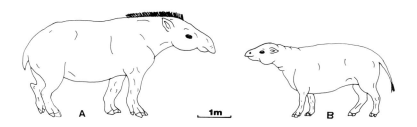

Fig. 1. (A) Recent South American *Tapirus bairdi* (modified from Grzimek 1972). (B) Middle Eocene North American tapiroid *Heptodon posticus* (reconstructed from skeletal reconstruction in Radinsky 1965b).

Pleistocene immigrant from the North American tapir stock, in which case it would be more closely related to the New World species than if it represented a relict of the tapirs of the Eurasian Tertiary.

Little is known of the behavior and ecology of tapirs in the wild, probably because of the difficulty of studying solitary forest-dwelling animals. They apparently feed on relatively low cellulose content foliage, such as leaves, fresh sprouts, and small branches, and the Central American tapirs have been observed to take a good proportion of fruit in their diet (Janzen, personal communication). Their distinctive bilophodont molars, with wear on the cross lophs only, appear to be specialized for dealing with this type of low-fiber, folivorous diet, which mainly requires slicing of soft material without the need for extensive pulping or grinding (Janis, in preparation). The lowland-dwelling

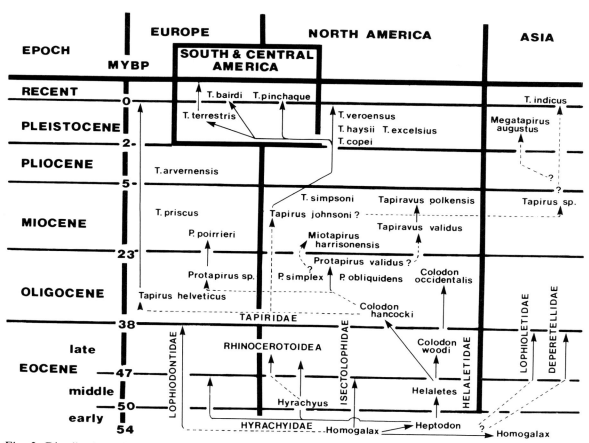

Fig. 2. Distribution of the Tapiroidea in space and time. (Family Tapiridae in detail.) (Species of *Tapirus* included to show diversity of genus and temporal position, and are not meant to indicate ancestor–descendant relationships.)

tapirs appear to be water dependent, taking a proportion of aquatic plants in the diet and defecating in the water in a manner similar to hippos. They are also reportedly good swimmers (Grzimek 1972).

In their anatomy, tapirs are primitive for ungulates in having an unreduced dental formula and short legs with the retention of a complete ulna and fibula. The manus is tetradactyl and the pes tridactyl. This is similar to the condition seen in the first member of the perissodactyls, *Hyracotherium*. They appear to be closer to the original therian condition than do more derived perissodactyls, such as equids, in that they depend more on olfaction than on vision, and their cerebral hemispheres are relatively smaller. They are also primitive for ungulates in various aspects of their behavior. They are solitary and are not territorial, although they may have a fairly circumscribed home range. When kept together in pens, they appear largely to ignore other individuals and do not possess the complicated repertoire of dominance and submission displays associated with maintaining rank order within a group hierarchy that are seen in more derived ungulates such as equids and ruminant artiodactyls. They have no distinct breeding season, as is typical of mammals living in nonseasonal tropical conditions, and have a gestation period of approximately 400 days, which seems rather lengthy for an animal of this body size. They have a single young at a time, which is a derived ungulate condition, but the mother and young both lie down for the young to suckle, which is a primitive mammalian type of behavior seen also in suoids but not in more derived ungulates. The young have little contact with peers and do not display complex play behavior (Grzimek 1972). In lying down, tapirs first adopt a position of sitting on the haunches, a behavior typical of primitive ungulates and rarely seen in equids or ruminant artiodactyls, which support themselves on the carpus when adopting this position (Zannier-Tanner 1965).

The superfamily Tapiroidea first appeared in the early Eocene of North America, in the genus *Homogalax* (Radinsky 1963). *Homogalax* was a small generalized perissodactyl with a body weight of approximately 10 kg and was probably derived from the slightly smaller equid genus, *Hyracotherium* (Radinsky 1969), which was first found in the late Paleocene. Although

Homogalax has been classified as a member of the family Isectolophidae, it was generalized enough in its anatomy to represent the ancestral tapiroid type from which all other lineages could be derived (Radinsky 1969). Its molars were of a generalized lophodont type and there was no molarization of the premolars, but studies of dental wear indicate a trend toward the bilophodonty of many later tapiroid lineages, and it was the most purely folivorous in its diet of all the contemporaneous early Eocene perissodactyls in North America (Janis, in preparation).

Figure 2 illustrates the diversity of tapiroids in space and time. It can be seen that they were diverse worldwide during the Eocene, but since the Oligocene the Tapiridae has been the sole surviving family (with the exception of the extension of the helaletid genus *Colodon* into the Oligocene of North America). Tapiroids were common during the Eocene, forming a dominant component of the fauna in North America, Europe, and Asia. After the Eocene they became extremely rare, although they remained in persistent element of the fauna on all these continents. No cladistic analysis exists for the Tapiroidea. The suggested cladogram in Fig. 3 represents a compilation of some of the discussions by Radinsky (1963, 1965a), but it is not the result of a detailed character analysis, even though it may serve as a general guide.

The Isectolophidae were almost exclusively a North American radiation and appear to have been closest to the original primitive perissodactyl type (Radinsky 1963, 1969). The Lophiodontidae were apparently an early specialized offshoot, more or less limited to the European Eocene, and were primitive in that they retained a good deal of ectoloph shear in the molars (Radinsky 1965a). The Tapiridae are fairly clearly derivable from the Helaletidae, probably from an early member (with a tetradactyl manus) of the late Eocene and early Oligocene genus *Colodon* (Radinsky 1963, 1965b). They share with the later helaletids the derived character of a retracted nasal incision, indicating the possession of a proboscis. The Hyrachyidae appear to be derived from earlier helaletids and were probably the stock from which the Rhinocerotidae were derived (Radinsky 1969). A problem arises in where to place the late Eocene Asian families

Fig. 3. Cladogram of the Tapiroidea within the order Perissodactyla.

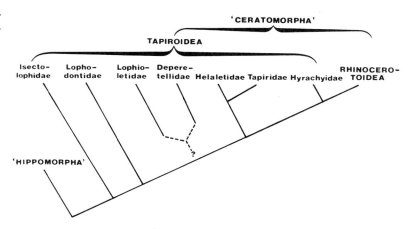

Lophioletidae and Deperetellidae. Radinsky (1965a) claims that they were more similar to helaletids than to lophiodontids or isecto-lophids, but does not enter into a detailed dis-cussion of character states.

The sister-group to the Tapiridae is thus the Helaletidae, a predominantly North American Eocene and Oligocene radiation of tapiroids. Radinsky (1965b) compares the skeleton of *Heptodon* (an Early Eocene helaletid) with that of *Tapirus* and shows that the differences be-tween them are slight. The main differences are that *Tapirus* is about 40% larger than *Heptodon* and has a relatively larger head (see Fig. 1), these facts alone accounting for most of the de-tailed differences in the postcranial skeletons of the two genera. *Tapirus* has relatively shorter and wider cervical vertebrae, with less flexibil-ity between the vertebrae in both neck and trunk region than in *Heptodon*. *Tapirus* also has a more expanded iliac blade and shorter and broader metapodials, features typical of a more graviportal type of animal than *Heptodon* and presumably related to the larger body size of the extant genus. However, *Tapirus* retains the primitive perissodactyl condition seen in *Hep-todon* of a tetradactyl manus and an unreduced fibula and ulna (although the ulna is fused to the radius at proximal and distal ends). The limb bones are in general more robust, with a greater prominence of the tuberosities for muscle at-tachment and a greater extent of the lateral epi-condyles at the joints, but this is all in keeping with the larger body size of *Tapirus*. The fore-limb of *Tapirus* shows a greater degree of curso-rial adaptation than that of *Heptodon*, with a

reduction of the acromium of the scapula (asso-ciated with the loss of the clavicle), a wider lateral epicondyle of the humerus, partial fusion of the radius and ulna, and a relatively wider and shorter radio-carpal articulation. However, this may not so much reflect a greater cursorial capacity in *Tapirus* as the need to maintain the same relative degree of cursoriality in an animal of larger body size . Thus the postcranial skele-ton of *Tapirus* is basically unchanged from that of an early Eocene perissodactyl.

A greater degree of modification is seen in the skull and dentition of *Tapirus* from the condi-tion in *Heptodon*. The most significant differ-ence is the retraction of the nasal incision and the shortening of the nasals, together with a higher position of the nasals and the develop-ment of the frontal sinus. This is a reflection of the possession of a mobile proboscis in *Tapirus*. The nasal diverticulum is displaced from its po-sition inside the nasal incision (as seen in living horses and rhinos) to a groove lying lateral to the incision, running along the maxilla and fron-tals and terminating with an anteroventral curl at the posterior border of the nasals. The brain of *Tapirus* shows a relative increase in the size of the cerebral hemispheres over the condition in *Heptodon*. The molar cusp pattern is similar, but the wear is restricted to the cross lophs and the premolars are almost completely molari-form, as opposed to the condition in *Heptodon*, where a certain amount of ectoloph shear ex-isted and the premolars were not molarized. Correlated with this increase in anterior extent of the occlusal surface in *Tapirus* is a forward shift in the area of origin of the masseter muscle, moving from a position over the middle of the second molar in *Heptodon* anteriorly to above the first molar in *Tapirus*. A specialized feature of *Tapirus* is the atrophy of the upper canine and the enlargement of the third upper incisor into a caniniform tooth (see Radinsky 1965b).

Protapirus, of the North American Late Oligocene, was intermediate in these conditions between *Heptodon* and *Tapirus,* although closer to *Tapirus,* but itself was too late in time to have been the direct ancestor of *Tapirus* (Radinsky 1965b). The first appearance of *Tapirus* was in the species *Tapirus helveticus* of the middle and late Oligocene of Europe (Schaub 1928). This was almost identical in anatomy to the living species, except the premolars were not as molarized. Thus the modern genus *Tapirus* is essentially unchanged since the middle Oligocene. *Tapirus* persisted as a rare, conservative genus in both Old and New World faunas until the Pleistocene, when it became restricted to its present geographic range. The Miocene of North America saw a slightly greater diversification of tapirs with the presence of *Miotapirus* and *Tapiravus*. However, these genera were of the same body size and general morphology as living tapirs. The only divergence from the conservative tapir type was the Asian genus *Megatapirus,* found in the Pleistocene of China, which had a skull some 18 in. in length and must have had a body weight approaching 800 kg. These animals may have been occupying a hippopotamus type of niche.

The sister-group of the Tapiroidea is the Rhinocerotoidea. Rhinocerotoids first appeared in the late Eocene and diversified and radiated during the Oligocene (Radinsky 1969), in direct contrast to the pattern of the evolutionary radiation of the tapiroids. They obviously represented the real adaptive solution of the tapiroid lineage to the problems caused in the late Eocene by climatic changes and artiodactyl competition. The rhinocerotoids differed from the tapiroids in the possession of molars that were higher crowned, with the enhancement of the ectoloph shear, which presumably reflected a difference in diet and foraging strategy. (It is of interest to note that similar changes in the molars occurred independently in the late Eocene in the tapiroid families Lophiodontidae and Lophioletidae, indicating a general adaptive

trend at this time towards the evolution of a rhinocerotoid ecological type [Radinsky 1969].) Rhinocerotoids also tended toward large body size, although they have displayed a diversity in body size throughout their evolutionary history. Most of them were larger than tapiroids, but the diceratherine rhinocerotids were about the same size as a living tapir, and the hyracodontids considerably smaller. However, large body size, with body weights ranging from 800 kg to 2500 kg, appears to have been the predominant feature of their evolution, and presumably was contributory to their ability to withstand competition from ruminant artiodactyls, since the ruminant digestive strategy is only efficient at body weights of under approximately 1000 kg (Van Soest 1981). Rhinocerotids were phylogenetically diverse and individually numerous among the faunas of North America, Europe, Asia, and Africa in the middle and late Tertiary, with the presence of hypsodont grazing forms as well as the less-derived brachyodont browsing genera. This is in direct contrast to the Tapiridae, which were conservative in morphology and feeding behavior, and low in diversity and absolute numbers during this period.

The extinction of most of the lineages of the Tapiroidea at the end of the Eocene was probably due to a combination of climatic change and artiodactyl competition, as suggested by Radinsky (1965a). Despite contention that there is no evidence for competition and ordinal replacement between artiodactyls and perissodactyls (Cifelli 1981), the differences in biology between the two orders make them differentially sensitive to changes in climate and vegetational types. It was not so much that the ruminant artiodactyls, when they evolved, were "better" than the tapiroids and so outcompeted them in the folivorous niche, but rather that the global changes in climate at the end of the Eocene (see Wolfe 1978) favored the ruminant artiodactyl type of foraging strategy over that of the tapiroids in what were then more temperate, seasonal latitudes. Tapiroids appear to have been specialists among the perissodactyls in the middle fiber range of foliage during the Eocene, whereas equids took higher fiber content foliage, and brontotheres and chalicotheres took foliage of lower fiber content (Janis, in preparation). The perissodactyl hindgut site of cellulose fermentation means that these animals require a greater daily intake of food than ruminant artiodactyls, which have a forestomach site of fermentation. Ruminants are limited by the quality of food that they can eat in a day but are better able to make maximal use of a limited amount of food than perissodactyls (see Janis 1976, 1979).

In a nonseasonal tropical forest habitat, the foliage is in general of poor quality, and the distribution of cellulose between the different parts of the plant is fairly uniform (Deinum and Dirvan 1972; Deinum 1973). A folivore in such a habitat would have to have a high daily intake of food in order to meet its nutritional requirements, and a perissodactyl type of feeding strategy would be the most adaptive. The artiodactyls of the early and middle Eocene were not folivorous, being either omnivorous or highly selective browsers taking only the nonfibrous growth parts of the plants (Janis 1976; in preparation). However, in the more seasonal conditions of the post-Eocene, the overall protein content of the herbage would have been greater, with a greater differentiation between fiber content of leaf and stem. Such vegetational conditions would now favor the foraging strategy of ruminant artiodactyls, which could then select the low-fiber parts of the plants and survive on a small total volume of food than a perissodactyl of similar body size.

Tapirs today live primarily in tropical forests, where there is still a paucity of folivorous ruminant artiodactyls, presumably because the high daily intake required for a diet of tropical foliage is not compatible with the limited intake capacity of ruminants. However, tapirs survived in limited numbers throughout the temperate latitudes during the Tertiary, becoming extinct in the Pleistocene in these areas presumably because of the disappearance of mesic woodland in more northern latitudes at this time (see Leopold 1968). The middle and late Tertiary tapirs may have been able to maintain themselves at low numbers in seasonal woodland, despite ruminant competition, by adopting the strategy of selecting only nonfibrous herbage. Here they would have an advantage over ruminant artiodactyls because they would not be constrained by having to ferment all ingested material prior to its entrance to the rest of the digestive tract. This process in ruminants destroys many of the other nutrients in the diet and is only advantageous if the diet contains a considerable amount of cellulose (Janis 1976; Van Soest 1981). The possession of a mobile proboscis in the Tapiridae suggests that they adopted a more selective mode of feeding, which may have enabled them to adopt this proposed foraging strategy at this time. The retraction of the nasal incision was seen in Helaletidae, but since it was not immediately accompanied by a reduction in the size of the nasals it cannot initially have reflected the presence of a mobile proboscis (Radinsky 1965b). Decrease in the size of the nasals was first seen in the early Oligocene genus *Colodon hancocki,* which is thought to be ancestral to the Tapiridae (Radinsky 1965b). Radinsky suggests that this modification of the nasal incision initially provided space for an enlarged nasal diverticulum and then served as an anatomical preadaptation in allowing the hypertrophy of the musculature of the upper lip and the subsequent formation of a mobile proboscis. (The function of the nasal diverticulum in perissodactyls is obscure, but its enlargement may have somehow been related to an increased sensitivity in the sense of smell, allowing tapirs in the late Eocene to search out more succulent food items, such as underground roots, at times of seasonal scarcity.) If the Helaletidae and their derivatives, the Tapiridae, were indeed the only tapiroid lineage with this fortunate preadaptation of the nasal region, allowing for the rapid development of a more selective mode of foraging in the face of the environmental conditions of the Oligocene, it would explain why the members of this lineage were the only ones able to survive past the end of the Eocene.

Thus, the genus *Tapirus* can be seen to be a relict of the Eocene radiation of folivorous perissodactyls, which selected herbage in the middle fiber content range, as do most folivorous ruminant artiodactyls of the present day. By virtue of the fortuitous fact that its ancestors had modified the nasal area in the retraction of the nasal incision, the Tapiridae were able to evolve a mobile proboscis in the early Oligocene and so modify their feeding strategy to survive in the face of climatic change and artiodactyl competition. However, the postcranial skeleton of *Tapirus* remains essentially unchanged from that of an early Eocene perissodactyl, and its skull and dentition are essentially unchanged since the first appearance of the genus in the middle Oligocene. Living tapirs live almost exclusively in tropical forest environments and presumably have had a style of feeding and social behavior essentially unchanged from that of middle Eocene perissodactyls. I have summarized my conclusions on the status of tapirs as living fossils in conjunction with my treatment of tragulids (Chapter 8, this volume). Both are herbivorous ungulates whose evolutionary radiations and distributions can be seen to be related to changes in global climate and vegetation throughout the Tertiary period.

Literature

Cifelli, R. L. 1981. Patterns of evolution among the Artiodactyla and Perissodactyla (Mammalia). Evolution 35(3):433–440.

Deinum, D. 1973. Preliminary investigations on the digestibility of some tropical grasses grown under different temperature regimes. Surinamse Landbouw 2:121–126.

Deinum, D., Dirven, J. G. P. 1972. Climate, nitrogen and grass. 5. Influence of age, light intensity and temperature on the production and chemical composition of Congo grass (Brachiana ruziziensis, Germain et Everard). Neth. J. Agric. Sci. 20:125–132.

Grzimek, B. 1972. Animal life encyclopedia, Vol. 13 (Mammals IV). New York: Van Nostrand Reinhold.

Janis, C. 1976. The evolutionary strategy of the Equidae, and the origins of rumen and cecal digestion. Evolution 30: 757–774.

Janis, C. 1979. Aspects of the evolution of herbivory in ungulate mammals. Ph.D. Diss., Harvard U.

Leopold, E. B. 1968. Late Cenozoic palynology, pp. 377–438. In: Tschudy, R. H., Scott, R. A. (eds.), Aspects of palynology. New York: Wiley.

Radinsky, L. B. 1963. Origin and early evolution of North American Tapiroidea. Bull. Peabody Mus. Nat. Hist. 17:1–106.

Radinsky, L. B. 1965a. Early Tertiary Tapiroidea of Asia. Bull. Amer. Mus. Nat. Hist. 129(2):185–263.

Radinsky, L. B. 1965b. Evolution of the tapiroid skeleton from Heptodon to Tapirus. Bull. Mus. Comp. Zool. 134(3):69–106.

Radinsky, L. B. 1969. The early evolution of the Perissodactyla. Evolution 23(2):308–328.

Schaub, S. 1928. Der Tapirschädel von Haslen. Ein Beitrag zur Revision der oligocänen Tapiriden Europas. Abh. Schweiz. Palaeont. Gesell. 47:1–28.

Van Soest, P. 1981. The nutritional ecology of the ruminant. Corvallis, Or: O. & B. Books.

Wolfe, J. A. 1978. A paleobotanical interpretation of Tertiary climates in the Northern Hemisphere. Amer. Sci. 66:694–703.

Zannier-Tanner, E. 1965. Vergleichende Verhaltensuntersuchungen über das Hinlegen und Aufstehen bei Huftieren. Z. Tierpsychol. 22:636–723.

8
Tragulids as Living Fossils

Christine Janis

Division of Biology and Medicine, Brown University, Providence, RI 02912

Tragulids, also known as chevrotains or mouse deer, belong to the family Tragulidae of the order Artiodactyla (even-toed ungulates). They are small, hornless ungulates and, as the term "mouse deer" suggests, look rather more like large rodents or rabbits than like true deer (see Fig. 1). The order Artiodactyla also includes suoids (suids, i.e. pigs and tayassuids, i.e. peccaries), deer, camels, bovids, and giraffes. The family Tragulidae is classified within the infraorder Tragulina in the suborder Ruminantia (see Fig. 2). The sister-group to the Tragulina is the infraorder Pecora, which in the present day represents the most numerous, diverse, and geographically widespread group of ungulate mammals.

Tragulids are found today in west and central Africa and in Asia. The Tragulidae has apparently always been an exclusively Old World family, although other families within the Tragulina, which are now extinct, were found in North America and Asia (see Fig. 3). Two living genera of tragulids are currently recognized, with 4 species and 56 subspecies. The largest of these is the African water chevrotain, *Hyemoschus aquaticus*, which has a body weight of 10–15 kg. The Asian genus, *Tragulus*, is smaller than this. The largest species is *Tragulus napu*, the larger Malay mouse deer, which has a body weight of 5–8 kg. The lesser Malay mouse deer, *Tragulus javanicus*, has a body weight of 2–2.5 kg. There are 28 subspecies of both Malayan species of mouse deer. Finally, there is the spotted mouse deer, *Tragulus meminna*, which is found in India and has a body weight of 2.25–2.70 kg. However, Groves and Grubb (in press) claim that the Indian tragulid should not be included in the genus *Tragulus*, as it is at present, but be assigned to a separate genus of its own, *Moschiola*.

Tragulids lack bony horns, which are prominent features of pecoran ruminant artiodactyls. The male is slightly smaller than the female and possesses large saberlike upper canines, while the female possesses much smaller canines. They are also primitive for ruminant artiodactyls in their possession of short legs and a short neck, with a tetradactyl manus and pes, and the presence of a gall bladder and appendix. The scrotum in the male is not sharply defined. Tragulids are in general nocturnal or crepuscular animals. They become sexually mature at 9–20 months and may live for 8–12 years. One young is produced at a time in *Hyemoschus* and two in *Tragulus* (see Grzimek 1972; Kingdon 1978).

Hyemoschus differs in its choice of habitat from *Tragulus*, being semiaquatic, whereas the species of *Tragulus* live in relatively drier habitats. The third and fourth metacarpals are not fused to form a distinct cannon bone in *Hyemoschus*, whereas they are in *Tragulus*. *Hyemoschus* also lacks a chin gland, and the intermaxillary bone does not reach the nasal, in contrast to *Tragulus* where the intermaxillary

87

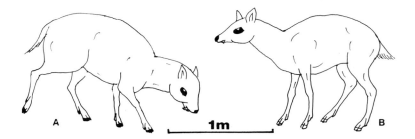

Fig. 1. (A) Recent African water chevrotain *Hyemoschus aquaticus* (modified from Kingdon 1978). (B) Oligocene North American *Leptomeryx,* a supposedly more open-country traguline.

and nasal are in contact. A chin gland is present in both Malayan species of *Tragulus* but is lacking in *Tragulus meminna.* For this reason, and also for the reason that the Indian mouse deer shares a spotted and striped coat with the African chevrotain, as opposed to the plain brown coat found in the Malayan mouse deer, it has been supposed that *T. meminna* is more closely related to the genus *Hyemoschus* than are *T.*

napu or *T. javanicus,* and it is sometimes placed in a separate subgenus *Moschiola* (see Grzimek 1972; Groves and Grubb, in press).

Tragulids are highly selective browsers, taking items of herbage that are high in protein content and that require minimal fermentation, such as fallen fruit, young leaves, buds, berries, and the young parts of some grasses. In addition to plant material, they have also been ob-

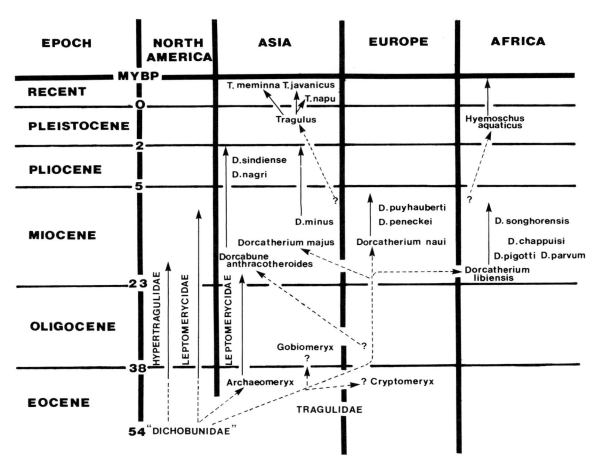

Fig. 2. Distribution of the Tragulina in space and time (family Tragulidae in detail). (Species of *Dorcatherium* and *Dorcabune* included to show diversity of genus and temporal position, and are not meant to indicate ancestor-descendant relationships).

Fig. 3. Cladogram of the Tragulina within the order Artiodactyla (modified from Webb and Taylor 1980).

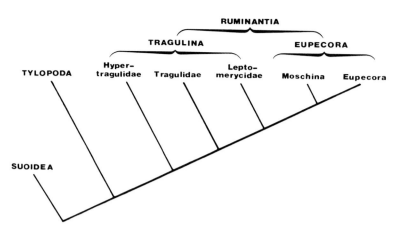

served to eat invertebrates, fish, small mammals, and carrion (Dubost 1963, 1975; Kingdon 1978). In their herbivorous habits they are dependent on the production of the young growth, nonfibrous parts of the plants, since ruminants of this small body size (or at least those under 5 kg in weight) are unable to cope with the larger intake necessitated by a diet of more fibrous vegetation (see Janis 1976). They are thus restricted in their distribution to habitats where such a diet is available year round. Tragulids have buno-selenodont molars and long, sectorial, nonmolariform premolars. Their molar morphology and wear is very similar to that of the tragulines of the Late Eocene and Early Oligocene, and presumably their diet and foraging stragies have changed little since then (Janis 1979; in preparation). Like other members of the Ruminantia, tragulids have a forestomach enlarged into a rumen as a chamber for cellulose fermentation. But their rumen is small, and primitive in that it resembles the rumen of a fetal, rather than an adult, pecoran. The third of the traditional four stomachs of ruminants, the omasum, is small or absent altogether in tragulids (Langer 1974). Tragulids search for their food with their snouts, like pigs, and they are incapable of rising up on their hind legs to browse, as are pecoran artiodactyls (Dubost 1975).

Tragulids are not only primitive for ruminant artiodactyls in aspects of their feeding and digestive behavior, but are also apparently primitive in their social behavior. They are predominantly asocial animals and avoid other individuals. The females have home ranges that are apparently without overlap, and the males have larger home ranges that encompass those of several females. However, they show no evidence of the complex territorial behavior seen in pecorans. They do not mark out their home ranges, and feces and urine are deposited anywhere rather than being used to demarcate boundaries, although they may use the interramal gland occasionally for marking twigs (Dubost 1975). Their fighting behavior is ritualized, as in pecorans, but the males fight in the primitive mammalian antiparallel stance, rather than in the head-to-head posture adopted by more advanced ungulates, and they slash at each other's neck and sides with the long canines (Ralls et al. 1975). The dense, tough skin over the dorsal area probably protects them from injury during these fights, as well as protecting them from abrasion in the dense undergrowth of their habitat (Kingdon 1978). Fighting and competition between males appears to be intense, as the sex ratio of males to females in the wild decreases markedly with age (Kingdon 1978). Despite the fact that fights occur between both males and females, there is no demonstration of rank order or obvious displays of aggression and submission typical of more advanced ungulates in captive groups of chevrotains (Dubost 1975).

The courtship of tragulids also lacks many of the derived features seen in the higher ruminants. The male finds the female by olfaction (Dubost 1975). The male tests the female's urine to see if she is in estrus, but he does not perform the "Flehmen" lip curl characteristic of more advanced ungulates (Ralls et al. 1975). Other primitive features of courtship behavior include rhythmical vocalizations by the courting male and the pressing of the chin of the male on the back of the female in order to test her sexual receptivity. The nursing posture of the female is also primitive for ungulates (Ralls et al. 1975). The play of the young is simple, and there is little contact with peers (Dubost 1975).

Tragulids rely little on vocal communication, and do not employ the use of facial expressions or ear signaling in communication, as seen in more advanced ungulates. They have a bold black and white patch of coarse hair under the throat that may be important in communication, in conjunction with the orientation of the neck and carriage of the head and the general body posture (Kingdon 1978). They lie down by means of an intermediate, sitting-on-the-haunches position, which is assumed to be primitive for ungulates (Zannier-Tanner 1965), as is the lying-down position with the back arched and the legs folded underneath the body (Dubost 1975; Kingdon 1978). No mutual grooming is seen between adults, as with pecorans (Dubost 1975). They rely on freezing and cryptic coat coloration to avoid predation and may "play dead" if caught (Kingdon 1978). However, this type of behavior is common to a number of small forest-browsing ruminants (Jarman 1974). Dubost (1975) considers tragulids to be much more similar to suids in their overall behavior than they are to higher ruminants and claims that *Hyemoschus* is more suidlike in its behavior than *Tragulus*. However, because tragulids have a different type of diet, habitat preference, and overall life style from most suids it is unlikely that this similarity in behavior is due to convergence, and anatomical evidence mitigates against tragulids being more closely related to suids than to pecorans. It seems more probable that both suids and tragulids have retained elements of the original primitive behavior of the mammalian stock ancestral to both lineages. (Suids are the least modified in their anatomy from the primitive mammalian condition, and apparently this retention of primitive characters also extends to their behavior. However, living species of suids are all derived members of the suoid lineage, and are not representative of the primitive condition of the group in the way that tragulids are apparently a primitive offshoot from the ruminant artiodactyl lineage.)

Tragulids have long been regarded as "living fossils." The first complete skeleton to be discovered of an early primitive ruminant artiodactyl was that of *Archaeomeryx*, of the late Eocene of Mongolia (Matthew and Granger 1925). It was claimed by Colbert (1941) that this animal differed little from present-day tragulids. However, for a long while the exact systematic position of *Archaeomeryx*, and the information that is really presented about the primitive nature of tragulids, remained unclear. A large number of small, apparently primitive, hornless ruminants existed during the early and middle Tertiary in both North America and the Old World, and their interrelationships and the exact systematic positions of fossil and living groups have only recently been subjected to any rigorous form of systematic analysis (Webb & Taylor 1980). Figures 2 and 3 reflect the phylogeny of Webb and Taylor, who show that tragulids are in fact early offshoots of the ruminant artiodactyl lineage, and genuinely primitive with regard to numerous morphological features.

Tragulids are clearly united with other members of the Ruminantia by the fusion of the cuboid and navicular bones in the tarsus, which is a unique specialization of the suborder. However, they remain primitive in a number of features, which they hold in common with the family Hypertragulidae, and which apparently reflect the primitive artiodactyl condition. These include numerous features of the basicranium and petrosal area; the broad lateral exposure of the mastoid, with a small, posteriorly positioned stylohyoid vagina; the short and peg-like odontoid process of the axis (as opposed to the spoutlike shaped process of higher ruminants); the short limbs with unfused full-length lateral digits II and IV, and with the central metapodials unfused or partly fused; the incomplete distal keel on the metapodials; and the elongate and relatively narrow astragalus, with the distal ginglymi medially deflected. However, some features of the Tragulidae show them to be less primitive than the Hypertragulidae, and unite them with the higher ruminants, including *Archaeomeryx*, which has now been ascertained to be a leptomerycid (Webb and Taylor 1980). These include the fusion of the trapezoid and magnum bones in the carpus; the absence of the trapezium in the carpus; a distinct fibular malleolar bone and the loss of the first metacarpal; and the confluence of the jugular foramen with the posterior lacerate foramen. The Tragulidae also possess a number of uniquely derived features, which primarily involve the areas of the postorbital bar and the auditory bullae, and stem from the fact that the postorbital bar was apparently closed and the auditory bullae enlarged in a manner parallel to, but independent from, the acquisition of this condition by higher ruminants. Recent tragulids are also modified in the fusion of the cubonavicular with the ectomesocuneiform in the tarsus (Webb and Taylor 1980).

Thus tragulids must have branched off from the ruminant artiodactyl lineage by the middle Eocene, because the first leptomerycids are known from the late Eocene, despite the fact that there is no fossil record of the family prior to the early Miocene. Although tragulids have independently acquired a few features seen in the higher ruminants, such as the closure of the postorbital bar and the greater degree of fusion of the tarsal elements, in most respects they are essentially unmodified from the morphological condition seen in the Tragulina of the late Eocene and early Oligocene.

Chromosomal studies of the two Malayan species of *Tragulus* shows that they have a diploid number of 32, which is lower than that of most higher ruminants (Yong 1973; Todd 1975). Living cervoids and bovoids have the karyological distinction of an X-autosome translocation-fusion, or evidence of this having been the case at some time in their evolutionary history (Todd 1975). The X chromosome of living suids and camelids is of the simplex primitive mammalian type, and that of *Tragulus*, while there is no translocation-fusion, is not of the simplex type but shows a degree of modification that would be expected to preceed a karyological event of this nature (Todd 1975). This suggests that tragulids are related to pecorans, but were isolated from this group prior to the incorporation of this feature. However, there are certain problems with Todd's analysis and interpretation of chromosome configuration in ruminant artiodactyls (see Scott and Janis, in press).

Duwe (1969) interpreted his immunological studies on skeletal muscle antigens as showing that *Tragulus javanicus* has more features in common with suoids than with higher ruminants. However, as pointed out by Webb and Taylor (1980), such studies only show that tragulids share primitive artiodactyl features with suoids, and neither support nor deny the concept of a special phylogenetic relationship between them, although they do testify to the antiquity of the divergence of the tragulids from the lineage leading to present-day pecorans. As far as I am aware, no studies of this nature have been done comparing living species of tragulids with one another.

The first undoubted fossil tragulid was *Dorcatherium*, from the early Miocene of Africa and Eurasia. However, earlier fragmentary material has been referred to the Tragulidae by various workers, for example *Gobiomeryx* from the Oligocene of Mongolia (Trofimov 1958; Musakulova 1963), and *Cryptomeryx* from the Late Eocene of Bavaria (Schlosser 1886). *Dorcatherium* is virtually indistinguishable from *Hyemoschus*, differing only in the greater variety of body sizes (ranging in weight from approximately 5 kg to 50 kg), the partial persistence of the lower first premolar, the contact of the premaxilla with the nasal, and the retention of occasional primitive dental characters such as the fact that the premolar row tends to be slightly longer relative to the molar row (Whitworth 1958; Gentry 1978).

Dorcatherium was a fairly widespread genus during the middle Miocene, with a moderate amount of specific diversity, the species differing from each other primarily in the feature of body size. These species all had buno-selenodont molars, but some displayed a greater degree of development of selenodont crests, suggesting a more folivorous diet. In addition to *Dorcatherium*, the genus *Dorcabune* was also present in Asia during the Miocene and Pliocene. This was a large tragulid, which had very bunodont molars and would have weighed close to 100 kg. Its molar wear showed a greater amount of tip crushing than seen in other tragulids, suggesting a more suoidlike omnivorous diet. This genus is known only from dental material, and though Colbert (1935) claims that it is clearly a tragulid, its status has been questioned by Gentry (1978), who thinks that it may actually be an anthracothere (an extinct type of suoid). *Dorcatherium* disappeared from the fossil record of Europe and Africa during the late Miocene, although it persisted in Asia until the late Pliocene. The first fossil record of the living genera of tragulids is from the Pleistocene (Romer 1966), with the exception of the doubtful *Tragulus sivalensis,* known from a single molar from the middle Siwaliks of Pakistan (Colbert 1935). Whereas *Hyemoschus* is very similar to *Dorcatherium,* and may in fact represent the same genus (Gentry 1978) *Tragulus* appears to be somewhat more specialized from the tragulid lineage in aspects of both morphology (Grzimek 1972) and behavior (Dubost 1975).

Thus tragulids had a moderate radiation in the present-day subtropical and temperate latitudes during the middle Miocene, but following the climatic and vegetational changes in the northern hemisphere of the Old World in the late Miocene (see Leopold 1968; Wolfe 1978), they became restricted to more equatorial regions. Their molar morphology suggests that they have always been limited to habitats where a year-round supply of nonfibrous browse was available, and the fossil record evidence suggests that they retreated back toward the equator together with this type of vegetation during the deterioration of the climate during the Late Tertiary. Other members of the Tragulina, the hypertragulids and leptomerycids, may have been more open-country living animals than the forest-dwelling tragulids. In North America, taphonomic evidence exists to show that *Leptomeryx* lived in open-plains areas (Clark et al. 1967), and North American tragulines had longer legs and more selenodont molars than recent and fossil tragulids. They may have had a foraging strategy similar to that of present-day small, open-country bovids, such as Thomson's gazelle (*Gazella thomsoni*), which are highly selective grazers, taking the seeds and young growing parts of the grass plants and including a good deal of dicotyledonous material (see Bell 1969). However, North American tragulines also appear to have been highly sensitive to climatic changes, since they had all disappeared by the end of the early Miocene.

The sister-group to the Tragulina is the Pecora, which includes the Moschina and Eupecora of Webb and Taylor (see Fig. 2). (However, see Scott and Janis [in press] for a more recent analysis of the systematic position of the Moschidae within the Pecora.) Moschines were of similar body size to the larger tragulines, and also lacked bony horns and possessed large canines, although their molar morphology and wear indicates that their diet was more folivorous. The one surviving species, *Moschus moschiferous* (the musk deer), has a digestive system that appears to be similar to that of other pecorans, and moschines appear to have been better adapted to deal with the leafy diet available in more temperate habitats than were tragulines. The fossil evidence suggests that moschines took over the traguline niches in the more seasonal conditions of the later Tertiary in both Eurasia (Flerov 1971) and North America (Webb 1977) and were possibly also present in Africa during the Miocene (Scott and Janis, in preparation). However, their success was relatively short-lived, and the surviving *Moschus* is restricted to central and east Asia.

However, considering the Pecora as a whole to comprise the sister-group to the Tragulina, the Eupecora have undergone an outstandingly successful radiation since the Middle Tertiary, with the Bovidae and Cervidae being the most numerous and geographically widespread ungulate families of the present day. The Eupecora are in general of considerably larger body size than the Tragulina, although some small genera exist. Most pecorans weigh between 70 and 200 kg, although some genera may weigh up to 1000 kg. The metapodials are fused in all pecorans, and many genera have highly elongated limbs, with complete loss of the side toes and elongated necks. The molars are selenodont and the premolars are partially molarized in all families, and the cheek teeth are hypsodont in the Bovidae and Antilocapridae. In addition, all living eupecorans (with the exception of the cervid *Hydropotes*) have bony horns or hornlike organs in the males (with horns also present in the females of some species), and they display complex behavior in their social interactions, with the maintenance of territories and the establishment of rank hierarchies among individuals. They also display complex and ritualized courtship behavior (see Leuthold 1977). Eupecorans have diversified into a variety of feeding types, in correlation with the elaboration of the dentition and digestive system (with the enlargement and cornification of the rumen and the increase in size and complexity of the omasum), with many genera sustaining themselves partially or entirely on a diet of grass.

In short, the Eupecora represent an adaptive radiation that took advantage of the climatic changes of the middle and late Tertiary, radiating into a wide variety of open-country or temperate woodland forms that could utilize a diet of high fiber content and seasonal availability. In contrast, the Tragulidae have stayed essentially unchanged in morphology, diet, and social behavior since the end of the Eocene, and have become restricted in their individual numbers, diversity, and geographic range along with the restriction in availability of their original type of nonseasonal forest habitat.

Summary: Evidence for Genuine Living Fossils Among Living Ungulates

I have chosen the families Tapiridae and Tragulidae as examples of living fossils, because they both appear to represent relict groups that were diverse and widespread during the Eocene, when there was a broad extent of tropical forest throughout northern latitudes (Wolfe 1978). Apparently, these groups have remained in essentially the same habitat since these times and so have become "frozen" in their evolution. The evidence for this viewpoint comes from four different areas:

1. The fossil record shows that contraction in geologic range and diversity of the families is broadly coincident with the decreasing availability of suitable habitats throughout the latter parts of the Tertiary.
2. Studies of dental wear of living and fossil forms suggest that the diets of these animals have remained essentially unchanged since the end of the Eocene.
3. Studies of the morphology of both living and fossil forms show that little post-Eocene change has occurred.
4. Studies of the behavior of the living animals reveal the lack of many of the derived features characteristic of most ungulates. Many features of their behavior may merely be correlated with a forest-dwelling, browsing habitat rather than being truly "primitive," such as the lack of territoriality and the solitary behavior (Jarman 1974). However, both tapirs and tragulids lack the elements of communication and display seen in most ungulates, which are associated with the maintenance of dominance hierarchies in social

groups, and also show simple behavior with regard to their courtship and reproductive activities. The complex behavior in other ungulates appears to have been developed independently in the different families, in response to the increased frequency of social interactions in more open environments (Leuthold 1977). If tapirs and tragulids had secondarily adapted to their current habitats, it would be expected that they would show evidence of the complex behavior typical of other ungulates. The absence of this suggests early divergence from the other families of ungulates in their orders and the retention of the original ungulate habitat type. Both families also retain other apparently primitive mammalian types of behavior, such as the postures adopted when lying down or when suckling their young.

Various other extant ungulates might be thought to represent examples of living fossils, although none of these represent such a clearcut case as tapirs or tragulids, The three living genera of hyraces resemble Eocene perissodactyls in many aspects of their dental and postcranial anatomy, but the family Procavidae appears to be a fairly late and specialized offshoot of the order Hyracoidea (Meyer 1978). Although they retain many morphological features characteristic of primitive members of other ungulate orders (Meyer 1978), living hyraces appear to be specialized in many aspects of their physiology and behavior (Janis 1979, 1983).

Among the Perissodactyla, the rhinocerotoid genus *Dicerorhinus* has been in existence since the late Oligocene (Romer 1966), and the living Sumatran rhino, *Dicerorhinus sumatrensis* may well be a genuinely primitive example of a relatively small, browsing, forest-dwelling rhino. Among the Artiodactyla, the musk deer, *Moschus moschiferous,* as already mentioned, is the sole surviving member of the Moschina (see Webb and Taylor 1980) and appears to be very similar in morphology to Miocene moschids. The okapi, the rare forest-living giraffid, appears based on morphological evidence to be a very primitive member of the giraffoid radiation, although the living genus lacks a fossil record (Hamilton 1978). Since chromosomal evidence indicates that giraffids are the most primitive of the Eupecora (Todd 1975), *Okapia johnstoni* may well represent a true relict example.

Literature

Bell, R. H. V. 1969. The use of the herb layer by grazing ungulates in the Serengeti, pp. 111–128. In: Watson, A. (ed.), Animal populations in relation to their food resources. Blackwells: Symp. Brit. Ecol. Soc. Oxford.

Clark, J., Beerbower, J. R., Kietzke, K. K. 1967. Oligocene sedimentation, stratigraphy, paleoecology and paleoclimatology in the Big Badlands of South Dakota. Fieldiana Geol. Mus. 5.

Colbert, E. H. 1935. Siwalik mammals in the American Museum of Natural History. Trans. Am. Phil. Soc. 26:1–401.

Colbert, E. H. 1941. The osteology and relationships of *Archaeomeryx*, an ancestral ruminant. Amer. Mus. Novit. 1135:1–24.

Dubost, G. 1963. Un ruminant a régime alimentaire partiallement carné: le chevrotain aquatique. C. R. Acad. Sci. Paris. 256:1359–1360.

Dubost, G. 1975. Le comportment du chevrotain Africain. *Hyemoschus aquaticus.* Z. Tierpsychol. 37:403–501.

Duwe, A. E. 1969. The relationship of the chevrotain, *Tragulus javanicus,* to the other Artiodactyla based on skeletal muscle antigens. J. Mam. 50(1):137–140.

Flerov, C. C. 1971. The evolution of certain mammals during the Late Cenozoic, pp. 479–491. In: Turekian, K. K. (ed.), The Late Cenozoic glacial ages. New Haven: Yale U. Press.

Gentry, A. W. 1978. Tragulidae and Camelidae, pp. 536–539. In: Maglio, V. J., Cooke, H. B. S. (eds.), Evolution of African mammals. Cambridge: Harvard U. Press.

Groves, C. P., Grubb, P. In press. Relationships of living Cervidae. In: Wemmer, C. (ed.), Biology and management of the Cervidae. Symp. Nat. Zool. Park. Washington, DC: Smithsonian Institution Press.

Grzimek, B. 1972. Animal life encyclopedia, Vol. 13 (Mammals IV). New York: Van Nostrand Reinhold.

Hamilton, W. R. 1978. Fossil giraffes from the Miocene of Africa and a revision of the phylogeny of the Giraffoidea. Phil. Trans. Roy. Soc. London B. 283(996):165–229.

Janis, C. 1976. The evolutionary strategy of the Equidae, and the origins of rumen and cecal digestion. Evolution 30:757–774.

Janis, C. 1979. Aspects of the evolution of herbivory in ungulate mammals. Ph.D. Diss. Harvard U.

Janis, C. 1983. Muscles of the masticatory apparatus in two genera of hyraces (*Procavia* and *Heterohyrax*). J. Morph. 176:61–87.

Jarman, P. J. 1974. The social organization of antelope in relation to their ecology. Behaviour 48:213–267.

Kingdon, J. 1978. East African mammals, Vol. IIIB. London: Academic.

Langer, P. 1974. Stomach evolution in the Artiodactyla. Mammalia 38(2):295–314.

Leopold, E. B. 1968. Late Cenozoic palynology, pp. 377–438. In: Tschudy, R. H., Scott, R. A. (eds.), Aspects of palynology. New York: Wiley.

Leuthold, W. 1977. African ungulates. Berlin: Springer-Verlag.

Matthew, W. D., Granger, W. 1925. New mammals from the Shara Marun Eocene of Mongolia. Amer. Mus. Novit. 196:1–11.

Meyer, G. E. 1978. Hyracoidea, pp. 284–314. In: Maglio, V. J., Cooke, H. B. S. (eds.), Evolution of African mammals. Cambridge: Harvard U. Press.

Musakulova, L. T. 1963. *Gobiomeryx* from the Paleogene of Kazakhstan. Mat. Ist. Fauny e Flory Kazakhstan 4:201–203.

Ralls, K., Barasch, C., Minkowski, K. 1975. Behaviour of captive mouse deer, *Tragulus napu.* Z. Tierpsychol. 37:356–378.

Romer, A. S. 1966. Vertebrate paleontology 3rd edn. Chicago: U. Chicago Press.

Schlosser, M. 1886. Beitrage zur Kenntris der Stammesgeschichte der Hufthiere und Versuch einer Systematic der Paar- und Unpaarhufer. Morph Jb. Bd. 12:1–136.

Scott, K. M., Janis, C. In press. Phylogenetic relationships of the family Cervidae. In: Wemmer, C. (ed.), Biology and management of the Cervidae. Symp. Nat. Zool. Park. Washington, DC: Smithsonian Institution Press.

Todd, N. B. 1975. Chromosomal mechanisms in the evolution of artiodactyls. Paleobiology 1(2):175–188.

Trofimov, B. A. 1958. New Bovidae from the Oligocene of central Asia. Vert. Palasiatica 2(4):243–247.

Webb, S. D. 1977. A history of savanna vertebrates in the New World. Part 1. North America. Ann. Rev. Ecol. Syst. 8:355–380.

Webb, S. D., Taylor, B. E. 1980. The phylogeny of hornless ruminants and a description of the cranium of *Archaeomeryx*. Bull. Amer. Mus. Nat. Hist. 167(3):121–157.

Whitworth, T. 1958. Miocene ruminants of east Africa. In: Fossil mammals of Africa Brit. Mus. (Nat. Hist.) Pub. 15.

Wolfe, J. A. 1978. A paleobotanical interpretation of Tertiary climates in the northern hemisphere. Amer. Sci. 66:694–703.

Yong, H. S. 1973. Complete Robertsonian fusion in the Malaysian lesser mouse deer (*Tragulus javanicus*). Experientia 29(3):366–367.

Zannier-Tanner, E. 1965. Vergleichende Verhaltensuntersuchungen über das Hinlegen und Aufstehen bei Huftieren. Z. Tierpsychol. 22:636–723.

9
Conceptual and Methodological Aspects of the Study of Evolutionary Rates, with some Comments on Bradytely in Birds

Joel Cracraft

Department of Anatomy, University of Illinois, Chicago, IL 60680

The analysis of evolutionary rates has received scant attention within ornithology. The primary reason would seem to be the nature of the avian fossil record: If an understanding of rates depends upon having a time dimension, which most paleontologists believe can only be extracted from fossil data, how can we hope to study rates using the notoriously poor record of birds? No one would deny that the avian record is less complete than other vertebrates or many groups of nonvertebrates, yet this cannot be the entire story, for there are easily over a thousand paleospecies of birds known, some of which provide information about rates. Another contributing factor, probably, is our relatively poor knowledge of avian phylogenetic relationships. All assessments of rates, whether absolute or relative, depend upon some hypothesis about the phylogenetic relationships of the taxa being studied.

Simpson (1944, 1949, 1953) pioneered the analysis of evolutionary rates based on the fossil record, and it is fair to say that even today most discussions of rate are still greatly influenced, conceptually and methodologically, by his important contributions. Although the mathematical analysis of evolutionary rates has been refined considerably since the time Simpson wrote, surprisingly little attention has been paid to a critical assessment of the systematic and evolutionary assumptions underlying Simpson's approach to rates or those of his successors. Most workers have perceived rates as a straightforward problem of measurement, taken of course within the context of an incomplete fossil record. But is the analysis of rates primarily a problem of measurement (i.e., finding the fossils, dating them, and then comparing quantitative and qualitative traits)? That question will be the subject of the first part of this paper. Following that, I will turn to the specific theme of this volume and discuss bradytelic (slow) evolutionary rates within birds.

The Analysis of Evolutionary Rates

Simpson's Classification of Rates

Simpson (1953) recognized three major groupings of rates: genetic, morphological, and taxonomic. Although he considered (1953:4) that genetic change within ancestral–descendant populations would be the "ideal" rate of measure, he devoted little discussion to this, obviously because his focus was paleontological. Thirty years later, of course, much has been published about the rates of nucleotide and amino acid substitutions for many different proteins, both within and between lineages. For their part, paleontologists have spent much time (until recently) measuring morphological rates, but especially in the last decade, with the emergence of macroevolutionary analysis and the study of Phanerozoic diversity patterns, there has been a shift toward the measurement of taxonomic rates.

Because Simpson's classification of evolutionary rates, his methodological approaches, and his underlying assumptions continue to have such a profound effect on the analysis of rates, it is important to review that classification. In the next section, I will examine some of the systematic and evolutionary assumptions of this work and discuss their implications for future investigations of evolutionary rate.

Simpson (1953) specified three kinds of morphological rates. The first described a quantitative rate change for a chosen character. Ancestral–descendant sequences of fossil taxa are identified, a character is measured, and from the stratigraphic distribution of these taxa a rate is calculated. The classic example, and one presented by Simpson himself, describes the change in tooth size within the lineage of horses (1953:11–17). Simpson's second type of morphological rate involves a change in "character complexes." Rather than measure change in an individual character, this rate attempts to measure the degree of change of entire organisms (or taxa). Probably the most well-known example, also discussed in detail by Simpson (1953), is Westoll's (1949) assessment of evolutionary change in fossil lungfish (Dipnoi). Westoll scored discrete character variability in terms of a primitive-derived scale and then obtained a composite score of "primitiveness" for each dipnoan genus from the Devonian to the Recent. Finally, Simpson (1953) identified a third form of morphological rate, lineage allomorphic rate, which is simply the amount of change in one character relative to another. The measured value is usually taken to be the allometric coefficient calculated from the allometric power function.

Within his categorization of taxonomic rates, Simpson (1953) proposed two subdivisions, phylogenetic taxonomic rates and taxonomic frequency rates. Phylogenetic taxonomic rate is itself divisible into "phyletic" and "group" rates. The concept of a "phyletic" rate was not defined precisely but was said to be measured either by the reciprocal of the generic durations for a particular group, or by dividing the number of genera by the total duration of the group as a whole to yield a "rate" of number of genera per unit time (usually per million years). Underlying Simpson's use of the term "phyletic" is the idea that the taxa are "successive" (ancestral–descendant?) to one another; he notes (1953:32), for example, that to qualify for this type of analysis, genera should "be essentially monophyletic in origin, that they have a significant extension in time, and that they be horizontally divided from preceding and following units of the same rank (even though such division is usually an artifact of taxonomy)." "Group rates," to Simpson, are simply the average durations of genera within a higher taxon. Plotted as survivorship curves, comparisons from group to group express differences in "generic turnover."

Simpson's second subdivision of taxonomic rate was that of "taxonomic frequency." The basic data are the relative frequencies of various taxa (of whatever rank) at a particular point in time or time interval. The slope of the frequency curve through time measures rate, but, as noted by Simpson, this simply reflects the rate of origination and extinction. As such, this form of taxonomic rate is not distinct from the preceding.

Assumptions of Rate Measurement

All measurements of evolutionary rate entail the use of certain systematic or evolutionary assumptions (Cracraft 1981a; Eldredge 1982; Novacek and Norell 1982). In the majority of cases, however, these assumptions are rarely, if ever, discussed and probably have been overlooked. Assumptions about the integrity of the stratigraphic record, on the other hand, are generally recognized by all workers and need not be discussed here (see Dingus and Sadler 1982).

Although the measurements of morphological and taxonomic rates share similar conceptual and methodological problems, each has some unique assumptions. Measurements of both kinds of rate must include a systematic assumption about the phylogenetic relationships of the taxa being investigated, and both must also make an assumption about the ontological status of those taxa. In addition, virtually all studies of morphological rates include an evolutionary assumption about ancestry and descent, whereas this is generally not a component of studies examining taxonomic rates. The latter, on the other hand, are exceedingly sensitive to the ''comparability'' of the taxa being counted, and this is a function not only of which concept of relationships has been adopted, but also of the ontological status of the taxa themselves.

Morphological rates. Quantitative rate change in single characters is usually calculated from measurements of selected taxa whose absolute temporal position can be estimated accurately. For this rate to be valid, the taxa must be of specific rank and must also be a true ancestral–descendant sequence. Morphological rates can be calculated, of course, within a temporal sequence of specimens assigned to a single species, in which case a prior hypothesis of conspecificity must have been accepted. The fact is that a large number of studies examining morphological rates do not use specific-rank taxa but instead calculate rates for a sequence of supraspecific taxa, usually genera (e.g., Patterson 1949; Westoll 1949). But unless all these genera are monotypic, which is rarely the case, such sequences are not legitimate evaluations of rates: Supraspecific taxa do not have any ontological status as ancestral evolutionary units (Wiley 1979, 1981; Eldredge and Cracraft 1980). The notion that a genus can give rise to another genus (or higher taxon) merely because the ancestral species is said to be a member of the former genus is simply taking the evolutionary status of species and reifying it to higher taxa. By definition, an ''ancestral'' genus is paraphyletic because the ancestral species is more closely related to its descendants than to other species within that ''genus''; the ancestral genus, therefore, does not have an ontological status as a discrete historical entity but is instead a taxonomic artifact.

Once it is determined that the taxa of a study of rates are species, then the hypothesis that they form an ancestral–descendant sequence must be addressed. Unfortunately, few investigations of evolutionary rate include a detailed character analysis to support a linear ancestral-descendant hypothesis. The systematic methods to construct such hypotheses have been presented elsewhere (Eldredge and Cracraft 1980; Wiley 1981), but their consequences for the analysis of evolutionary rates have been little discussed. Consider Fig. 1 in which four species (A–D) are represented by fossil samples at single points in time (to have each sample distributed over time would only serve to make the example more complicated and yet would not detract from the argument made here): in Fig. 1A each species is assumed to be defined by a combination of characters, yet each of these characters is more primitive than their homologues in the species immediately above it stratigraphically. Thus, a direct ancestral–descendant sequence (A–B–C–D) is postulated. Assuming that the rate of character change does not fluctuate greatly when viewed over narrow increments of time, an average rate of change between successive species probably approximates the real rate measured at any point in time.

Figure 1B reflects a different systematic pattern: Each of the four species is again defined by a unique combination of characters, but each has one or more derived characters, thus ruling out direct ancestry and descent. We must postulate, instead, three ancestral species (a, b, c) that are as yet unknown; consequently their po-

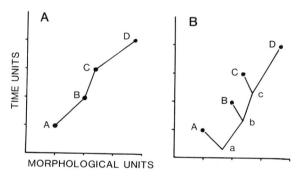

Fig. 1. Four diagnosably distinct species (A–D) with narrow distributions in time and some measurable aspect of morphological space. (A) Each species is primitive in all respects to that species occurring immediately above it in the stratigraphic column. Thus, a direct ancestral–descendant relationship is postulated, and an interspecific rate of change can be calculated. (B) Each species has one or more derived characters, and systematic analysis yields a phylogenetic hypothesis of common ancestry. Morphological rates now become much more difficult to measure because the position of the common ancestors (a, b, c) in time and morphological space is unknown. See text.

sitions in time and morphological space are also unknown (although the latter could be estimated by phylogenetic analysis; see below). What can be said about evolutionary rates now? Probably not much, unless one is willing to accept a number of *ad hoc* assumptions regarding the position of the ancestors in time and the direction of change from them to their descendants (i.e., to the species of our sample).

Absolute morphological rates, then, are not easy to measure unless one has sufficient systematic data to postulate ancestry and descent. The example of Fig. 1A refers to cases in which species are defined as being discrete by character analysis and thus have status as real evolutionary units; they are not artifacts of taxonomy. There is no morphological "transition" to be seen in this example; nevertheless, character analysis indicates that a hypothesis of ancestry and descent is plausible. Situations in which a transition is said to exist between two species— and this is frequently reported in the literature—present special methodological problems. No interesting questions are solved if species are defined as arbitrary segments of morphological continua (Wiley 1979). In fact, to define species that way and then measure the rate of morphological change within the continuum merely creates an artificial rate between artificial species. These lineages are not without significance for rate analysis, however, because a lineage itself may be definable as a discrete taxonomic unit (species) on the basis of character analysis, with rates then being determined *within* that unit.

Absolute morphological rates, at least in some sense of the term, can be measured in those cases in which the phylogenetic pattern is one of common ancestry. These rates are not necessarily calibrated with respect to time (although they might be) but to the amount of divergence from common ancestors and are derived directly from the phylogenetic analysis itself. Any cladistic analysis of discrete characters, for example, yields a hypothesis showing the amount of character change along each lineage (the amount of information retrieved is, of course, related to the amount of data entered into the analysis). In one sense, this would be a measure of absolute change (from the common ancestor); in another, it is also a measure of relative change when sister-lineages are com-

pared. Continuous quantitative change can be incorporated in two ways, by "superimposing" that change on a phylogenetic hypothesis derived from discrete characters, or it can be examined by direct analysis. In the latter case, numerical cladistic techniques such as distance Wagner analysis (Farris 1972) can produce parsimonious trees and assign an amount of change to each branch. The branch lengths, then, would represent an overall measure of relative rate for the data included in the study.

In summary, then, phylogenetic analysis represents a powerful analytical tool to assess the amount of morphological differentiation within and among lineages. Absolute measures of that change, in terms of time, can be calculated only once the time of branching is known. Fossils alone cannot provide that, but they can be employed to establish upper and lower temporal limits to branch points depending on the position of fossil taxa within the phylogenetic hypothesis. The time of branch points also can be estimated if the phylogenetic hypothesis can be related to historical biogeographic patterns that themselves can be dated by independent geological criteria (Rosen 1978; Cracraft 1982a, 1983a). This method has been used, for example, to establish the time of lineage separation of various vertebrate taxa on the southern continents by correlating that separation with the time of continental fragmentation (see below). Yet another method of dating branch points would be to "date the organisms," so to speak, by assuming the existence of a "molecular clock" by which the number of estimated nucleotide replacements can be translated into a measure of time.

Taxonomic rates. Perhaps the only evolutionarily meaningful taxonomic rates are those that measure either the number of speciation events per unit time for a lineage or the relative number of speciation events between sister-taxa. We are faced with an epistemological problem, however: How can we ever discover whether we are actually measuring the true rates? Not only does the patchiness of the vertical and horizontal extent of the stratigraphic record ensure that some species will go unrecorded, but in those cases in which rates are being estimated from the Recent biota, undetected extinction produces an underestimation of speciation rate.

The analysis of taxonomic rates is fraught

with methodological difficulties, and it is not an exaggeration to say that most of these are generally not discussed within the literature. This is particularly true for paleontology (Cracraft 1981a), but many of the same problems exist when analyzing diversity patterns within the Recent biota.

The first significant problem is the use of supraspecific groups to measure taxonomic rates. Only specific-level taxa represent true evolutionary units. It should be noted here that the use of a "biological species" concept (Mayr 1963, 1970) will tend to underestimate the number of evolutionary units in a sample. That concept frequently treats diagnosably distinct units as the "same" species either because of evidence of interbreeding or, if allopatric, the assumption of being capable of interbreeding. A phylogenetic species concept (Cracraft 1983b, which attempts to recognize all evolutionary units, would be more appropriate for the analysis of taxonomic rates. The use of genera to measure the "amount of taxonomic evolution" will be accurate only to the extent that each genus has, on average, the same number of species. That is clearly not the case.

The use of genera or other supraspecific taxa in rate analysis has three additional potential difficulties, related to their monophyly, the criteria used for deciding their rank, and their historical equivalency (Cracraft 1981a). To be used in rate analysis, supraspecific taxa must be strictly monophyletic. A paraphyletic or polyphyletic taxon has no ontological status as a real historical entity; they are taxonomic artifacts. Given an understanding of the phylogenetic relationships of the species within the lineages being studied, the problem created by nonmonophyletic taxa need not arise.

Potentially much more troublesome is the problem introduced by criteria used in taxonomic ranking (many workers have referred to the "equivalency" of genera from group to group, usually without any detailed discussion). If we had, for example, a corroborated phylogenetic hypothesis for a very large, monophyletic group of species, we could arrange sets of species in many different combinations to produce variable numbers of strictly monophyletic genera both within and between clades of the group as a whole. The number of species per genus would obviously vary from classification to

classification, and it is easy to see why counts of genera derived from any one classification may not necessarily reflect the underlying amount of taxic (species) evolution. This example, it must be emphasized, represents the *best* situation, since we at least have specified that all the genera are strictly monophyletic. In real-world classifications, based as they generally are on the principles of evolutionary systematics, paraphyletic taxa are common, which only increases the problems when using them in rate analysis.

The third potential difficulty, which is related to the preceding one, is that of their historical equivalency. I have already discussed this in some detail elsewhere (Cracraft 1981a:462–463; see also Novacek and Norell 1982). Comparisons of classes, orders, or families in terms of their relative diversity of genera or species are not especially meaningful unless they are postulated to have shared an immediate common ancestor relative to other taxa of equivalent rank (i.e., unless they are sister-taxa). Sister-taxa, by definition, have had the same age of origin, therefore differences in relative diversity directly reflect differences in speciation and/or extinction rates. Comparisons among higher taxa that are not each other's sister-group inevitably contain an inherent bias of not being the same age. If one assumes, for example, that order A is the sister-group of both orders B and C, a comparison of relative diversity between A and B, or A and C, will generally be of little value because the taxa are not historically equivalent; taxon A is older than either B or C. If such comparisons are undertaken within the context of a phylogenetic hypothesis—unfortunately, they rarely are—then not all comparisons will necessarily be uninteresting (if A, despite being older, has much less diversity than B or C, for example). Given the fact that most contemporary classifications often have substantial numbers of paraphyletic taxa, workers should exercise caution when making comparisons among supraspecific taxa in the absence of a phylogenetic (cladistic) hypothesis.

Finally, it is clear that standing diversity at any point in time is related not only to speciation rates but also to extinction rates (omitting from consideration variation introduced by the stratigraphic record itself). Paleontologists since Simpson (1944, 1953) have sought to recognize this fact when examining rates of evolution by constructing survivorship curves, the rate of evolution (or taxonomic turnover) supposedly being reflected in the structure of the curve. Novacek and Norell (1982) have pointed out, on the other hand, some of the difficulties of relying solely on the fossil record—i.e., the first and last occurrences of taxa—without taking into account the probable times of divergence based on a phylogenetic analysis. They show, for instance, significant differences in the shapes of the survivorship curves of primates calculated over the last 30 million years when estimating the age of origin from the fossil record versus basing estimates on information from a phylogenetic analysis. Novacek and Norell also stress that primates have one of the better fossil records and are perhaps better understood phylogenetically than any other group; survivorship curves for all other groups are likely to have increased margins of error and difficulties of interpretation simply because of inadequacies in the data.

The above comments are not meant to discourage the analysis of taxonomic rates. Quite the contrary, for investigations into speciation rates (and "background" as compared to "mass" extinction rates) are essential if we are to describe patterns of diversity through time and investigate their potential causes. Indeed, the analysis of speciation and extinction rates is central to the field of macroevolution (e.g., Stanley 1979; Eldredge and Cracraft 1980; Cracraft 1982b). Awareness of some of the difficulties of rate analysis should help us better investigate and understand these types of problems.

Bradytelic Evolution in Birds

Birds apparently originated sometime in the Middle to Late Jurassic. Our knowledge of their Mesozoic history is poor, at least in terms of the fossil record, but some of that history can be inferred from phylogenetic studies of higher taxa in which fossil species are also included (e.g., Cracraft 1982c), or when those relationships can be correlated with Mesozoic continental fragmentation (Cracraft 1974, 1982d). These studies thus indicate, even though tentatively, that a substantial number of the major avian lineages have been in existence since the Early Cretaceous.

Combining what little we know about avian paleontology with systematic analysis of Recent taxa does permit some preliminary comments to be made about bradytelic (slow) evolutionary rates in birds. In this short summary, a distinction will be made between bradytelic morphological rates and taxonomic rates. Although there must be some general correlation between the two—after all, new taxa manifest new phenotypes by which they are recognized—it is by no means exact. Lineages with low speciation rates can have very distinct, apomorphic species, and species within lineages showing high speciation rates can be relatively similar morphologically (the relationship between the two is discussed also in Cracraft 1982b).

Bradytelic Morphological Rates

Two methods are available to infer bradytelic morphological rates in birds. The first is to document the longevity of avian genera as revealed by the fossil record. Avian paleontologists have traditionally accepted the notion that avian species do not extend beyond the Pleistocene (Brodkorb 1971:48–49); consequently older forms are usually described as new paleospecies. Whether this assumption is true or not—and Pliocene species should be examined to check this—it will not be addressed further here. Many genera, on the other hand, extend deep into the Tertiary. Within avian paleontology, generic-level taxa typically do not encompass much morphological variability. Species within a fossil genus usually differ only in size or show minor shape variation; these genera, then, provide a basis to examine the longevity of basic avian morphologies through time.

Several problems arise when assigning fossil species to modern genera. Most avian fossils are described from isolated bones or bone fragments, and because of this, accurate systematic placement is often difficult. Within different lineages it is clear that different portions of the skeleton have changed more than others. Thus, if a fossil happens to be known only from a "conservative" part of the skeleton, even if other portions did change considerably, that fossil will likely be placed in a modern genus. In any sample of fossil taxa, therefore, the "true" numbers of modern genera for any time period will actually be less than indicated by the fossils.

A contrasting situation also exists. Many well-defined modern genera often show exceedingly little intergeneric variation in skeletal anatomy, the genera being defined on soft-part anatomy or behavior. Consequently, fossil material of these groups might be lumped into a single generic-level taxon, and thus the "true" numbers of genera would be more than that estimated from fossil material alone.

Finally, another problem exists when interpreting lists of fossil taxa and their stratigraphic longevity. Many of the fossil species in the avian record were first found in the extensive Tertiary deposits of Europe, particularly France. Most of these fossils were described in the last century or early part of this century. Avian paleontologists working then generally had a much broader concept of the genus and had much less modern comparative material than is currently available. Thus, many small species of ducks were placed in *Anas*, species of geese in *Anser*, hawks in *Buteo*, owls in *Bubo*, and so on. Some of these fossils have been redescribed as new fossil genera or placed in other modern genera, but the systematic position of many has not yet been reassessed.

With the above in mind, Table 1 presents a list of the Recent genera whose time of first occurrence is prior to the Pliocene (about 10 million years before the present). The beginning of the Pliocene was chosen as the upper limit to enhance identification of those genera presumably exhibiting slow evolutionary rates. Not unexpectedly, large land birds and those with aquatic habits predominate; only three nonpasseriform genera of small species (*Apus*, *Collocalia*, *Pterocles*) and two passeriform (songbird) genera (*Corvus*, *Lanius*) are included. All of these genera belong to families and/or orders that undoubtedly originated in the Cretaceous or very Early Tertiary. Some of these genera are known from more than one specimen and more than one species. Hence, the list is probably a reasonable approximation of those genera that have not changed much morphologically over substantial periods of time.

A second method that can be employed to identify slow morphological rates is to examine the results of phylogenetic analyses of taxa in which fossil evidence is available or whose biogeographic patterns can be used to date branch points. Shared derived characters of the different lineages signify, then, similarities that have been conserved over the two lineages since branching. With respect to my own work on avian relationships, three examples in particular identify slow rates of change in certain morphological characters of specific lineages.

Figure 2 presents three phylogenetic hypotheses based on cladistic analysis of skeletal anatomy. The living ratite birds (Fig. 2A) are large, flightless forms distributed widely on the southern continents (a ratite of still uncertain affinity has also been found in the Quaternary of New Caledonia, but it probably is related to the moas

Table 1. Times of first occurrence of extant avian genera recorded prior to the Pliocene (Brodkorb 1963–1978).

Late Miocene:	*Struthio, Oceanodroma, Ardea, Cygnus, Anser, Nettion, Hieraaetus, Haliaeetus, Aquila, Sterna, Aethia, Corvus*
Middle Miocene:	*Diomedea, Morus, Hypomorphnus, Falco, Cyrtonyx, Numenius, Charadrius, Vanellus, Cercorhinca, Uria, Otus*
Early Miocene:	*Podiceps, Pelecanus, Phoenicopterus, Eudocimus, Querquedula, Aythya, Milvus, Erolia, Larus, Strix, Apus, Collocalia, Lanius*
Late Oligocene:	*Anas*
Middle Oligocene:	*Puffinus, Buteo*
Early Oligocene:	*Phalacrocorax,*[a] *Sula*
Late Eocene:	*Limosa, Totanus, Pterocles,*[b] *Bubo,*[b] *Asio*[b]

[a] Perhaps Middle Oligocene.
[b] Perhaps Early Oligocene.

and kiwis of New Zealand; Cracraft, unpublished observations). The phylogenetic relationships are documented by large numbers of shared-derived characters (see Cracraft 1974 for details). Adopting the parsimonious hypothesis that these taxa differentiated in concert with the fragmentation of Gondwanaland, the shared characters are evidence for slow morphological change. Thus, the common ancestor of the ostrich in Africa and rheas in South America became isolated sometime prior to 85–90 million years ago (branch point 1); all the characters they share are at least that old, if not considerably older (the reader can consult Cracraft 1974 for a list of those characters).

Likewise, within the order Gruiformes (Fig. 2B), another set of transantarctic relationships reveals slow morphological change (Cracraft 1982d). The New Zealand–New Caledonian continental block separated from west Antarctica (and thus South America) prior to about 80 million years ago. Consequently, characters shared between the Eurypygidae on the one hand and the Rhynochetidae + Aptornithidae on the other (branch point 1) are 80 million years or older. Those characters shared by all the taxa (branch point 2) are older still, and presumably date from sometime near the middle part of the Cretaceous (perhaps 100 million years or more). In an earlier paper, I listed a large number of characters for both branch points (Cracraft 1982d).

Finally, when fossil taxa can be interpolated into a phylogeny of Recent taxa, the age of the branch points can be estimated. Figure 2C shows a phylogenetic hypothesis for a lineage of diving birds whose lower branch points almost certainly are Early Cretaceous in age (Cracraft 1982c). Both the baptornithids and hesperornithids are found in Upper Cretaceous deposits. Another fossil taxon, *Enaliornis* from the Lower Cretaceous, is closely related to them, or to them plus loons (gaviids) and grebes (podicipedids). This implies that branch point 2 may be Early Cretaceous in age, which would make branch points 3 and 4 even older. All of these diving-bird taxa share a number of characters (described in Cracraft 1982c), some of which may be as old as 125 million years. Add to this the probable close relationship of the penguins (Spheniscidae) and the pelecaniform–procellariiform birds, and it is not difficult to see that some aspects of avian anatomy have evolved at very slow rates indeed. Even if the details of these relationships are altered by subsequent work, any pattern of interrelationships of these taxa to one another, or to other major lineages of birds, probably will not alter these conclusions about rates significantly.

Bradytelic Taxonomic Rates

The fossil record of birds is not, in general, sufficiently dense to determine the first and last

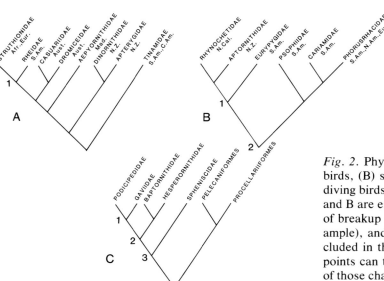

Fig. 2. Phylogenetic hypotheses for (A) ratite birds, (B) some gruiform birds, and (C) some diving birds. The age of the branch points in A and B are estimated by biogeography (the time of breakup of the southern continents, for example), and in C by the age of fossil taxa included in the analysis. The age of the branch points can then be used to investigate the age of those characters shared at different levels of the hierarchy. See text.

Table 2. Some examples of sister-taxa showing significant differences in diversity probably related to tachytelic and bradytelic rates of evolution (speciation).

Tachytelic taxon	Bradytelic taxon
Pelecaniformes	
other pelecaniforms (59)[a]	Phaethontidae (3)
Ciconiiformes	
Ardeidae (62)	Balaenicipitidae (1)
Anseriformes	
Anatidae (147)	Anhimidae (3)
Falconiformes	
Accipitridae + Falconidae (279)	Pandionidae (1)
Galliformes	
Cracidae + Phasianidae (256)	Megapodiidae (12)
Gruiformes	
Rallidae (142)	Heliornithidae (3)
Columbiformes	
Columbidae (306)	Pteroclididae (16)
Cuculiformes	
Cuculidae (129)	Musophagidae (18)
Strigiformes	
Strigidae (135)	Tytonidae (11)
Piciformes	
Picidae (204)	Indicatoridae (16)
Passeriformes	
all other suboscines (1097)	Acanthisittidae (4)
Furnariidae (218)	Dendrocolaptidae (52)
Laniidae (74)	Vangidae (13)
Corvidae (106)	Grallinidae (4)
Estrildidae (136)	Bubalornithidae (2)
Estrildini (113)	Viduini (10)

[a] Number of species; from Bock and Farrand (1980).

occurrences of species or genera with any accuracy. Hence, rates of taxonomic turnover within families or orders are not available at this time. It is necessary, therefore, to examine taxonomic bradytely using the Recent biota.

Taxonomic bradytely can be inferred from comparisons of sister-taxa in which the two lineages have significantly different diversities. All one has to assume is that, on average, extinction rates have been the same in each lineage, that is, the low diversity lineage has not been subject to high extinction. Differences in diversity must therefore be due to differences in speciation rate subsequent to divergence from a common ancestor.

It is my purpose here to record some probable cases of taxonomic bradytely. Possible explanations for these particular patterns, and those of taxotely (Raup and Marshall 1980) in general, are beyond the scope of this contribution. Table 2 presents 16 pairs of presumed sister-taxa in which bradytely is evident. Information about most of these hypothesized relationships is summarized in Cracraft (1981b); the numbers of species are from Bock and Farrand (1980). There is no present evidence from the fossil record to suggest that any of the identified bradytelic lines underwent significant extinction relative to their sister-group. Indeed, the data were chosen to represent extreme differences in diversity in order to avoid this possibility. Thus, these taxa almost certainly reflect real differences in speciation rate.

Many such examples, at all levels of the taxonomic hierarchy, could be given provided phylogenetic hypotheses are available. This type of comparison has rarely been used within neontology or paleontology (see Vrba 1980; Eldredge and Cracraft 1980), but it has much to offer those interested in searching for the general causes of large-scale patterns of diversity and macroevolution (Cracraft 1982b).

Acknowledgments: This paper was funded by NSF grant BSR-7921492.

Literature

Bock, W., Farrand, Jr., J. 1980. The number of species and genera of Recent birds: a contribution to comparative systematics. Amer. Mus. Novit. 2703:1–29.

Brodkorb, P. 1963–1978. Catalogue of fossil birds. Parts 1–5. Bull. Florida State Mus. 7:177–293 (1963); 8:195–335 (1964); 11:99–220 (1967); 15:163–266 (1971); 23:139–228 (1978).

Brodkorb, P. 1971. Origin and evolution of birds, pp. 19–55. In: Farner, D. S., King, J. R. (eds.), Avian biology, Vol. 1. New York: Academic.

Cracraft, J. 1974. Phylogeny and evolution of the ratite birds, Ibis 116:494–521.

Cracraft, J. 1981a. Pattern and process in paleobiology: the role of cladistic analysis in systematic paleontology. Paleobiology 7:456–468.

Cracraft, J. 1981b. Toward a phylogenetic classification of the Recent birds of the world (Class Aves). Auk 98:681–714.

Cracraft, J. 1982a. Geographic differentiation, cladistics, and vicariance biogeography: reconstructing the tempo and mode of evolution. Amer. Zool. 22:411–424.

Cracraft, J. 1982b. A nonequilibrium theory for the rate-control of speciation and extinction and the origin of macroevolutionary patterns. Syst. Zool. 31:348–365.

Cracraft, J. 1982c. Phylogenetic relationships and monophyly of loons, grebes, and hesperornithiform birds, with comments on the early history of birds. Syst. Zool. 31:35–56.

Cracraft, J. 1982d. Phylogenetic relationships and transantarctic biogeography of some gruiform birds. Geobios. Spec. Mem. 6:393–402.

Cracraft, J. 1983a. Cladistic analysis and vicariance biogeography. Amer. Sci. 71:273–281.

Cracraft, J. 1983b. Species concepts and speciation analysis, pp. 159–187. In: Johnston, R. F. (ed.), Current Ornithology, Vol. 1. New York: Plenum.

Dingus, L., Sadler, P. M. 1982. The effects of stratigraphic completeness on estimates of evolutionary rates. Syst. Zool. 31:400–412.

Eldredge, N. 1982. Phenomenological levels and evolutionary rates. Syst. Zool. 31:338–347.

Eldredge, N., Cracraft, J. 1980. Phylogenetic patterns and the evolutionary process. New York: Columbia U. Press.

Farris, J. S. 1972. Estimating phylogenetic trees from distance matrices. Amer. Nat. 106:645–668.

Mayr, E. 1963. Animal species and evolution. Cambridge: Harvard U. Press.

Mayr, E. 1970. Populations, species, and evolution. Cambridge: Harvard U. Press.

Novacek, M. J., Norell, M. A. 1982. Fossils, phylogeny, and taxonomic rates of evolution. Syst. Zool. 31:366–375.

Patterson, B. 1949. Rates of evolution in taeniodonts, pp. 243–278. In: Jepsen, G. L., Mayr, E., Simpson, G. G. (eds.), Genetics, paleontology, and evolution. Princeton: Princeton U. Press.

Raup, D. M., Marshall, L. G. 1980. Variation between groups in evolutionary rates: a statistical test of significance. Paleobiology 6:9–23.

Rosen, D. E. 1978. Vicariant patterns and historical explanation in biogeography. Syst. Zool. 27:159–188.

Simpson, G. G. 1944. Tempo and mode in evolution. New York: Columbia U. Press.

Simpson, G. G. 1949. Rates of evolution in animals, pp. 205–228. In: Jepsen, G. L., Mayr, E., Simpson, G. G. (eds.), Genetics, paleontology, and evolution. Princeton: Princeton U. Press.

Simpson, G. G. 1953. Major features of evolution. New York: Columbia U. Press.

Stanley, S. M. 1979. Macroevolution: pattern and process. San Francisco: Freeman.

Vrba, E. S. 1980. Evolution, species and fossils: how does life evolve? S. Afr. J. Sci. 76:61–84.

Westoll, T. S. 1949. On the evolution of the Dipnoi, pp. 121–184. In: Jepsen, G. L., Mayr, E., Simpson, G. G. (eds.), Genetics, paleontology, and evolution. Princeton: Princeton U. Press.

Wiley, E. O. 1979. Ancestors, species, and cladograms. Remarks on the symposium, p. 211–225. In: Cracraft, J., Eldredge, N. (eds.), Phylogenetic analysis and paleontology. New York: Columbia U Press.

Wiley, E. O. 1981. Phylogenetics. The theory and practice of phylogenetic systematics. New York: Wiley.

10
Crocodilians as Living Fossils

Eugene R. Meyer

The Johns Hopkins University, Department of Earth and Planetary Sciences, Baltimore, MD 21218

Introduction

Nile crocodiles and American alligators belong to a group of reptiles called broad-nosed crocodilians. In the warmer parts of the world, broad-nosed crocodilians are the largest predators to walk on land. They are living fossils in the sense that they resemble ancient forms in the shapes and the ruggedness of their heads and bodies.

Their ecology and evolution is far better understood today than a few years ago. It is now possible to show which natural history patterns are universal and which are confined to one species, and to test explanations as to why certain crocodilians are living fossils. In this chapter I will outline their natural history and point out where new information has caused major changes in perspective.

It is now clear that only one group of crocodilians, the broad-nosed group, is at all stable. Species change, predators and prey change, but the common broad-nosed shape is perpetually renewed. The question arises: What processes make them living fossils?

Phyletic constraints (inability to produce major new forms) are an unlikely explanation in view of new fossil evidence. The broad-nosed group has given rise to many offshoots, including forms with delicate narrow snouts, forms with duckbills, and forms with strange tall snouts. Some tall snouted species were so ter-

restrial that they had hooves! The broad-nosed group now appears as a central core in crocodilian evolution, able to innovate on many levels.

An ecological explanation for the stability of the core group is that the ancient design is selected by combat and by dispersal in avoidance of combat. Combat and dispersal are important in three processes: predation of small crocodilian species by larger species, cannibalism of subadults by adults, and sexual competition. Because these three conflicts are interactions among members of the group, the shape of individuals is partly insulated from extrinsic selective pressures from other groups, such as predators and prey.

Divergence from the core group can occur when a population gives up one or more conflicts. It can then replace the combat-ready design with a design better suited to catching small aquatic prey or to catching terrestrial prey. The divergent population may change further when it contacts broad-nosed crocodilians which eat or drive the new specialist from habitats where it can be caught.

Analogous conflicts between life stages within a species, between species of similar shape, and sexual competition may stabilize shape in other groups of animals and in plants.

I outline below the ecology, geography, relationships, variation, fossil record, and diversity of crocodilians, arranged in sections so that the reader may follow his or her own interests. The

Table 1. Diversity of head shapes in the broad-faced crocodilians of North and Central America. Every published occurrence of species known from good comparative material is shown.

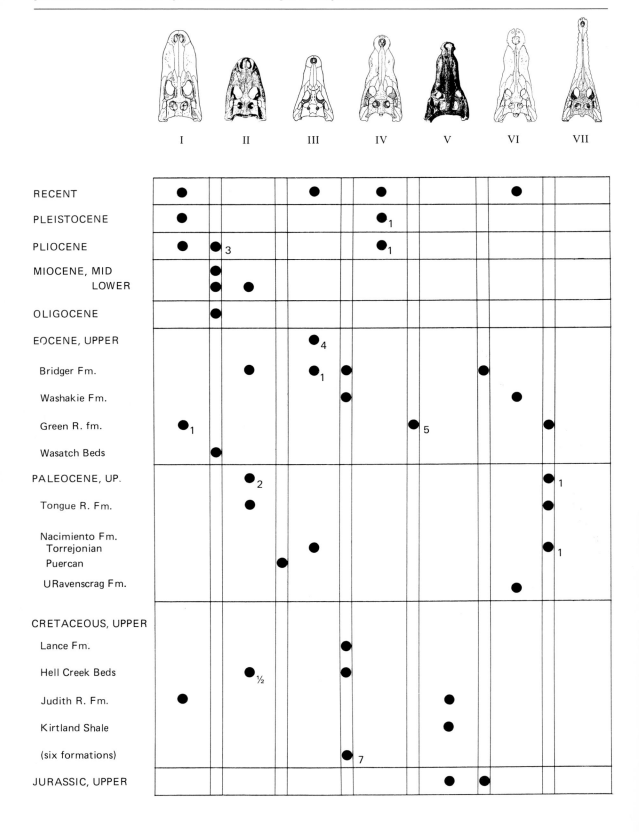

data consist of my observations of American alligators in natural habitats and the primary literature. Generalizations are built around examples from one continent and time; and to avoid oversimplification brief note is made whether exceptions are known from other continents and times. We begin with the age of the broad-nosed shape and its common traits.

Relative Constancy of Shape

Many living broad-nosed crocodilians are so similar to extinct species that if they were exhibited side-by-side in zoos, the public and many herpetologists would be hard pressed to distinguish them. Examples of the first known forms are *Goniopholis lucasii* of North America and *G. simus* of Europe (shown as outline V in Table 1). Although they lived in the Upper Jurassic, more than 140 million years ago, their skulls had only a small difference in outline from several living crocodiles: The notches in the snout that receive the large canine teeth of the lower jaw are somewhat deeper in the Jurassic forms. The external similarity may not reflect direct ancestry of the living species, because the vertebrae and the position of the major air passages in the skulls of the ancient species show that they belong to an extinct suborder.

The common features of broad-nosed species are an amphibious head profile, a massive head in adulthood, a relatively wide snout, and a distinctive body shape described below. Some di-

versity is apparent in head outlines when viewed from above.

The amphibious head profile has nostrils, eyes, and ears higher than the rest of the head (Schmidt 1944; Cott 1975). Iordansky (1973:244) suggested that this profile makes a crocodile "inconspicuous to its terrestrial prey . . . when floating". This profile may be more important in concealment—crypsis—from other crocodilians in predation, sexual competition, and cannibalism. It may have little role in concealment from many underwater prey and predators.

The adult skull is massive, and the jaw gape is large in all species. The head is used with an unusual technique in combat, and in dismembering very large prey. The crocodilian seizes a limb or other convenient part in its jaws, then rolls over and over, tearing the part off (Schmidt 1944; Cott 1961; photos in Root and Root 1971). Broad-nosed species generally lack shearing teeth (Langston 1965; many others), and so cannot carve their meat, unlike the extinct terrestrial crocodilians and other terrestrial predators.

The external differences among broad-nosed species often seem minor because of the amphibious outline and sturdy construction of the head, and because head outline as seen from above changes during early growth. Good clean skulls can be assigned to species by the sizes and shapes of major bones, and aspects of their taxonomy based on anatomy have been confirmed by chromosome data in Cohen and Gans (1970). Some of the most useful traits in keys to

---

◁ • Head shapes are repeated from the Upper Cretaceous to the present, with little progression or divergence among amphibious forms.
 • Shapes intermediate between categories occur repeatedly.
 • Each dot represents a different species, except for two Recent species, as shown in Appendix 1.
 [a] Head profiles:

I	*A. mississippiensis*	V	*Goniopholis simus*
II	*Alligator mcgrewi*	VI	*C. acutus*
III	*Caiman sclerops fuscus*	VIII	*C. johnstoni:* narrow-snouted, shown for comparison only.
IV	*Crocodylus moreleti*		

Subscripts in the Table are:
 1 Shape is inferred from partial material.
 2 Snout has a constriction near the base deeper than in II.
 3 Shape shares traits with I, II, and III.
 4 Rear of skull is narrower than in III, resembling *Paleosuchus trigonatus*.
 5 Shape is intermediate between V and III.
 7 This giant skull has an unusual outline in Colbert and Bird's 1954 restoration, based on fragmentary material.
 The head profiles of living species are for adults of moderate size for that species. The profiles are approximate because all change with growth and some change with geography (see text). Profile II is from Schmidt (1941), V is from Hulke (1878:Fig. 3) and may be distorted at the quadrates in preservation, and all others are from Wermuth and Mertens (1961). Species names and references are in Appendix 1.

living specimens are skin color, eye color, and scale arrangements (Brazaitis 1973; Wermuth and Mertens 1961).

Body shapes are quite similar throughout history in the broad-nosed group. Their major limb bones are short and almost slender, the feet are always plantigrade and are never paddles, and the tail is long and flat.

Habitats

With very few exceptions, the meeting place of land and water has always been home for the crocodilians.

(Minton and Minton 1973:15)

Alligator and crocodile habitats range from quite stable to very unstable. Their way of life or niche is broad as defined by habitat variety and by predators and prey.

The American alligator lives in almost every wetland category in eastern Georgia, from flooded forest to the Atlantic Ocean (Meyer 1975; in preparation). Its broad ecological tolerance is indicated by the large differences in available sunlight, temperature, and visibility between habitats such as flooded forests and flooded marshes. Tolerance is also known by alligator occurrence at seasonally dry sites and at manmade places such as canals and dammed ponds. Use of a wide range of habitats is also seen in Nile crocodiles (Cott 1961) and Australian estuarine crocodiles (Messel et al. 1981).

Use of water offers advantages in temperature regulation and energy savings, in addition to concealment. Spotila et al. (1972:1100) found that use of water "enables the alligator to survive a wide range of temperatures . . . Without water the alligator is severely restricted during the day and at night." Swimming is energetically less costly than walking for marine iguanas (Gleeson 1979) and I infer the same for amphibious crocodilians, for whom dispersal over long distances is important.

Subadult alligators use some sites where water is lacking for several months of the year. Subadults often live in marginal habitats lacking large adults. Habitat shift of subadults away from adult sites is best documented for American alligators at natural sites (Meyer 1975; in preparation). (*Natural* is defined in Appendix 2.) It may also occur in estuarine crocodiles and

in Nile crocodiles, but most of those subadults simply disappear from sight, often not reappearing until they are approximately 1.5 m long (Messel et al. 1981; Cott 1961). Day/night shifts of subadult activity also may occur at some sites (Cott 1961; Watson et al. 1971). Cannibalism, discussed in the next section, appears to be the driving force of habitat shift.

Deep open ocean is rarely occupied by crocodilians at any lifestage today, although they often live on shorelines and in estuaries. They disperse across deep ocean, but rarely feed and never breed there (Meyer, in preparation). On warm continents there are few wet areas not occupied by crocodilians. These few have very rough water (Modha 1967; Messel et al. 1981; personal observation), or cold springs. They also include certain large African lakes after a local faunal extinction (Beadle 1974; many others), and a section of the Mekong River running in a very deep gorge (G. Davis, personal communication). Habitats and ecology are discussed further in three reliable introductions with citations: Minton and Minton (1973), Webb (1977), and the beautifully illustrated Cott (1975).

Prey and Predators

The prey and predators of broad-nosed crocodilians span an enormous range of sizes and orders, and vary with geography and time. The most universal predation on this group is cannibalism. Large adults are almost immune from predation.

Large prey is usually taken by drowning (Colbert 1962). Single Nile crocodiles have been seen seizing and drowning adult male giraffes, full-grown African buffalo bulls, and an adult male lion (Pienaar 1966, 1969). An American alligator was seen drowning a feral boar weighing more than 227 kg (over 500 lb) (McIlhenny 1935). Adult Nile crocodiles also seize mammals on land by waiting next to game trails (Cott 1961). In contrast, I find no record of crocodiles in Australia or caimans in South America taking large native mammals.

Diets differ radically from site to site, and small prey comprise almost the entire diet at some sites. Consider the contrast between the primary diets of Nile crocodiles at three localities: (1) snails, by crocodiles of many size classes, including the largest, up to 3.68 m (12 ft) long (Cott 1961); (2) insects, frogs, and cane rats (Cott 1961); and (3) a single species of fish, by crocodiles from less than 1 m to more than 4 m long, at a lake where available invertebrates are rare (Graham 1968). American alligator diets may diverge even more by locality: at one site, subadults take primarily apple snails (Fogarty and Albury 1967); at another, almost exclusively muskrats, while nearby alligators took snakes, fishes, and crabs (McIlhenny 1935); at a third site, a mixed primary diet of crayfish, turtles, and muskrats (Giles and Childs 1949): and finally, entirely vertebrates—herons, turtles, and garfish (McIlhenny 1935).

The idea that crocodilians are opportunists and generalists, eating the most locally abundant prey, is not supported by Chabreck's (1971) work with an alligator subadult population. In fresh water, they took primarily crayfish, birds, and a few fishes, but ignored abundant mammals and frogs. In nearby saline water, they took less food by volume, primarily crabs and crayfish, but no birds or mammals. The opportunist and generalist notion is supported by diets of some individuals, e.g., a wild estuarine crocodile eating crabs, an eel, a green sea turtle, and apparently pigs (Allen 1974), and occasionally by eating introduced species—livestock and humans.

Foraging goes beyond ambush predation to include search hunting. For example, American alligators repeatedly steal eggs from Canada goose nests at one site (Chabreck and Dupie 1976); juvenile alligators and Nile crocodiles act as terrestrial insectivores under some conditions (Cott 1961; Meyer, in preparation); and adults scavenge carcasses in water and on land.

Predators on crocodilians vary greatly by continent, and except for cannibals many are modern in aspect. To emphasize crocodilian survival in the face of such diverse predation, I list the species involved. Black bears and raccoons often eat American alligator eggs, and otters and fishes take the young at times (Dietz and Hines 1980; Joanen 1969; Metzen 1977; others). In contrast, native rodents and varanid lizards eat Australian estuarine crocodile eggs (Webb 1977). Olive baboons often rob Nile crocodile nests, and a wide array of adaptive types take eggs at various sites, including:

varanid lizards, hyenas, jackals, mongooses, warthogs, and maribou storks (Cott 1961; P. Shipman, personal communication). Birds of six families, including eagles, take juveniles and lions kill the smaller adults (Cott 1961). Eggs are drowned when nests are flooded (Magnusson 1982); abiotic causes of mortality are not otherwise very important.

Cannibalism is the only predation common to all continents. Nile crocodiles are "much addicted to cannibalism" based on "(i) injured specimens, (ii) direct observation, and (iii) stomach contents" (Cott 1961). The larger a Nile crocodile is, the more likely it is to eat another (Cott 1961). Cannibalism is also known in estuarine crocodiles (Messel et al. 1981; Webb 1977) and the medium-sized spectacled caiman (Staton and Dixon 1975). Nichols et al. (1976) noted that "this mortality source is probably the major density dependent factor operating on Louisiana alligator populations."

Cannibalism is limited to conflicts where the prey is unlikely to inflict much damage on the predator. The cannibalism cited above involves adults eating subadults. I have circumstantial evidence that subadults eat unprotected juveniles: When a subadult moved into their burrow at a natural site during a drought, most juvenile alligators vanished, and a thorough search failed to find them. (Size classes are defined in Appendix 2.) At some sites, eggs are eaten by adults, including females who did not nest that season, cited under female competition below.

Where there are strong clues that cannibalism might interfere with an individual's genetic contribution to the next generation, cannibalism is not reported in crocodilians. I find no record of a male killing a nesting female of his own species at any natural site, even though males are often larger. I find no eyewitness report of adults cannibalizing juveniles in *natural* conditions in any species; instead, adult response to juvenile distress calls is well known (e.g. Romero 1983). At nursery sites in zoos or disturbed situations, subadults may be chased away or bitten gently, if they can then escape. This behavior has been reported for Nile crocodiles, (Pooley 1977), Morelet's crocodiles (Hunt 1977), American alligators (Hunt and Watanabe 1982) and marsh crocodiles (Whitaker 1974).

Very large crocodilians are almost immune from predation from any source (Dowling, personal communication; Cott 1961). The evidence I accept for near-immunity is (1) the maximum size known killed is much less than the maximum observed body size. (The largest individual reported killed by any predator was a lion-killed Nile crocodile 3.53 m (11 ft 7 in.) long, 0.5 m above minimum male breeding size in some populations [Cott 1961].) (2) A very long life is required to reach the great sizes documented even in the 20th century, because adult growth rates in nature (McIlhenny 1935; Cott 1961; Chabreck and Joanen 1979) are so low; and (3) many adult crocodilians are larger than their potential predators.

Comparing body sizes of predators walking on land indicates that at least one crocodilian species on every warm continent grows large enough to be almost free from predation. A wild alligator or crocodile 4.6 m (15 ft) long weighs more than 400 kg, heavier than any sympatric terrestrial predators: tigers, lions, bears of warm climates, and giant snakes. Lengths greater than 15 ft have been measured in the wild, although weights are rarely taken. American alligators have been measured at 5.84 m and 5.64 m (19 ft 2 in. and 18 ft 6 in., McIlhenny 1935), estuarine crocodiles at 6.1 m (20 ft) with larger estimates (Australia and India; Webb and Messel 1978a), Nile crocodiles at 5.5 m (18 ft) and a likely 6.4 m (21 ft) (Africa, see ref. in Cott 1961), and Orinoco crocodiles measured at 6.78 m (22 ft 3 in.) and estimated at 7.3 m (24 ft) (South America; Humboldt 1876).

Great size and probable near-immunity from predation also occurred in the past. The largest predators of all times to walk on land were the dinosaur *Tyrannosaurus* and the crocodile *Deinosuchus* (*Phobosuchus*) (Kurten 1978). Both lived in the Upper Cretaceous. Colbert and Bird (1954) estimate the giant crocodile's length in life as up to 50 ft (15.2 m). I estimate the likely boundaries of its size are 11 m (36 ft) and 6 metric tons, up to a less likely 15.2 m (50 ft) and 18.8 tons. This size range reflects the need to reevaluate its skull shape (W. Langston, personal communication) and the problems of size estimates suggested by Webb and Messel (1978a) for the estuarine crocodile. Based on co-occurrence of fossils in deposits in six states, Baird and Horner (1979) suggest that the giant crocodile, rather than *Tyrannosaurus*, was a major predator on duckbilled dinosaurs.

Distribution in Space

Broad-nosed crocodilians occur widely in tropical and warm temperate zones. They have unusual dispersal ability, moving easily to new river systems and to isolated wetlands. There is a clear pattern of endemism: The same species is not found in the interior of more than one continent.

Crocodilians live on every tropical continent and at least six species live in subtropical and temperate areas (Meyer, in preparation; Honegger 1975; see Dowling and Duellman 1978). The primary physical limit to ranges today is temperature, with all species limited to the zone between 40° north and 40° south of the equator (modified from limits of 35° north and south, suggested by Ostrom 1969). Cool climates have often set limits to distributions in the past (Sill 1968). The geography of the higher living taxa was thoughtfully reviewed by Dowling and Duellman (1978) and by Darlington (1957).

The Nile crocodile has the largest land distribution of any crocodilian. Its range includes all of Africa except the northwest, and even some oases in the Sahara desert in historical times (Minton and Minton 1973; others). An extinct African species also had a very large range (Tchernov 1976). Ranges of 200,000 sq km are moderate to small for most species in the interiors of continents.

Species with much smaller ranges are usually confined to islands or an isthmus by a very large crocodile. Examples are the Cuban crocodile (maximum length 3.7 m) and Morelet's crocodile (maximum approximately 2.9 m), both of which are confined by the American crocodile (maximum 6? m), (Schmidt 1924, 1932, 1944). This confinement presumably stems from direct predation, which Medem (1971) describes as the limit to sympatry. The small range of the Chinese alligator may result from elimination by humans (Honneger 1975; Wermuth and Mertens 1961). I find no indication that the range of any broad-nosed species is limited primarily by a mammal or bird group, e.g., by distinct allopatry with a warm-blooded predator.

Dispersal

Subadult and adult crocodilians have the unusual dual ability to travel easily overland and by water. This enables species to cross barriers between drainage systems. It also allows crocodilians to occupy isolated wetlands, come into chronic conflict with each other, and to leave areas of intense conflict.

The most numerous dispersing groups are subadults and young adults. They are potentially abundant in species of large body size because clutches are large, and growth time to size of first reproduction is long (9 to 15 yr) (data in Cott 1961; McIlhenny 1935; Graham 1968; Chabreck and Joanen 1979). Subadult alligators are "consistently more active over a wider range of environmental conditions" than adult alligators (McNease and Joanen 1974:499). In view of Terpin et al.'s findings that "a very small alligator is more closely tied to the limits of its climate space" (1979:311), I suggest that they live closer to their physiological limits than adults do, especially when moving overland distant from water.

Overland travel is extensive for the Nile crocodile (Pienar 1966), American alligator (McIlhenny 1935; my observations), and three caiman species (Medem 1971; Gorzula 1978). My data indicate that the sizes walking on dry land range from subadult to large adult, because females from isolated sites are courted by adult males, and their progeny later disperse from those sites. The importance of overland travel is underscored by Zug's (1974) report that crocodilians are the only living reptiles to gallop.

The best documented long distance swimming records are for estuarine crocodiles. A radiotagged male 3.2 m (10.5 ft) long moved "at least 130 km, some 80 km being around the sea coast" of Australia (Webb and Messel 1978b). A 3.8 m (12.5 ft) male lived at Ponape Island, Eastern Caroline Islands, 1360 and 2400 km from the two nearest crocodile populations (Allen 1974). In Louisiana marshes, with their network of manmade canals, tagged bull alligators moved an average of 0.75 km each summer day. A large adult and a subadult each moved 52+ km in less than 9 months (Joanen and McNease 1972; McNease and Joanen 1974). These record distances were set where no other crocodilian species was present, and all were set by males.

Endemism

All fossil broad-nosed species from North America were endemic (compare Appendix 1 with taxa listed in Steel 1973; Meyer, in preparation). No broad-nosed crocodilian species occurs in the interior of more than one continent today or in history, even when the continents were closer together in the Late Jurassic and Early Cretaceous. Where a species occurs on two continents or on many islands, its populations seem restricted to within 300 km of the seacoast on the second land mass. Examples are the spectacled caiman and American crocodile (my interpretation of Central American maps of Smith and Smith 1977, C8 and C10), and estuarine crocodile (Australian maps and data in Messel et al. 1981). Exceptions to precise coastal limits may occur in major rivers. Endemism determines the minimum number of species worldwide.

Variation Within Species

There is substantial variation in anatomy and ecology in the few species carefully surveyed.

It is surprising to find, among a conservative group like the crocodiles, evidences of rapid evolutionary change and faunal discontinuities, as is demonstrated among Neogene and Quaternary taxa in eastern Africa.

(Tchernov 1976:370)

The Nile crocodile has changed substantially in time: A fossil jaw from Olduvai is more robust than that of the modern form (1.5 to 1.9 times as high, and broader). A related species shows little geographic variation in the Miocene, but underwent a "local adaptive radiation" into "four significantly different, readily distinguishable populations" in the Pliocene and Pleistocene (Tchernov 1976).

Several living species show considerable geographic variation. For example, one of Medem's subspecies of the living spectacled caiman is strikingly different in the taper of its snout from other *Caiman sclerops* (1955, 1960). In the Nile crocodile, living populations vary greatly from each other in head shape and in maximum body size (Kaelin 1933; Cott 1961). (Seven subspecies are suggested by Fuchs et al. (1974), but the adequacy of the material is doubted by Friar and Behler (1983).) The American alligator was not divisible into subspecies in a thorough study of scale patterns, but there was much variation in those patterns within populations (Ross 1979). Dodson (1975:350) found that variation in a growth series of American alligator skeletons "compared favourably with values of *V* for taxonomically homogeneous samples of mammals."

Enzyme polymorphism data are published only for the American alligator. For three populations, Adams et al. (1980) found the proportion of polymorphic loci, $P = 0.15$, similar to values in other vertebrates. However, other work with single populations found $P = 0.06$ (Gartside et al. 1977), and $P = 0.045$ (Menzies et al. 1979). All three studies found heterozygosity low: respectively, $H = 0.022, 0.021$, and 0.0086. Adams et al. review possible causes of low H.

Lack of subspecies and low heterozygosity in the American alligator may be related to two peculiarities of its range: (1) It was probably reduced to the edges of its continent several times in the Pleistocene; and (2) it lacks inland sympatric species and the many isolated drainages found in Africa and South America (Meyer, in preparation).

Relationships and Origins

There is general agreement about evolutionary relationships among broad-nosed crocodilians, suggested by evidence from anatomy, geography (Meyer, in preparation), karyotypes (Cohen and Gans 1970), and biochemistry (Densmore and Dessauer 1979). There is much less agreement about relationships among narrow-nosed or terrestrial groups. One certainty emerges: Convergence and parallelism have been unusually important in crocodilian history (Langston 1973; Buffetaut 1979).

Crocodilians have strongly braced skulls, with complex sutures joining the bones (Langston 1973). The skulls are highly derived (S. B. McDowell, personal communication), and have sixteen ordinal skull traits (Langston 1973). Two groups are proposed as the first crocodiles. The Middle Triassic *Proterochampsa,* which clearly has an amphibious profile, is the first crocodile in Sill's view (1968, and others), but in Romer's (1971) view it is only a parallel form, and the first crocodilians are *Protosuchus, Orthosuchus,* and their Late Triassic relatives of perhaps more terrestrial habits. *Protosuchus* is discussed in detail by Colbert and Mook

(1951). Until ancestry is better known, statements about the nearest relatives and evolutionary rates of the earliest crocodilians will be quite speculative. Olson (1971) outlines five alternate classifications that differ primarily in which major groups are recognized as separate families and Mook (1962) adds a sixth.

Discovery of evolutionary paths to the modern crocodile suborder "furnished a stronger support to the hypothesis of evolution than even that of *Hipparion* and the Horse" (Duncan, in Huxley 1875). Modern crocodilians are called eusuchian. Compared with extinct mesosuchians, they have the internal nares (ends of the major air passage) placed further back on the palate, and have ball and socket rather than flat articulations between the vertebrae. Eusuchians were diverse by the Late Cretaceous, and several intermediate genera from the Early Cretaceous are known. The intermediate genera have small body sizes, fitting Cope's Rule as discussed by Stanley (1973). The new suborder apparently rose by mosaic evolution, and a key trait, placement of the internal nares, changed slowly (Joffe 1967; Buffetaut 1979).

All living species and their extinct close relatives are placed here in one family for brevity, the Crocodylidae s.l. (sensu lato). In the Late Cretaceous the crocodylids appear with three fully formed subfamilies: Alligatorinae, Crocodilinae, and Tomistominae. The caimans may be a later group. Each of the major broad-nosed groups—alligators, crocodylines, and caimans—has had its own radiation into a variety of head shapes. Each has tended to form parallel morphological assemblies, leading to repeated treatments of each as a separate family. These radiations and parallelisms are another indication of evolutionary vigor.

The origins of one broad-nosed genus demonstrate an evolutionary pattern. One species is so intermediate between the living *Alligator* and the extinct *Allognathosuchus* that Patterson (1931) was "somewhat undecided" to which genus it should be assigned, finally designating it *Allognathosuchus riggsi.* Simpson (1933) implied it should be placed in *Alligator* with the earlier *Alligator prenasalis,* which he thought was closer to *Allognathosuchus* than to the living alligators. (Simpson also noted similarities of *A. prenasalis* with three South American caimans, but in view of their karyotypes (in Cohen and Gans 1970), those similarities now appear convergent.) This chain of intermediate forms is so subtle that experienced taxonomists disagree where one genus ends and the next begins.

Wide snouts (I and II in Table 1) appear first in the Upper Cretaceous. Their late appearance may be an advance made possible by the novelty of overlapping tooth rows, or it may be an artifact of the fossil record, because there are so few pre-Late Cretaceous skulls. (This paucity of pre-Late Cretaceous material is true of many other reptile groups [Ostrom 1969]). The only major dental change in the broad-nosed amphibious group since its origin is the periodic appearance and loss of bulbous rear crushing teeth. These were probably "independently acquired in several groups of small crocodilians . . ." (Buffetaut and Ford 1979).

Distribution in Time

The record of broad-nosed amphibious species in North America is shown in Table 1 as a range of head shapes, and in Appendix 1 as a list of species. North America's record covers the most time, has the longest well-sampled time interval, and has the most broad-nosed species of any continental record.

Species turnover is more rapid in the best sampled interval from the Uppermost Cretaceous through the Eocene than I had expected for a living fossil group (see Appendix 1). Most broad-nosed species are known from good diagnostic material from a single epoch or stage, counting Pleistocene/Recent appearances as one. This is also the rule in South America, which I interpret from Langston's (1965) accounts, and in the rest of the world, from Steel's (1973) summary, with a few exceptions below. Generic longevity is bimodal, another unexpected pattern.

Part of the restriction of species in time may result from a sampling problem, and part is real. The sampling problem is that only one good skull has been described in print for most species, even when several good skulls are known. As yet, teeth are often not identifiable to species. Few taxa have been carefully reviewed for change over time and in space (see Tchernov 1976 for such a review).

Three lines of evidence indicate that time restriction is real. (1) There are no published accounts of a broad-nosed species reappearing after a long absence, say 15 million years, from the record in any part of the world. (2) Change over time is substantial in two well-sampled African species (Tchernov 1976; variation). (3) There is little indication that a continent could have many similar species, each with a small range, so that ancient taxa could long persist undetected. In the interiors of continents, almost all species today have large to very large geographic ranges, even in complex drainages like the Amazon and the Orinoco.

Estimates of species turnover and faunal complexity depend on taxonomic reliability. Only species known from good comparative material are included in Table 1 and Appendix 1. Taxonomy of broad-faced taxa known from good skulls has been quite stable on the species level since 1910, despite the discovery of many new fossils and the naming of new species. Systematic practices in this group have improved greatly since the time of Cope and Marsh, in part owing to the examples set by occasional crocodilian contributions from C. W. Gilmore, B. Patterson, and G. G. Simpson.

There are a few apparently long-lived species: (1) the Nile crocodile, which was present but had more robust jaws in an early Pleistocene fossil from Olduvai, perhaps 1.8 million years ago (Tchernov 1976; Leakey 1971); (2) the living *Crocodylus porosus*, to which a partial snout is referred from the Pliocene of Australia, 4 to 4.5 million years ago (Molnar 1979); (3) the living marsh crocodile (*C. palustris*) of India, which is quite similar to the ?Pliocene *Crocodylus sivalensis* (Lydekker 1886); (4) the living *A. mississippiensis*, which occurs in the Late Pleistocene (Holman 1978), and which may be represented in Pliocene Florida skulls that have not been carefully studied; the Miocene Florida species is the extinct *A. olseni* (Auffenberg 1967).

Generic longevity is curiously bimodal. Genera commonly appear for either less than approximately 10 million years or for more than 20 million years. Genera of intermediate survival time are strikingly rare in North America (Appendix 1). This pattern is confirmed in South America (species accounts in Langston 1965), and in Europe (ranges in part from Russell et al.

Fig. 1. Adult American alligator (*Alligator mississippiensis*) basking in the sun in the Okefenokee.

1982; and Steel 1973), and the rest of the world. It is as if those passing a stringent selective test persist for many millions of years. Perhaps only the Holarctic *Asiatosuchus* lasts an intermediate time, although taxonomic changes could produce others, e.g., by assigning *Allognathosuchus riggsi* and *Alligator*. This pattern is linked to the effect of two major extinctions (at the end of the Cretaceous and at the end of the Eocene). Each may be related to loss of many short-lived genera, but not to loss of many long-lived genera. It does not appear that the bimodal pattern of "generic longevity" has been demonstrated in other reptiles.

Diversity

I suggest here that diversity has been relatively stable for broad-nosed crocodilians since at least the Upper Cretaceous in three ways: (1) the number of species alive at the same time, (2) the presence of four or fewer species in each well-studied fauna, and (3) variety of head shapes.

There are 16 to 21 broad-nosed species alive today—the total varies with recognition of a few forms as species or subspecies. A count ranging from 10 to 30 species at any one time may be typical of the group, after the time sequence of fossils becomes clear in the Late Cretaceous. This count is in accord with endemism patterns and geographic ranges discussed above, and with ecological processes and faunal structures, and the list of species known from good fossils in Appendix 1. (Species diversity before the mid-Cretaceous is difficult to analyze because the time relationships and taxonomy of earlier taxa are uncertain and the sample size is small.)

Each well-studied fauna has only one to four broad-nosed species—at any time since the origin of the broad-nosed group. Fauna here means all of the species in a drainage system at the same time. The presence of few species in particular faunas was recognized by Schmidt 1924, 1928; Patterson 1936; Langston 1965; and especially Medem 1971. Possible exceptions are discussed below. Even the faunas in South America today have few species, although it has the widest range of crocodilian morphology of any continent today.

Predation among crocodilian species maintains sympatry at low levels in the great drainage basins of South America, even in the vast Amazon Basin (Medem 1971). For example, large crocodiles eat the medium-sized spectacled caimans, which in turn feed on *Pa-*

leosuchus caimans in other habitats. After hide hunters eliminated the huge *Crocodylus* from the Atrato River, the spectacled caiman became abundant there for the first time. It had been confined to lagoons, creeks, and marshes previously by *Crocodylus* (Medem 1971). Similarly, in Central America the large American crocodile (*C. acutus*) displaces the smaller Morelet's crocodile into "small and vegetated bodies of water, and microsympatry does not occur" (del Toro, in Smith and Smith 1977).

I have found several size classes of alligators together, but not the smallest and the largest in a few localities, usually large bodies of water disturbed by man. Because cannibalism between alligator size classes is similar to predation between crocodilian species, I suggest that two crocodilian species may cohabit where the smaller individuals can evade the larger.

Another proposed mechanism for low levels of sympatry is continuous growth with changes in diet, causing each crocodilian species to pass through a series of potential niches (Dodson 1975). Continuity of growth is a new insight, but current data show that diet does not predictably change from small subadults to large adults at some sites. A predictive model of faunal structure is described in the section on forces renewing an ancient shape.

In every fauna today, no two species share the same head shape and body size. Examples are South America today and the North American fossil record. Faunas are indicated in Appendix 1 as horizontal groups of species in the same state or physiographic province, excluding the Jurassic taxa whose time relationships are not understood. In contrast to the similarity of faunas and body shapes of broad-nosed crocodilians through time, large predatory mammals today have much different head and limb proportions than their counterparts in the Orellan (29–32.5 million years ago) (Van Valkenburgh 1982).

Diversity in physiology may correlate with patterns of body size in faunas, with smaller species having unusual capacities that may allow them to live in habitats free of larger species, or to escape pursuit by them. For example, the dwarf caiman survives exposure to low temperatures fatal to the sympatric, middle-sized spectacled caiman (Medem 1971). The latter, in turn, has a tidal lung volume seven times

greater than that of the larger allopatric Nile crocodile although values for baseline ventilation are similar (Glass and Johansen 1979). In another case, the two smaller crocodiles of the Congo Basin have broader nesting tolerances than the Nile crocodile (Lang, in Schmidt 1919). However, more data comparing species are needed (Smith 1979). A major study of American alligator metabolism (Coulson and Hernandez 1983) has laid the foundation for detailed comparative work.

Head-shape diversity is indicated in Table 1. Almost the entire historical range is shown, with each species represented on the continuum of adult shape. Diversity changes little from the Late Cretaceous to today; short-nosed species are now absent from North America, but remain in South America. Three unusual groups occurred, all offshoots of the broad-nosed group. The first group (*Brachychampsa* and *Ceratosuchus*) had more pinched snout bases than II; in addition, *Ceratosuchus* had small horn-like squamosals (Appendix 1). A more unusual shape group contains two duckbill clades with longer and more delicate snouts than I (*Mourasuchus* confined to the South American Miocene and ?Pliocene, Langston 1965, 1966; and two(?) genera confined to part of the Late Cretaceous of Africa, referenced in Steel 1973). The third unusual group is a miscellany comprised of species for which published plates show skulls difficult to interpret or terrestrial rather than broad-nosed amphibious in design.

Unusual head shapes appear in a very large fauna in the paleontologist's broadest sense, reviewed by Langston (1965). The Colombian La Venta fauna occurs in deposits some 700 m thick. It was resolved into four broad-nosed amphibious forms, an aberrant duckbill, two narrow-nosed species, and *Sebecus*, which many workers regard as highly terrestrial. The most complex lithologic unit, "the Monkey unit, with six crocodilian-producing localities" has probably no more than five species (Langston 1965), of which I believe only two are broad-faced amphibious forms. Further work on Central and South American fossils (Ferrusquia-Villafranca 1978) shows the same pattern of presence of few broad-nosed species in any local fauna.

The "Bridger" crocodilians are a possible exception to the presence of few species in each fauna (Appendix 1), but the vague locality descriptions in Cope (1883) and others do not allow definite assignment of many crocodilians to that formation, or to the Bridgerian land mammal age as defined by Berggren et al. (1978). The Bridger formation is 695 m thick (Bradley 1964), and I suggest that "Bridger" crocodilians include several successive faunas in the zoologist's sense of contemporary species in a drainage system. Their taxonomy is chaotic, with similar species named from isolated teeth and fragments, a practice long since abandoned by crocodilian workers.

Behavior and Mortality

Crocodilians have the most advanced behavior of any living reptile. Only two aspects of behavior are discussed here, both related to injury and mortality: sexual competition and parental defense. The reader interested primarily in evolutionary consequences of behavior may prefer to read the next section on forces renewing shape first.

One of Darwin's first examples of sexual selection was alligator "fighting, bellowing and whirling around" (1859:88). Injuries from sexual combat in male alligators and American crocodiles are described by McIlhenny (1935) and Hornaday (1875). C. B. Cory, Curator of Ornithology at the Field Museum, observed a fight between two male alligators in a Florida pond. The battle ended when the larger seized "his opponent by the upper jaw and immediately rolled over and over, breaking his opponent's jaw close to his head, killing him instantly" (Cory 1896:68).

Male Nile crocodiles patrol territories at breeding grounds and fight with other males (Modha 1967; Pitman 1941; Pooley 1977). Frequent deaths of male Nile crocodiles from fighting in the mating season were recorded in the 1930s (Pitman 1941). I find no indication that adult Nile crocodiles fight outside the seasons of courtship and nesting. Combat is rarely observed in nature today because the leather trade has decimated most populations and made the survivors wary of humans. In a popular book, Neill (1971) denies the existence of combat, cannibalism, and other natural history facts reported by competent observers (Cott 1972; King 1972; and especially Fogarty 1972).

The injury rate increases with size to 63% in the largest male size class and 41% in the largest female size class in Nile crocodiles (Cott 1961), and with size in estuarine crocodiles and spectacled caimans of unstated sex (Webb and Messel 1977; Gorzula 1978). Crocodilians have great ability to recuperate from all but the most serious injuries (Gorzula 1978; Brazaitis 1981). At sites or in seasons where secondary factors increase the lethality of small injuries—such as the piranha attack of wounded caimans seen by Roosevelt (1924)—the risks of battle may be so high that disputes are settled with visual signals (Meyer, in preparation). Crocodilians of three continents do use a wide variety of visual signals, described by Modha (1967), Garrick, Lang, and Herzog (1978), Staton and Dixon (1977), and clearly illustrated by Cott (1975). Combat can uncover deception and poor judgment, inevitable when mere display controls important resources. Male narrow-nosed gharials may have a special organ of display, but unlike Darwin's (1859) other examples of sexual selection all broad-nosed species lack them.

Female Nile crocodiles fought to death at one nesting area (Pitman 1941) but communal nesting was documented at two other sites (Cott 1961, 1975). Nests of estuarine crocodiles are close to each other in some habitats in Australia (Magnusson 1980). American alligator females are not known to fight in the wild and their natural injury rate seems low today. Yet there does appear to be sexual competition and territoriality, as indicated by the following observations: (1) Only one adult with juveniles was resident at each *natural* isolated nesting and nursery site in Georgia (Meyer 1975, 1977); (2) nests and nurseries were generally well spaced in the vast Okefenokee marshes, and occurred in three different wetland categories, including apparently suboptimal breeding grounds, and at highly disturbed sites (in preparation; nest data in Metzen 1977); (3) alligator eggs or shells were found in 13% of stomachs of adult females, including those who had not nested that season (McNease and Joanen 1977), and eggs were found in stomachs of Nile crocodiles of unstated sex (Welman and Worthington 1943). I infer that nest protection from conspecifics can be important. Here sexual competition and parental defense merge.

If an alligator can modify the topography of an isolated site in ways useful for hiding and defense, it may then have a great advantage over an intruder in choosing a place and time to fight. Alligators deepen and burrow in their wetlands extensively, and in isolated sites such changes in bottom contour are more extensive than those made by any other animal group (my data). Young females may either avoid small isolated nesting sites that are already occupied, or die in the attempt to invade. This adds to the obvious thermal advantages of deepening ponds and digging burrows and creates a perennial area of open water for combat. The need to defend ideal nesting sites could select for ability to modify sites, to make decisions quickly, and to learn local topography. Parental defense of nest and juveniles may also favor sturdiness, crypsis, and stealth (in preparation).

Selection for learning ability in defense of eggs and young by female alligators was suggested by Kushlan and Kushlan (1980). After ingenious experiments on defense, they concluded "alligators distinguish between types of potential predators . . . Such behavioral plasticity and attendant ability to learn quickly from experience is obviously highly adaptive." Dietz and Hines (1980:257) found that in response to human disturbance, female alligators "are able to modify an innate behavior such as nest defense without altering other behaviors such as nest maintenance or liberation of young." Alligators who protected their nests and juveniles most actively from human researchers had by far the lowest rates of nest loss to bears in Okefenokee (Metzen 1977 and my data). Nest defense is not common in some areas today, probably because hide hunters lured adults with juvenile calls (personal communication from hide hunters).

Parental attendance in natural habitats lasts up to two years in alligators at natural sites, comparable to care length in eagles and large predatory mammals (Meyer 1977). It depends on weather in some habitats (Chabreck 1965). Hatchling Australian estuarine crocodiles disperse within days after emergence in some habitats, but remain together in others (Webb and Messel 1978b). Parental behavior is complex in all species investigated (e.g., Pooley 1977; Meyer 1978), and may even include provisioning the young by the adult female (McIlhenny 1935). Vocalizations are varied in character, frequency shift, and pulse rate shift (Herzog 1974; Herzog and Burghardt 1977; Meyer 1977). I see little indication that parental behavior is stereotyped.

Territoriality for food seems unlikely at first glance, since low energy requirements would not seem to require feeding territories at today's population densities. However, densities are artificially low now, and the concept should be re-examined. Spacing and group occurrence of alligators outside of reproduction are suggested by Meyer (1977; in preparation) and spacing of equal-sized estuarine crocodiles with tolerance of individuals in certain smaller size classes, by Messel et al. (1981).

Renewing an Ancient Shape

To inflict such terrible injuries on the armor plating of another crocodile, those jaws must be among the strongest in nature.

(Myers 1972:151)

I propose combat and its avoidance select for the classic shapes and construction of broadnosed crocodilians. Combat and dispersal are common to three processes: predation between species, cannibalism, and sexual battle.

Combat would select for sturdy construction of the head and torso and for amphibious shapes offering concealment at the water's surface. Avoiding battle by hiding at a moment's notice would again select for amphibious head shapes. Avoidance by travel to isolated sites where large crocodilians are uncommon places a premium on walking overland and swimming long distances. The limbs and tail necessary for such dispersal are less ruggedly made than the head and torso, and are often severely injured in battle. The resulting compromise is the ancient design of broad-nosed crocodilians.

Based on data from five continents, the three processes suggested above could be more consistent in time and space than predator/prey relations. The data indicating that each conflict is a significant source of mortality are outlined in the relevant sections above. Other explanations are outlined in the following section. For the sake of brevity, these processes are called *conflicts* within crocodilians—*competition* seems too vague a word when competing individuals are torn limb from limb.

I outline below how habitat shift and dispersal join the three conflicts, the evolutionary consequences of each conflict, the evolutionary result if a deme gives up cannibalism and sexual combat, and why these conflicts preserve shape in crocodilians more effectively than in some predatory mammals.

Habitat shift and dispersal intensify conflict in broad-nosed crocodilians. The most numerous dispersing groups are the subadults and young adults of large species, fleeing cannibalism and sexual competition. Habitat change and dispersal overland and by water cause each species to spread quickly in one drainage system, and to cross from one stream system to the next. This brings every broad-nosed species into prolonged conflict with other crocodilians in a large region.

Predation between species causes the habitat boundaries between species to be sharp, and sympatry to be low (a maximum of 4 species) (Medem 1971, discussed in Diversity above). Medem found that the larger species eat the smaller. The result inferred here is that any species or deme not fit for combat, e.g., having weak heads and teeth, will be eliminated from every habitat in which it cannot evade typical crocodilians.

The evolutionary consequences of predation and of habitat shifts can be predicted to be strong for small species. Even if they live in habitats quite different from those of adults of large species, they will be exposed to an influx of subadults and young adults of large species. Adults of smaller species would need to be better protected in battle and better equipped for avoidance of battle. They should be more robust, more terrestrial, or more agile than larger species, or they should be physiologically specialized for difficult environments avoided by larger species.

I believe this explains why several small living and fossil species have such formidable canine teeth and so much bony armor. They carry the broad-nosed morphotype almost to extremes. In performance, Medem (1958) notes the *Paleosuchus* caimans are agile jumpers and

fast swimmers compared with the larger caimans. In behavior, he notes their aggressiveness. In physiology, Medem (1971) notes the unusual temperature tolerance of the dwarf species.

Sexual combat and cannibalism may stabilize shape if only one broad-nosed species occupies a region. The frequency of injury and death from sexual combat at some sites (discussed in Behavior above), suggests that these conflicts are not trivial, and that the forces exerted on the skull in battle must be enormous. Busbey (1977) describes in detail the skull structures that resist maximum stress; Busbey (personal communication) and I have concluded independently that the skull design is well suited for maximum loads from rolling with an object between the jaws. I believe that the general source of maximum loads while rolling is another crocodilian, not prey, because large prey are not taken on several continents today and may have often been missing from local and continental faunas in the past.

Cannibalism of subadults and young adults may maintain skull sturdiness and dispersal ability. Cannibalism in crocodilians occurs when the predator is unlikely to be severely injured (as mentioned under Prey and Predators). A small increase in skull sturdiness, making a subadult more formidable as potential prey, would be rewarded by increased freedom from cannibalism.

If a deme gives up cannibalism and sexual combat to avoid injury, it can specialize in anatomy for a new diet. It risks losing ecological breadth and evolutionary longevity. For example, by evolving delicate feeding structures, a deme could improve feeding efficiency upon small abundant animals, especially fishes. It could gain in crucial aspects of life history theory: increased growth rates, clutch size, and hatchling size; decreased time to sexual maturity; and decreased feeding time with its exposure to predation.

At first, such a deme may survive if out of contact with other broad-nosed crocodilians, including typical demes of its own species. When contact with typical forms is resumed, such a deme will be confined to habitats in which it can escape being eaten by typical forms of the same body size—the fate of narrow-nosed crocodiles today. Confinement to fewer habitats may reduce variety of selective pressures, setting the stage of further divergence in anatomy and further confinement in habitats.

Interactions among large crocodilians are different from those among large carnivorous mammals for four reasons: (1) Large crocodilians produce an order of magnitude more young annually per mother (20 to 80) than do large predatory mammals or birds. (2) Growth patterns are different. The young grow very slowly to *size* of first reproduction (9 to 15 yrs) in large species and even more slowly to maximum size (20 to 50 yrs). Hatchlings are tiny compared to adults: their weight ratio is 1 : 800 to 1 : 5000+ (Meyer, in preparation). Continuity of growth and niche change (Dodson 1975) are discussed under Diversity. (3) Almost all individuals return daily to shallow water near land, and thus can meet at a common setting. (4) Reptiles differ from mammals in biomass and feeding efficiency in ways that increase the frequency of encounters between individuals of similar shape.

Ectothermy (being cold blooded) allows potential crocodilian biomass to rise to a very high level, while allowing great bursts of activity. Recent reptiles and amphibians "are able to devote a very large portion of their ingested energy to producing new biomass". Their "net long-term conversion efficiencies . . are many times greater than those of birds and mammals" (Plough 1980:102). Ability to survive long fasts and to fast rather than to feed actively when food is scarce also increases potential reptile biomass by reducing food competition at crucial times. The result I infer is that crocodilians become unusually abundant for top predators and can encounter each other frequently in many kinds of wet habitats. The capacities of small reptiles for burst activity are similar to those of small mammals, although varying with temperature and species (Bennett and Ruben 1979; Bennett 1980). High burst activity could be crucial in the relations of crocodilians with mammals and when crocodilians encounter each other.

The construction of the legs, feet, and the distal half of the tail is surprisingly light in broad-nosed crocodilians. The limbs and tail are often severely injured in fights (injury lists are in Cott 1961 and Webb and Messel 1977).

Their continued slender construction indicates that avoidance by dispersal may be quite important as a selective process, as indicated in Distribution in Space. After the hatchling stage, long distance travel is useful in reaching water and prey which vary with geography and seasonal rains, in avoiding cannibalism, interspecific predation, and sexual conflict, and in finding isolated mates.

The body plan of typical crocodilians may thus be shaped primarily by combat and by avoidance of combat. In ecological time these processes operate to structure faunas. In geological time, these processes may produce rapid convergence to basic crocodilian shapes and rugged construction, when a large slow-growing terrestrial predator enters the amphibious life.

Phyletic Constraints and Alternate Explanations

Other explanations for preservation of ancient shapes are simpler. They seem at first glance to be plausible, but are contradicted by the natural history data. Phyletic constraints—inability to produce major innovations—are discussed first, with an outline of the major divergent lines produced by the broad-nosed group. Following that are brief outlines of theories of large population sizes, unchanging environments, and the small area of the earth's fresh waters. These explanations have a common pattern: each is a simple, vague extension of an idea that is valid when limited to individuals or to parts of an ecosystem.

Phyletic constraints are a possible explanation for stasis. For example, the embryology of a group might be so integrated that major new features cannot appear. Thus failure to innovate can result from developmental constraints rather than from selection (Gould and Lewontin 1979). To test this idea, the record can be examined to see if members of a group have moved into entirely new ways of life, or have originated morphological oddities.

In fact, several new ways of life and odd morphologies have been produced by broad-nosed crocodilians, while the stable central shape group has always remained. They produced the following divergent forms, which share one ecological feature: They live in habitats in which the broad-nosed forms are rare.

1. Highly terrestrial crocodiles appeared in several lineages. A lineage closely related to the living family Crocodylidae s.l., the pristichampsids, survived into the Eocene in North America and Europe. *P. vorax* had flattened serrated teeth, a high skull, and a round tail (Langston 1975; Busbey 1977, in preparation). The related *P. geiseltalensis* of Europe (see Kuhn 1968) had hooves!

2. Duckbilled caimans with delicate snouts and many slender teeth appear in the South American Miocene (*Mourasuchus,* Langston 1965, 1966). They may have gathered algae or small animals by a straining technique, or grubbed in the mud. Their remains are scarce in sediments having remains of other eusuchians, suggesting habitat separation from them. The giant duckbilled crocodiles of Mid-Cretaceous Africa (*Stomatosuchus*) were "even more aberrant" and were independently derived (Langston 1965).

3. Narrow-faced crocodiles have arisen from broad-faced groups several times. There are two living *Crocodylus* of medium size with very slender snouts, one in Australia (*C. johnstoni*) and the other in Africa (*C. cataphractus*). Each is separately derived from typical *Crocodylus* (Longman 1925), and their karyotypes (in Cohen and Gans 1970) are different. Snout elongation occurred over time in populations of three African *Crocodylus* species but was never reversed (Tchernov 1976). Jurassic goniopholids also diversified into narrow-faced and broad-faced genera (Buffetaut and Ingavat 1980). (Habitat separations are outlined in the next section.)

4. Peculiar variants have appeared in body parts even of the broad-nosed species. For example, the keeled plates of bone that armor the backs of crocodiles and alligators have had strange forms, including unique spikes and blades on a Paleocene alligator (O'Neill et al. 1981), and smooth overlapping plates on a Paleocene crocodile (Erickson 1976). The scutes of some European Early Cretaceous goniopholids had lateral pegs projecting under adjacent scutes (Owen 1878). The scutes of a North American relative had both pegs and odd grooves (my data).

Thus predictions based on the hypothesis that phyletic constraints did not allow divergence by broad-nosed crocodilians, are contradicted by the record. New work in embryology indicates that alligator embryos do respond to experimental manipulation (Ferguson 1979). From this I infer that alligator development is not too canalized for change in one generation, and that phyletic constraints operating over many generations are unlikely from a biological point of view.

A second possible explanation is that some modes of speciation are prevented by large population sizes. I believe that modes of change occurring in small isolated populations can operate in some crocodilians, because dwarf caimans occur as small isolates in some areas (Medem 1971), and African dwarf crocodiles can be predicted from the few field data, and from the theory advanced here, often to live in small isolates. They could evolve in several of the transilience modes described by Templeton (1981) or in the allopatric model of Mayr (1954, 1975). Chromosomal evolution is obvious in crocodilians (karyotypes in Cohen and Gans 1970). Thus the data do not support confinement of crocodilians to a few modes of evolution.

A third idea—that unchanging environments are linked with stasis—is supportable if stated as: The persistence of certain physical aspects of wetlands has permitted the persistence of conflicts stabilizing amphibious crocodilians. The chain of logic is: Most wetlands have some open water at some time of the year; the borders of open water with dry land and with air have distinct physical qualities, which are ageless as ideals but depend locally on the presence of vegetation; at open water the amphibious head shape of crocodilians is easily concealed, and at banks the typical body plan allows rapid entry and egress. The idea that selection by combat and its avoidance favor key traits in crocodilians rests on the occurrence of open water and open banks for at least part of each year.

However, the idea that wetlands are unchanging is incorrect if taken literally. Wetlands change quickly; many dry up seasonally. Wetlands are quite diverse, and their availability changes over time. Wooded swamp, marsh, and river are quite different from each other in temperature, tree cover, and water flow. Their relative areas change with fluctuations in sea level, aridity, glaciation, topography, and plant evolution. The animals that live at the edges of wetlands—crocodilian predators and prey—also vary with time and geography. However, broad-nosed crocodilians on different continents (e.g., Africa/Australia) do not show major changes of shape with proven differences in their predators and prey. The unchanging environment idea is also incorrect if stated as: Living in wetlands guarantees success of amphibious groups. The fossil record shows that there have been many extinctions and radiations of amphibious predators, from labyrinthodont amphibians and phytosaurs to modern water snakes (e.g., *Natrix* s.l., *Nerodia*), and narrow-nosed crocodiles.

Finally, when Darwin coined the term *living fossils,* he suggested that the relatively small area of the earth's fresh waters has reduced rates of species origination, competition, and extinction (1859:107). Estes (1970) outlines the long history of many living freshwater lower vertebrates now in southeastern North America. However crocodilian ecology extends far beyond fresh water animals, to life and death interactions with terrestrial mammals. Also, at least seven Recent species including the American alligator occur in salt water (Meyer 1975; Messel et al. 1981; many others).

Narrow-Nosed Crocodilians

The fate of the narrow-nosed group supports the mechanism proposed for stasis of the broad-nosed group. Their snouts and teeth usually but not always lack the sturdiness needed for combat with broad-nosed crocodilians of the same body size. There are three very large species today. One, the tomistoma, is a living fossil in the sense that it is the last survivor of a subfamily. The second species is intermediate between narrow and broad nosed, the Orinoco crocodile, aptly named *Crocodylus intermedius.* It may cross a threshold during growth to reach sturdiness comparable to that of smaller broad-nosed species, becoming one of the dominant crocodilians in its fauna (see Medem's 1971 fau-

nal list). It is the last species in the New World to approach the narrow-nosed condition. The third large species, the gharial, is the last survivor of a once widespread rather young family. The two species of medium size are derived from broad-nosed *Crocodylus* (see above). They seem confined to regions in which large broad-nosed species are uncommon (my interpretation of Lang, in Schmidt 1919; Messel et al. 1981:459).

The upper jaw of a very large tomistoma or gharial is formidable in appearance, but large subadult heads seemed delicate in the museum specimens of tomistoma, gharial, and Orinoco crocodile that I examined. Ross (1974, in Whitaker and Rajamani 1974) observed gharial and marsh crocodiles together in a deep pool in a river, and cites aggression and avoidance but not biting. This is the only described instance of aggression without predation between any crocodilian species at a natural site. Gharials are highly derived, and as Buffetaut's (1978) family Gavialidae, may date only from the Eocene.

The narrow-nosed group has fluctuated in diversity and geographic range all through its history. Many adaptive types are extinct (teleosaurs, metriorhynchids, and dyrosaurids; see Buffetaut 1979, Romer 1966). Some adaptive types were so derived that they were not amphibious, but aquatic, e.g., the limbs were paddles and the tail had a reversed heterocercal bend in metriorhynchids. Convergence and parallelism led to taxonomic chaos in many narrow-nosed taxa, especially the dyrosaurids, until Buffetaut's recent work.

The Future

Crocodilians are now intensively killed for their skins to make vanity goods such as expensive purses. Most species are now threatened or endangered, as detailed by Honegger (1975). Crocodile farming was planned to reduce hide hunting in the wild, but instead it has worsened the conservation problem in Papua New Guinea by screening illegal killing of crocodiles in the wild, and it has failed to reduce captive juvenile mortality below levels in nature (Burgin 1980). The present partial bans on trade in skins are easily evaded—finished leather from endangered species is simply passed off as from legally exported species, covering up the smuggling of hundreds of thousands of skins annually. The threat of extinction for many species will not abate, until international bans on trade in skins apply uniformly to all crocodilians.

Public opinion often favors conservation of crocodilians, even though they are correctly seen as dangerous to humans and livestock under some circumstances. The impact of crocodilians on human fisheries is beneficial in certain areas, by fertilization of low-nutrient waters in the Amazon (Fittkau 1973), and by removal of predators on valuable fishes in Africa (Cott 1961; Campbell, personal communication). The impact of alligators on wetlands is beneficial, because alligators dig holes that provide the last remaining water for many animal species during drought (Kushlan 1974; Meyer, in preparation; others). In Florida, a poll found that the American alligator is widely perceived as ecologically important, interesting, and sometimes hazardous to people, and respondents strongly favored its conservation (Hines and Scheaffer 1977).

Concluding Remarks

Darwin (1859) recognized that the young enter a world filled with older individuals, and that selection acts on all life stages during growth. Crocodilians illustrate how selection on all life stages can stabilize a group. Similar kinds of selection may stabilize shape in other animals and in plants when individuals encounter each other frequently under common conditions, and drive rapid divergence in small isolates.

Three stabilizing processes can be outlined in terms applicable to many groups. (1) Predation between species: Adults of dominant species (usually of large body size) select for survivors among several life stages of less dominant species. A more subtle interaction may be just as important: Subadults of dominant species interact with adults of less dominant species, and their selection is mutual. (2) Sexual competition: Within demes, sexual competition can stabilize anatomy, curtailing specializations that allow faster growth to adult size but reduce ability to compete at the time of reproduction. (3)

Cannibalism: Within species, adults effectively select for survivors among the young and determine the conditions of their lives. Among the young, the older ones effectively select for survivors among the younger, and they select for aspects of parental treatment of the youngest.

This explanation for stability allows for variation within populations and within species of the stable group. It is compatible with a variety of genetic modes of evolution described by Templeton (1981), and the allopatric model of Mayr (1954, 1975). It does not rely on phyletic constraints or genetic stasis. There are certainly other causes of stability.

As a general hypothesis for conservation of shape, the degree to which these processes stabilize a species or a larger shape group is testable in some taxa and only inferable in others. For example, in eagles and arboreal mammals, it may be possible to estimate the ability of young to escape detection and chases by adults in ecological time, and to compare shapes and ways of life that are stable with those that are unstable in geological time. The inevitable gaps in the data base could be spanned by the judgment of field ecologists, and by noting differences in ecology and morphology on different continents, as done here for crocodilians. This theory may also account for the low diversity in general shape of the tyrranosaurs and other Cretaceous carnivorous dinosaurs, but lack of many kinds of data leaves the general hypothesis as but one of several possible explanations for a phenomenon that has received too little attention. For trees, testability is high. The majority of woody plant species in California are "15 to 50 million years" old (Stebbins 1982). Analogous processes could be evaluated as a means of stabilizing wood and leaf architecture within demes, for increasing physiological differences between species, and for founder and divergence effects.

These processes stabilize a "central" core, often while driving innovation into side groups. Neither stability nor innovation requires that each genetic species lasts a long time. One or more of the conflicts can be given up, and a deme can then change morphology and give rise to a specialized offshoot that lives in habitats free of the ancestral forms. Crocodilians have produced such specialists again and again.

Wilson (1980) describes the general origins of structured demes and of differences between demes.

These processes may cause divergence in small isolates. A divergent first generation could select for survivors among the next generations—an ecological founder effect. This could reinforce genetic mechanisms for change, without requiring populations to remain so small that they undergo genetic founder effects. When the isolate again contacts the stable core group, it may diverge further if the core group excludes it from all but a few habitats. Thus the processes stabilizing anatomy, ecology, and physiology are inevitably linked to those driving divergence.

For Appendix see p. 124

Appendix 1. The broad-faced crocodilians of North and Central America. Every published occurrence of species known from good comparative material is shown.

	Alligatorines		Crocodylines	Goniopholids: G
RECENT	*Alligator mississippiensis*	I	*Crocdylus acutus* = Amer. crocodile	
	Amer. alligator, southern U.S.			VI
	Caiman scleros fuscus	III	Florida to Antilles and S. America	
	Brown caiman, Mexico to Colombia		*C. rhombifer* = Cuban crocodile	IV
			C. moreleti, Mexico & Cent. America	
				IV
PLEISTOCENE	*A. mississippiensis,* southern U.S.	I	*C. rhombifer, C. antillensis,* Cuba	IV
			C. moreleti subsp., Guatamala 1	IV
PLIOCENE	*Alligator* sp., Florida	I	cf. *C. moreleti,* Baja Calif. 1	IV
	A. mefferdi, Nebraska	×II×III		
MIOCENE, MID	*A. thomsoni,* Nebraska	I×II		
LOWER	*A. mcgrewi,* Nebraska	II		
	A. olseni, Florida	I×II		
OLIGOCENE: South	*A. (Caimanoidea) prenasalis*	I×II		
Dakota	*Caimanoidea visheri*	U		
Titanotherium Beds	*Allognathosuchus riggi*	NO		
EOCENE, UPPER	*Procaimanoidea utahensis* 4	III		Terrestrial
				Pristichampsine
Bridger? Fm.	*Procaimanoidea kayi* U 1	III	*C. affinis* IV×III	*P. vorax* UUU
Wyoming	*Allognathosuchus polyodon*	II	*Brachyuranochampsa zangerli* VI×V	
	Diplocynodon stuckeri ID	NO	?*C. elliottii* IV	??*C. grinelli* NO
Washakie Fm.			*C. clavis* IV×III	*P. vorax* UUU
Wyoming			*B. eversolei*	VI
Green R. Fm.	*Alligator* ?n. sp., Wyoming	I×II	*Leidysuchus wilsoni,* Wyoming V×III!	
	of Grande 1980.		*C. acer,* Utah	VI×VII
Wasatch Beds	*Allog. heterodon*	U I?	*Orthogenysuchus olseni*	UU
Wyoming	*Allog. wartheni*	NO		
PALEOCENE,	*Ceratosuchus burdoshi,* Colorado 2		*L. riggsi,* Colorado 1 U	VI×VII
UPPER		II		
Tongue R. Fm.	*Wannaganosuchus brachymanus,*		*L. formidabilis,* N. Dakota	VI×VII
	N. Dakota	II		
Nacimiento Fm.	*Akanthosuchus langstoni,* N. Mexico		*L. multidentatus,* N. Mexico UVI×VII	
Torrejonian Faunal		? NO	*Navajosuchus novomexicanus*	III
Zone				
U Ravenscrag Fm.			*L. acutidentatus,* Saskatchewan	VI
Puercan Zone	*Allog. mooki*	II×III		
Lance Fm, near Cret./	*Prodiplocynodon langi,* Wyoming		*L. sternbergi,* Wyoming	IV×III
Paleocene	III×IV boundary			
UPPER	*Brachychampsa montana* 1, 02	II	*L. sternbergi,* Montana	IV×III
CRETACEOUS			*Pinacosuchus mantiensis* ?crocodile	
Hell Cr. Beds				NO
North Horn Fm.				
Judith R. Fm.	*Albertochampsa langstoni*	I	*L. canadensis*	V
Alberta			*L. gilmorei*	V
Kirtland Shale			*Goniopholis kirtlandicus,* G N.Mex.	
				V
"Greensand"	*Bottosaurus harlani,* New Jersey NO			
in six Fms.				
			Deinosuchus 7 U	?IV×III
			(giant crocodiles in six states)	
Dakota Sandst.			*Dakotasuchus kingi,* Kansas	NO
Mowrie Sh. Fm.?			*Coelosuchus reedii,* Wyoming	NO
UPPER JURASSIC			*Amphicotylus (G.) lucasii,* G. ColoradoV	
All Morrison			*G. gilmorei,* G. Wyoming	V×VI
Fm. in the			*G. stovalli,* G. Oklahoma	V×VI
broad sense			*Goniopholis felix,* G. Colorado V×VI	
			Eutretauranosuchus delfsi,	
			G. Colorado	U

Species references: *Akanthosuchus* (O'Neill et al. 1981)/*Albertochampsa* (Erickson 1972)/*Alligator* in Pliocene and Miocene Florida (Auffenberg 1967), *A. mississippiensis* in Pleistocene Florida (Holman 1978), assignment of Pleistocene midcontinent material uncertain to species (Preston 1979)/*A. mefferdi* head shape is closer to *M. niger* skull ca. 20 cm long, than to I×II./*Alligator* ?n. sp. is shown in oblique view in Grande 1980. The head shape assigned here is an approximation./*Allognathosuchus wartheni* (Bartels 1980 for formation)/*Bottosaurus* content and range (Gilmore 1911). *Caiman sclerops = Caiman crocodilus* following Medem (1955, 1971, etc.) as the acknowledged best taxonomy; its range in Central America and that of *C. acutus* (Smith and Smith 1977)/*Crocodylus:* assignment of Eocene species from North America to this genus is not certain, and *C. clavis* may be a junior synonym of *C. affinis* (Buffrenil and Buffetaut 1981, W. Bartels, personal communication 1983)/*C. acer* and *L. wilsoni* (Grande 1980 for formation)/cf. *C. moreleti* in late Pliocene (Miller 1980)/*C. johnstoni* rather than *C. johnsoni* is used by current workers on that species, (e.g., Messel et al. 1981)/*Dakotasuchus* (Vaughn 1956 for formation)/*Deinosuchus* treated as a genus here—the three named species are not well distinguished by their authors; Baird and Horner 1979 for presence in six states; W. Langston, personal communication that skull shape needs re-evaluation; D. Baird, personal communication that lateral profile of anterior snout is unusual/*Go-niopholis* (Mateer 1981 that perhaps North American species should be placed in *Eutretauranosuchus*)/*Hyposaurus* (G.) *natator* excluded from this list because it was narrow snouted (Troxell 1925) although included in list of broad-snouted species in Steel 1973/*Leidyosuchus* (Baird and Horner 1979 and D. Baird, personal communication of occurrence in the Late Cretaceous in North Carolina and Georgia)/*L. canadensis* and *L. gilmorei* (W. Langston, personal communication 1983, that designation of Judith River Fm. rather than Old Man Fm. is correct)/*L. formidabilis* (Erickson 1976)/*P. vorax* (Langston 1975; Busbey, in preparation)/*Wannaganosuchus* (Erickson 1982).

The times and formations listed are the most precisely known strata of good published material. Stratigraphic names prior to 1965 follow USGS practice in Keroher (1966) or the *Lexicon of Geologic Names in Alberta* (1954). There are three exceptions, where major changes in formation names make the location of described fossils uncertain: the Wasatch Beds, New Jersey Greensand, and Plateau Valley Beds. For all North and Central American species the primary literature was reviewed. For brevity, I refer the reader to Steel (1973) and Kuhn (1936) for the remaining citations, which are accurate. The taxonomy used throughout is the best published work. Assignments to subfamily are convenient but often problematical. Synonymizing of species names in two unpublished theses is not followed here, in accordance with Smith's (1981) rules of priority.

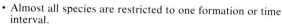

- Almost all species are restricted to one formation or time interval.
- Each potential fauna, a horizontal group in the same state or physiographic province, has very few species. Fauna is used in the zoologist's sense. (The apparent exception, the Bridger species, is discussed in the text. The time sequence and taxonomy of the Upper Jurassic species are uncertain.)
  Head profiles are given in Table 1. The unusual profiles, marked U, are discussed in the text. Fm. = formation.
  1 Shape is inferred from partial material.
  2 Snout has a constriction near the base deeper than in II.

3 Shape shares traits with I, II, and III.
4 Rear of skull is narrower than in III, resembling *Paleosuchus trigonatus*.
5 Shape is intermediate between V and III.
7 This giant skull has an unusual outline in Bird's restoration, based on fragmentary material.
? Taxonomic distinctness is uncertain.
NO No skull has been figured.
ID Species not based on comparative material, but it is distinct in its fauna.

Appendix 2. Definitions.

Hatchling: of size just emerged from egg.

Juvenile: potentially still in care of an adult, e.g., up to 80 cm for large species like the American alligator.

Subadult: smaller than common size of first breeding, e.g., up to 180 cm for female American alligators, larger for males.

Adult: of breeding size.

Natural habitat; a working definition for crocodilians is: the site is not modified by canals, water pumping, or other human influences on the presence of open water and of dry banks. Exotic plants are rare or absent from the wetland, especially plants offering cover for crocodilians, and wetland forest structure has no major alterations by logging or fires related to logging. Overland egress to other sites is not artificially restricted. Motorboats and airboats are rarely present (less than one day each three months), and harassment of crocodilians by humans rarely occurs.

Broad-nosed and narrow-nosed: these are equivalent to the respective terms, short-nosed (brevirostrine, often used as synonymous with alligators or broad-nosed crocodiles) and long-nosed (longirostrine, often used as synonymous with gharials and other narrow-nosed taxa). A glance at the head profiles in Fig. 1 shows the ambiguity in the short-nosed/long-nosed dichotomy, when used as descriptors of major shape groups.

Acknowledgments: Dedicated to the work of Hugh B. Cott, which has set a high water mark in vertebrate ecology.

Reviewers of various drafts included: G. Zug, B. Van Valkenburgh, S. M. Stanley, P. Shipman, C. A. Ross, P. W. Kat, W. Langston, K. W. Kaufmann, J. B. C. Jackson, T. H. Hughes, C. Crumley, A. B. Busbey, and D. Baird. The many suggestions by S. M. Stanley are appreciated.

Field work was greatly assisted by C. Carter, R. Curtis, H. G. Dowling, M. Hopkins, P. W. Kat, J. Stokes, and J. T. Woods.

Support from: the Society of Sigma Xi, the American Museum of Natural History and the Noble Foundation, New York University, and The Johns Hopkins University, Department of Earth and Planetary Sciences.

I thank the following for discussions: W. Auffenberg, D. Baird, R. Bakker, W. Bartels, A. B. Busbey, H. W. Campbell, C. Crumley, R. Curtis, D. Dietz, H. G. Dowling, who suggested the zoo comparison, T. H. Hughes, Jewett Hall, R. Inger, J. B. C. Jackson, P. W. Kat, K. W. Kaufmann, S. B. McDowell, F. Medem, K. Rose, C. A. Ross, J. Roze, P. Shipman, S. M. Stanley, B. Van Valkenburgh, A. Walker, M. Watanabe, and G. Zug.

Literature

Adams, S. E., Smith, M. H., Baccus, R. 1980. Biochemical variation in the American alligator. Herpetologica 36(4):289–296.

Allen, G. R. 1974. The marine crocodile, *Crocodylus porosus,* from Ponape, Eastern Caroline Islands, with notes on food habits of crocodiles from the Palau Archipelago. Copeia 1974(2):553.

Auffenberg, W. 1967. Fossil crocodilians of Florida. Plaster Jacket, Florida State Mus., No. 5.

Baird, D., Horner, J. R. 1979. Cretaceous dinosaurs of North Carolina. Brimleyana 2:1–28.

Bartels, W. S. 1980. Early Cenozoic reptiles and birds from the Bighorn Basin, Wyoming, pp. 73–79. In: Gingerich, P. D. (ed.), Early Cenozoic paleontology and stratigraphy of the Bighorn Basin, Wyoming. U. Mich. Papers Paleont. No. 24.

Beadle, L. C. 1974. The inland waters of Tropical Africa. London: Longman Group.

Bennett, A. F. 1980. The metabolic foundations of vertebrate behavior. BioScience 30(7):452–456.

Bennett, A. F., Ruben, J. A. 1979. Endothermy and activity in vertebrates. Science 206:649–654.

Berggren, W. A., McKenna, M. C., Hardenbol, J., Obradovich, J. D. 1978. Revised Paleogene polarity time scale. J. Geol. 86:67–81.

Bradley, W. H. 1964. Geology of Green River Formation and associated Eocene rocks in southwestern Wyoming and adjacent parts of Colorado and Utah. Geological Survey Prof. Paper 496-A, U.S. Gov't Printing Office.

Brazaitis, P. The identification of living crocodilians. Zoologica (NYZS) 1973:59–101.

Brazaitis, P. 1981. Maxillary regeneration in a marsh crocodile, *Crocodylus palustris.* J. Herpetology 15(3):360–362.

Buffetaut, E. 1978. Sur l'histoire phylogenetique et biogeographie des Gavialidae (Crocodylia, Eusuchia). C. R. Acad. Sc. Paris, t.287, Serie D, No. 10:911–914.

Buffetaut, E. 1979. The evolution of the crocodilians. Sci. Amer. 1979:130–144.

Buffetaut, E., Ford, R. L. E. 1979. The crocodilian *Bernissartia* in the Wealden of the Isle of Wight. Paleontology 22(4):905–912.

Buffetaut, E., Ingavat, R. 1980. A new crocodilian from the Jurassic of Thailand, *Sunosuchus thailandicus* n. sp. (Mesosuchia, Goniopholidae) and the paleogeographic history of Southeast Asia in the Mesozoic. Geobios 13(6):879–889.

Buffrenil, V. de, Buffetaut, E. 1981. Skeletal growth lines in an Eocene crocodilian skull from Wyoming as an indicator of ontogenetic age and paleoclimatic conditions. J. Vert. Paleont. 1(1):57–66.

Burgin, S. 1980. Crocodiles and crocodile conservation in Papua New Guinea, pp. 295–300. In: Morauta, L., Pernetta, J., Heaney, W. (eds.), Traditional conservation in Papua New Guinea: Implications for today. Inst. Applied Social and Economic Research, Boroko, Papua New Guinea.

Busbey, A. B. 1977. Functional morphology of the head of *Pristichampsus vorax* (Crocodilia, Eusuchia) from the Eocene of North America. M.A. thesis, U. Texas, Austin. Unpublished, cited with author's permission.

Chabreck, R. H. 1965. The movement of alligators in Louisiana. Proc. SE. Assn. Game and Fish Comm. 19:102–110.

Chabreck, R. H. 1971. The foods and feeding habits of alligators from fresh and saline environments in Louisiana. Proc. SE. Assn. Game and Fish Comm. 25:117–124.

Chabreck, R. H., Dupie, H. P. 1976. Alligator predation on Canada goose nests. Copeia 1976(2):404–405.

Chabreck, R. H., Joanen, T. 1979. Growth rates of American alligators in Louisiana. Herpetologica 35(1):51–57.

Cohen, M. M., Gans, C. 1970. The chromosomes of the order Crocodilia. Cytogenetics 9:81–105.

Colbert, E. H. 1962. The weights of dinosaurs. Amer. Mus. Novit. 2076:1–16.

Colbert, E. H., Bird, R. T. 1954. A gigantic crocodile from the Upper Cretaceous beds of Texas. Amer. Mus. Novit. 1688:1–22.

Colbert, E. H., Mook, C. C. 1951. The ancestral crocodilian *Protosuchus*. Bull. Amer. Mus. Nat. Hist. 97(3):147–182.

Cope, E. D. 1883/4. The Vertebrata of the Tertiary Formations of the West. Book 1. Washington, DC: U.S. Gov't Printing Office.

Cory, C. B. 1896. Hunting and fishing in Florida, including a key to the water birds. 2nd ed. Boston: Estes and Lauriat.

Cott, H. B. 1961. Scientific results of an inquiry into the ecology and economic status of the Nile crocodile (*Crocodylus niloticus*) in Uganda and northern Rhodesia. Trans. Zool. Soc. London 29:210–357.

Cott, H. B. 1972. Review of: W. T. Neill (1971). Nature 237:468.

Cott, H. B. 1975. Looking at animals: a zoologist in Africa. New York: Scribners.

Coulson, R. A., Hernandez, T. 1983. Alligator metabolism: studies on chemical reactions *in vivo*. Comp. Biochem. Physiol. 74(1):1–182.

Darlington, P. J. 1957. Zoogeography: the geographical distribution of animals. New York: Wiley.

Darwin, C. R. 1859. On the origin of species. (Facsimile reprint of the first edition, 1964). Boston: Harvard U. Press.

Dietz, D. C., Hines, T. C. 1980. Alligator nesting in North-Central Florida. Copeia 1980(2):249–258.

Densmore, L. D., Dessauer, H. C. 1979. Preliminary molecular evidence on relationships within the Alligatoridae, p. 24. In: 26th Ann. Meeting of the Herpetologist's League, 21st Ann. Meeting of the SSAR.

Dodson, P. 1975. Functional and ecological significance of relative growth in *Alligator*. J. Zool. London 175:315–335.

Dowling, H. G., Duellman, W. E. 1974/78. Systematic herpetology: a synopsis of families and higher categories. New York: HISS Publications.

Erickson, B. R. 1972. *Albertochampsa langstoni*, gen. et sp. nov., a new alligator from the Cretaceous of Alberta. Sci. Publ. Sci. Mus. Minnesota 2, No. 1:1–13.

Erickson, B. R. 1976. Osteology of the early eusuchian crocodile *Leidyosuchus formidabilis*, sp. nov. Sci. Mus. Minnesota Monogr. 22:1–61.

Erickson, B. R. 1982. *Wannaganosuchus*, a new alligator from the Paleocene of North America. J. Paleont. 56(2):492–506.

Estes, R. 1970. Origin of the Recent North American lower vertebrate fauna: an inquiry into the fossil record. Forma et Functio 3:139–169.

Ferguson, M. J. W. 1979. The American alligator (*Alligator mississippiensis*): a new model for investigating developmental mechanisms in normal and abnormal palate formation. Medical Hypotheses 5:1079–1090.

Ferrusquia-Villafranca, I. 1978. Distribution of Cenozoic vertebrate faunas in middle America and problems of migration between North and Central America. Bol. Inst. Geol. Univ. Nat. Auton. Mexico, 101:193–321.

Fittkau, E. J. 1973. Crocodiles and the nutrient metabolism of Amazonian waters. Amazonia 4(1):103–133.

Fogarty, M. J. 1972. Review of W. T. Neill (1971). J. Wild. Mgmt. 36:1370–2.

Fogarty, M. J., Albury, J. A. 1967. Late summer

foods of young alligators in Florida. Proc. SE. Conf. Game and Fish Comm. 1967:220–222.

Friar, W., Behler, J. L. 1983. Review of H. Wermuth und R. Mertens (1977). Liste der rezenten Amphibien und Reptilien. Testudines, Crocodylia, Rhynchocephalia. Das Tierreich, W. De Gruyter, New York. Herp. Rev. 14(1):23–25.

Fuchs, K., Mertens, R., Wermuth, H. 1974. Die unterarten des Nilkrokodils, *Crocodylus niloticus*. Salamandra 10:107–114.

Garrick, L. D., Lang, J. W., Herzog, H. A. 1978. Social signals of adult American alligators. Bull. Amer. Mus. Nat. Hist. 160(3):153–192.

Gartside, D. F., Dessauer, H. C., Joanen, T. 1977. Genic homozygosity in an ancient reptile (*Alligator mississippiensis*). Biochem. Genet. 15(7/8): 655–663.

Giles, L. W., Childs, V. L. 1949. Alligator management on the Sabine National Wildlife Refuge. J. Wild. Mgmt. 13:16–28.

Gilmore, C. W. 1911. A new fossil alligator from the Hell Creek Beds of Montana. Proc. U.S. Natl. Mus. 41, No. 1860:297–302 + 2 pl.

Glass, M. L., Johansen, K. 1979. Periodic breathing in the crocodile *Crocodylus niloticus:* consequences for the gas exchange ratio and control of breathing. J. Exp. Zool. 208(3):319–325.

Gleeson, T. T. 1979. Foraging and transport costs in the Galapagos marine iguana, *Amblyrhynchus cristatus.* Physiol. Zool. 52(4):549–557.

Gorzula, S. J. 1978. An ecological study of *Caiman crocodilus crocodilus* inhabiting savanna lagoons in the Venezuelan Guayana. Oecologia 35:21–34.

Gould, S. J., Lewontin, R. C. 1979. The spandrels of San Marco and the Panglossian paradigm: a critique of the adaptationist programme. Proc. Roy. Soc. London 205:582–598.

Graham, A. 1968. The Lake Rudolf crocodile (*Crocodylus niloticus*) population. Mimeo. report to the Kenya Game Dept.

Grande, L. 1980. Paleontology of the Green River Formation with a review of the fish fauna. Geol. Surv. Wyoming, Bull. 63.

Herzog, H. A. 1974. The vocal communication system and related behaviors of the American alligator. (*Alligator mississippiensis*) and other crocodilians. M.A. Thesis, U. Tennessee.

Herzog, H. A., Burghardt, G. M. 1977. Vocal communication signals in juvenile crocodilians. Z. Tierpsych. 44:394–404.

Hines, T. C., Scheaffer, R. C. 1977. Public opinion about alligators in Florida. Proc. Ann. Conf. SE. Assoc. Fish & Wildl. Agencies 31:84–89.

Holman, J. A. 1978. The Late Pleistocene herpetofauna of Devil's Den sinkhole, Levy County, Florida. Herpetologica 34(2):228–237.

Honegger, R. E. 1975. Red data book, Vol. 3: Amphibia and Reptilia. IUCN, Morges, Switzerland.

Hornaday, W. T. 1875. The crocodile in Florida. Amer. Nat. 9:1–6.

Hulke, J. W. 1878. Notes on two skulls from the Wealden and Purbeck Formations indicating a new subgroup of Crocodilia. Geol. Soc. London Quart. J. 34:377–383.

von Humboldt, A. 1876. Personal narrative of travels to the equinoctial regions of America. Ross, T. (tr.), Vol. 2. London: Bell.

Hunt, R. H. 1977. Aggressive behavior by adult Morelet's crocodiles *Crocodylus moreleti* toward young. Herpetologica 33:195–201.

Hunt, R. H., Watanabe, M. E. 1982. Observations on maternal behavior of the American alligator, *Alligator mississippiensis.* J. Herp. 16:235–239.

Huxley, T. H. 1875. On *Stagonolepis Robertsoni,* and on the evolution of the crocodilia. Geol. Soc. London Quart. J. 31:423–436. (Duncan's comments & discussion are on pp. 436–438.)

Iordansky, N. N. 1973. The skull of the Crocodylia, pp. 256–284. In: Gans, C. (ed.), Biology of the Reptilia, New York: Academic.

Joanen, T. 1969. Nesting ecology of alligators in Louisiana. Proc. SE. Assn. Game and Fish Comm. 19:141–151.

Joanen, T., McNease, L. 1972. A telemetric study of adult male alligators on Rockefeller Refuge, Louisiana. Proc. SE. Assoc. Game and Fish Comm. 1972:252–275.

Joffe, J. 1967. The "dwarf" crocodiles of the Purbeck Formation, Dorset: reappraisal. Paleontology 10(4):629–639.

Kaelin, J. A. 1933. Beiträge zur vergleichenden Osteologie des Crocodilidenschädels. Zool. Jahrbuch, Anatomie 57:535–714.

Keroher, G. C., and others. 1966. Lexicon of geologic names of the U.S. U.S. Geol. Surv. Bull. No. 1200. 3 Vol.

King, F. W. 1972. Review of W. T. Neill (1971). BioScience 22:119.

Kuhn, O. 1936. Fossilium Catalogus. 1: Animalia, Pars 75: Crocodilia. Gravenhage: W. Junk.

Kuhn, O. 1968. Die Vorzeitlichen Krokodile. Munchen: Verlag Oeben.

Kurten, B. 1978. The age of dinosaurs. New York: McGraw-Hill.

Kushlan, J. 1974. Observations on the role of the American alligator (*Alligator mississippiensis*) in the southern Florida wetlands. Copeia 1974 (4): 993–996.

Kushlan, J., Kushlan, M. S. 1980. Function of nest attendance in the American alligator. Herpetologica 36(1):27–32.

Langston, W. 1965. Fossil crocodilians from Colombia and the Cenozoic history of the Crocodilia in South America. U. Calif. Publ. Geol. Sci. 52:1–157 + 5 pl.

Langston, W. 1966. *Mourasuchus* Price, *Nettosuchus* Langston, and the family Nettosuchidae (Reptilia: Crocodilia). Copeia 1966:882–885.

Langston, W. 1973. The crocodilian skull in historical perspective, pp. 263–284. In: Gans, C. (ed.), Biology of the Reptilia, Vol. 4D. New York: Academic.

Langston, W. 1975. Ziphodont crocodiles: *Pristichampsus vorax* (Troxell), new combination, from the Eocene of North America. Fieldiana Geol. 33(16):291–314.

Leakey, M. D. 1971. Olduvai Gorge, p. 291. Vol. 3. Cambridge U. Press.

Lexicon of geologic names in Alberta and adjacent portions of British Columbia and N.W. Territories. 1954. Alberta Soc. Petr. Geol., Calgary.

Longman, H. A. 1925. *Crocodilus johnsoni* Krefft. Mem. Queensland Mus. 8(2):95–102.

Lydekker, R. 1886. Palaeontologica Indica. Mem. Geol. Surv. India, ser. 10, V. 2:209–240.

Magnusson, W. E. 1980. Habitat required for nesting by *Crocodylus porosus* (Reptilia: Crocodilidae) in Northern Australia. Aust. J. Wildl. Res. 7:149–156.

Magnusson, W. E. 1982. Mortality of eggs of the crocodile *Crocodylus porosus* in northern Australia. J. Herp. 16(2):121–130.

Mateer, N. J. 1981. The reptilian megafauna from the Kirtland Shale (Late Cretaceous) of the San Juan Basin, New Mexico. p. 59. In: Lucas, S. G., Rigby, J. K., Kues, B. S. (eds.), Advances in San Juan Basin paleontology. U. New Mexico Press.

Mayr, E. 1954. Change of genetic environment and evolution, pp. 155–180. In: Huxley, J., Hardy, A. C., Ford, E. B. (eds.), Evolution as a process. London: Allen and Unwin.

Mayr, E. 1975. The unity of the genotype. Biol. Zbl. 94:337–388.

McIlhenny, E. A. 1935. The alligator's life history. Boston: Christopher Publ. (Also reprinted by the SSAR, 1976).

McNease, L., Joanen, T. 1974. A study of immature alligators on Rockefeller Refuge, Louisiana. Proc. SE. Assn. Game and Fish Comm., 28:482–500.

McNease, L., Joanen, T. 1977. Alligator diets in relation to marsh salinity. Proc. Ann. Conf. SE. Assoc. Fish and Wildlife Agencies 31:36–40.

Medem, F. 1955. A new subspecies of *Caiman sclerops* from Colombia. Fieldiana Zool. 37:339–343.

Medem, F. 1958. The crocodilian genus *Paleosuchus*. Fieldiana Zool. 39:227–247.

Medem, F. 1960. Notes on the Paraguay caiman, *Caiman yacare* Daudin. Mitt. Zool. Mus. Berlin 36(1):129–142.

Medem, F. 1971. Biological isolation of sympatric species of South American crocodilia, pp. 152–158. In: Crocodiles. IUCN Publ. N.S. No. 52.

Menzies, R. A., Kushlan, J., Dessauer, H. C. 1979. Low degree of genetic variability in the American alligator (*Alligator mississippiensis*). Isozyme Bulletin 12.

Messel, H., Vorlicek, G. C., Wells, A. G., Green, W. J. 1981. The Blyth-Cadell rivers system study and the status of *Crocodylus porosus* in tidal waterways of Northern Australia. Vol. 1 of the series: Surveys of tidal river systems in the Northern Territory of Australia and their crocodile populations. Sydney: Pergamon.

Metzen, W. D. 1977. Nesting ecology of alligators on the Okefenokee National Wildlife Refuge. Proc. SE. Assoc. Fish and Wildl. Agencies, 31st Conf.

Meyer, E. R. 1975. Alligator ecology and population structure on Georgia sea islands. Amer. Soc. Ichth. Herp. 55th Meeting: p. 54.

Meyer, E. R. 1977. Alligator behavior, ecology, and populations in some Okefenokee marshes. Amer. Soc. Ichth. Herp. 57th Meeting Program Abstract.

Meyer, E. R. 1978. Female alligator opening her nest and carrying hatchlings to water. Color motion picture footage in: Dragons of paradise, a one-hour documentary distributed to U.S. public television, 1978–1983.

Miller, W. E. 1980. The late Pliocene Las Tunas local fauna from southernmost Baja California, Mexico. J. Paleont. 54:762–805.

Minton, S. A., Minton, M. R. 1973. Giant reptiles. New York: Scribners.

Modha, M. L. 1967. The ecology of the Nile crocodile (*Crocodylus niloticus* Laurent) on Central Island, Lake Rudolf. E. Afr. Wildl. J. 5:74–95.

Molnar, R. E. 1979. *Crocodylus porosus* from the Pliocene Allingham formation of North Queensland, Australia: results of the Ray E. Lemley Expeditions, Part 5. Mem. Queensland Mus. 19(3):357–365.

Mook, C. C. 1962. A new species of *Brachyuranochampsa* (Crocodilia) from the Bridger Beds of Wyoming. Amer. Mus. Novit., No. 2079:1–6.

Myers, N. 1972. The long African day. New York: Macmillan.

Neill, W. T. 1971. The last of the ruling reptiles, alligators, crocodiles, and their kin. New York: Columbia U. Press.

Nichols. J. D., Viehman, L., Chabreck, R. H., Fenderson, B. 1976. Simulation of a commercially harvested alligator population in Louisiana. Louisiana State U. Agric. Exper. Station, Bulletin No. 691.

Olson, E. C. 1971. Vertebrate paleozoology. New York: Wiley.

O'Neill, F. M., Lucas, S. G., Kues, B. S. 1981. *Akanthosuchus langstoni,* a new crocodilian from the Nacimiento formation (Paleocene, Torrejonian) of New Mexico. J. Paleont. 55(2):340–352.

Ostrom, J. H. 1969. Terrestrial vertebrates as indicators of Mesozoic climates. Proc. N. Amer. Paleont. Conv. 347–376.

Owen, R. 1878. Order. *Crocodilia.* Suppl. No. VIII to the: Monograph on the fossil Reptilia of the Wealden and Purbeck Formations. London: Palaeontographical Society.

Patterson, B. 1931. Occurrence of the alligatoroid genus *Allognathosuchus* in the Lower Oligocene. Field Mus. Nat. Hist., Geol. Ser. 4(6):223–226 + 1 pl.

Patterson, B. 1936. *Caiman latirostris* from the Pleistocene of Argentina, and a summary of South American Cenozoic Crocodilia. Herpetologica 1:43–54.

Pienaar, U. de V., FitzSimons, V. F. M. 1966. The reptiles of the Kruger National Park. Koedoe No. 1.

Pienaar, U. de V. 1969. Predator-prey relationships amongst the larger mammals of the Kruger National Park. Koedoe 1969:108–176.

Pitman, C. R. S. 1941. About crocodiles. Uganda Journal 9:89–114.

Pooley, A. C. 1977. Nest opening response of the Nile crocodile *Crocodylus niloticus.* J. Zool. London 182:17–26.

Pough, F. H. 1980. The advantages of ectothermy for tetrapods. Amer. Nat. 115(1):92–112.

Preston, R. E. 1979. Late Pleistocene cold-blooded vertebrate fauna from the mid-continental United States. U. Mich. Paper on Paleont. No. 19:1–53.

Romer, A. S. 1966. Vertebrate paleontology. (3rd ed.) U. Chicago Press.

Romer, A. S. 1971. The Chanares (Argentina) Triassic reptile fauna XI. Two new long-snouted thecodonts, *Chanaresuchus* and *Gualosuchus.* Breviora, MCZ, No. 379.

Romero, G. A. 1983. Distress call saves a *Caiman c. crocodilus* hatchling in the Venezuelan llanos. Biotropica 15(1):71.

Roosevelt, T. R. 1924. Through the Brazilian wilderness. New York: Scribners.

Root, J., Root, A. 1971. Mzima, Kenya's spring of life. Natl. Geogr. 140(3):350–373.

Ross, C. A. 1979. Scalation of the American alligator. U.S. Fish and Wildl. Serv., Special Sci. Rept.: Wildlife No. 225:1–8.

Ross, C. A. 1974. Gharial in Corbett National Park (U.P.), p. 8. In: Whitaker and Rajamani 1974.

Russell, D. E. and 32 coauthors. 1982. Tetrapods of

the Northwest European Tertiary Basin. Geologisches Jahrbuch, Reihe A, Heft 60.

Schmidt, K. P. 1919. Contributions to the herpetology of the Belgian Congo based on the collection of the American Museum Congo Expedition, 1909–1915. Bull. Amer. Mus. Nat. Hist. 39:385–624.

Schmidt, K. P. 1924. Notes on Central American crocodiles. Field Mus. Nat. Hist., Zool. Ser. 12(220):79–92.

Schmidt, K. P. 1928. Notes on South American caimans. Field Mus. Nat. Hist., Zool. Ser. 12(252):203–231 + 6 pl.

Schmidt, K. P. 1932. Notes on New Guinean crocodiles. Field Mus. Nat. Hist., Zool. Ser. 18(8):164–173.

Schmidt, K. P. 1941. A new fossil alligator from Nebraska. Field Mus. Nat. Hist. Geol. Ser. 8(494):27–32.

Schmidt, K. P. 1944. Crocodiles. Fauna 6(3):67–72.

Sill, W. O. 1968. The zoogeography of the crocodilia. Copeia 1968:76–88.

Simpson, G. G. 1933. A new crocodilian from the *Notostylops* beds of Patagonia. Amer. Mus. Novitates, No. 623.

Smith, E. N. 1979. Behavioral and physiological thermoregulation of crocodilians. Amer. Zool. 19:239–247.

Smith, H. M., Smith, R. B. 1977. Synopsis of the herpetofauna of Mexico, pp. 49–116. Vol. 5.

Smith, H. M. 1981. A suggested protective declaration for taxonomic dissertations. Herp. Rev. 12:98.

Spotila, J. R., Soule, O. H., Gates, D. M. 1972. The biophysical ecology of the alligator: heat energy budgets and climate spaces. Ecology 53(6):1094–1102.

Stanley, S. M. 1973. An explanation for Cope's Rule. Evolution 27(1):1–26.

Staton, M. A., Dixon, J. R. 1975. Studies on the dry season biology of *Caiman crocodilus crocodilus* from the Venezuelan Ilanos. Memoria Cien. Nat., Caracas, 35:237–265.

Staton, M. A., Dixon, J. R. 1977. Breeding biology of the spectacled caiman, *Caiman crocodilus crocodilus,* in the Venezuelan Ilanos. U.S. Fish and Wildl. Serv. Wildlife Res. Rept. 5:1–21.

Stebbins, G. L. 1982. Perspectives in evolutionary theory. Evolution 36(6):1109–1118.

Steel, R. 1973. Crocodylia, pp. 1–116. In: Encyclopedia of paleoherpetology, Part 16. Stuttgart–Portland: Gustav Fischer Verlag.

Tchernov, E. 1976. Crocodilians from the late Cenozoic of the Rudolf Basin, pp. 370–378. In: Coppens, Y. et al. (eds.), Earliest man and environments in the Lake Rudolf Basin. U. Chicago Press.

Templeton, A. R. 1981. Mechanisms of speciation: a population genetic approach. Ann. Rev. Ecol. Syst. 12:23–48.

Terpin, K. M., Spotila, J. R., Foley, R. E. 1979. Thermoregulatory adaptations and heat energy budget analyses of the American alligator, *Alligator mississippiensis*. Physiol. Zool. 52(3):296–312.

Troxell, E. L. 1925. *Hyposaurus,* a marine crocodilian. Amer. J. Sci. 10:489–514.

Van Valkenburgh, B. 1982. Evolutionary aspects of terrestrial large predator guilds. Third North Amer. Paleont. Conv. Proc. 2:557–562.

Vaughn, P. P. 1956. A second specimen of the Cretaceous crocodile *Dakotasuchus* from Kansas. Trans. Kansas Acad. Sci. 59:379–381.

Watson, R. M., Graham, A. D., Bell, R. H. V., Parker, I. S. C. 1971. A comparison of four East African crocodile (*Crocodylus niloticus*) populations. E. Afr. Wildl. J. 9:25–34.

Webb, G. J. W. 1977. The natural history of *Crocodylus porosus*. In: Messel, H. and Butler, S. T. (eds.), Australian animals and their environments. Sidney: Shakespeare Head Press.

Webb, G. J. W., Messel, H. 1977. Abnormalities and injuries in the estuarine crocodile, *Crocodylus porosus*. Austr. Wildl. Res. 4:311–319.

Webb, G. J. W., Messel, H. 1978a. Morphometric analysis of *Crocodylus porosus* from the north coast of Arnhem Land, Northern Australia. Austr. J. Zool. 26:1–27.

Webb, G. J. W., Messel, H. 1978b. Movement and dispersal patterns of *Crocodylus porosus* in some rivers of Arnhem Land, Northern Australia. Austr. Wildl. Res. 5:263–283.

Welman, J. B., Worthington, E. B. 1943. The food of the crocodile (*Crocodylus niloticus* L.). Proc. Zool. Soc. London 113:108–112.

Wermuth, H., Mertens, R. 1961. Schildkröten–Krokodile–Brückenechsen. Jena: Gustav Fischer Verlag.

Whitaker, R. 1974. Notes on behaviour, ecology, and present status of the marsh crocodile (*Crocodylus palustris*) in south India. Madras Snake Park Trust and Conservation Center, India.

Whitaker, R., Rajamani, V. C. 1974. Gharial survey report. Madras Snake Park Trust and Conservation Center, India.

Wilson, D. S. 1980. The natural selection of populations and communities. Menlo Park: Benjamin/Cummings.

Zug, G. R. 1974. Crocodilian galloping: an unique gait for reptiles. Copeia 1974:550–552.

11
Family Chanidae and Other Teleostean Fishes as Living Fossils

Colin Patterson

British Museum (Natural History), London SW7, England

Introduction

This article is something of a fraud. I began work on it with the intention of showing that at least one Recent teleost, *Chanos chanos* (Forskål), deserved to be called a living fossil. In the end, I failed to convince myself. The article is an account of that failure, with comments on other potential living fossils among teleosts.

Chanos chanos, the milkfish, is widespread in coastal waters of the Indian and Pacific Oceans, ranging from the Red Sea and east coast of Africa to the west coast of Mexico and southern California (Schuster 1960). It is euryhaline, entering fresh water freely, and having remarkable tolerance of hypersaline waters. *Chanos chanos* is the only living species of a genus, family (Chanidae) and, according to some, a suborder (Chanoidei) of teleosts. The Chanidae have a reputed fossil record extending back to the Early Cretaceous (Neocomian), a situation virtually unique amongst teleosts. The only other teleostean families with credible Early Cretaceous records are the Hiodontidae (Greenwood 1970; Chang and Chou 1976, 1977; Taverne 1979), Osteoglossidae or Arapaimidae (Taverne 1979:134), Elopidae and Megalopidae (Forey 1973; Taverne 1974a), and Argentinidae (Taverne 1982), but in each of these groups shared primitive characters play a large part in the assignment of the fossils to the families (on osteoglossoids and hiodontids see Greenwood 1973; Taverne 1979; Grande 1980; on elopids and megalopids see Greenwood, 1977; Patterson and Rosen 1977; on argentinids see Taverne 1982:16, 24). In contrast, the characters shared by Recent *Chanos* and Early Cretaceous fossils are said to be "readily recognizable . . . unique specializations of the skull" (Patterson 1975:168; also Taverne 1981:976). The main purpose of this article is to evaluate the fossil record of chanids; to anticipate, I find that the evidence for Early Cretaceous chanids is no more convincing than is that for equally ancient hiodontids, osteoglossids, elopids, and megalopids. Thus there are no teleostean families with reliable Early Cretaceous records.

In the early part of the Late Cretaceous (Cenomanian), almost 20 extant families of teleosts are recorded. Some of these records probably depend on shared primitive characters, but others appear to be genuine. The number of candidates is too great to select one or a few as living fossils.

Relationships of *Chanos* to Other Living Teleosts

The history of the classification of *Chanos* is reviewed by Greenwood et al. (1966) and Lenglet (1974). The order Gonorynchiformes of

Greenwood et al. (1966) contains the genera first united as a suborder Gonorhynchoidei by Gosline (1960): *Chanos* (Chanidae), *Gonorynchus* (Gonorynchidae, one sp., marine, Indo-Pacific), *Phractolaemus* (Phractolaemidae, one sp., freshwater, Africa), and Kneriidae *s.l.* (4 genera, approximately 12 spp., freshwater, Africa). Rosen and Greenwood (1970) extended the argument, first proposed by Greenwood et al. (1966), that Gonorynchiformes are related to the Ostariophysi, those fishes with a Weberian apparatus, the major group of freshwater teleosts. Rosen and Greenwood classified the Gonorynchiformes within the Ostariophysi as a series Anotophysi, sister-group to a series Otophysi, equivalent to Osta-

riophysi of previous usage. This notion of relationship between Gonorynchiformes and ostariophysans, criticized by Roberts (1973), Taverne (1974b, 1974c, 1981) and Gosline (1980) among others, receives strong support from Fink and Fink's (1981) cladistic analysis.

Within the Gonorynchiformes, Greenwood et al. (1966) and Rosen and Greenwood (1970) included *Chanos* with *Phractolaemus* and the Kneriidae in a suborder Chanoidei, sister-group of the Gonorynchoidei (*Gonorynchus* only). They did not characterize or subdivide the Chanoidei, but their classification implied that *Chanos* is the sister-group of Kneriidae plus Phractolaemidae, or of one of those families. Lenglet (1974) saw the Chanoidei in a more

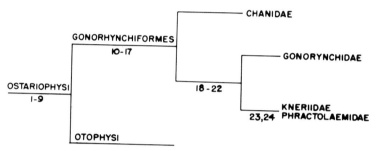

Fig. 1. A cladogram of Recent ostariophysan fishes. Numbered characters are as follows (modified from Fink and Fink 1981, whose character numbers are given in parentheses).

Characters of Ostariophysi: 1. Basisphenoid absent (1). 2. Sacculus and lagena situated more posteriorly and nearer the midline than in other primitive teleosts (8). 3. Dermopalatine absent (20). 4. Swim bladder divided into a smaller anterior and larger posterior chambers (54), anterior chamber with a silvery peritoneal tunic (55) which is attached to the first two ribs (56), and swim bladder suspended by a dorsal mesentery which is thickened anterodorsally (57). 5. Supraneural in front of neural arch of first vertebra absent (58). 6. Dorsomedial parts of first four neural arches expanded and abutting against one another, roofing the neural canal (63 *pars*). 7. Expanded part of first neural arch abuts against exoccipital (63 *pars*). 8. Alarm substance (117), nuptial tubercles with keratinous cap (118), adductor mandibulae with superficial ventral division (127). 9. Characters that are not unique to ostariophysans among primitive teleosts: (a) free neural arch between occiput and arch of first centrum absent (64); (b) hemal spines anterior to second pre-ural centrum fused to centra (111); (c) caudal skeleton with a compound element representing the first pre-ural and two ural centra and the first uroneurals (110).

Characters of Gonorynchiformes: 10. Orbitosphenoid absent, pterosphenoids small and widely separated (6). 11. Parietals very small, little more than canal-bearing ossicles (10). 12. Exoccipitals with prominent posterodorsal cartilaginous margin (14). 13. Quadrate condyle far forward, with elongation of quadratojugal process of quadrate, symplectic, interopercular and lower limb of preopercular (29). 14. Premaxilla thin and flat (38). 15. No teeth on fifth ceratobranchial (49). 16. Anterior neural arch especially large, with extensive, tight joint with exoccipital (65). 17. Characters that are not unique to Gonorynchiformes (within Ostariophysi): (a) no teeth in jaws (42); (b) no teeth on pharyngobranchials or basihyal (47, 48); (c) epibranchial organ behind fourth epibranchial (46); (d) no postcleithra (96); (e) one or two epurals (115); (f) no adipose fin (125).

Characters of Gonorynchidae, Kneriidae, and Phractolaemidae (Fink and Fink 1981:304): 18. Separation of palatine and ectopterygoid through reduction of latter. 19. Neural arches and parapophyses anterior to dorsal fin fused to centra. 20. No neural spine on first neural arch. 21. No ossified first basibranchial. 22. No ossified first pharyngobranchial.

Characters of Kneriidae and Phractolaemidae: 23. Supraoccipital with prominent cartilaginous margin (14). 24. Foramen magnum enlarged, bounded by cartilage above.

restricted sense, excluding *Cromeria* and *Grasseichthys* (both kneriids to Greenwood et al.). Lenglet listed four characters of the Chanoidei in her sense, but none is exclusive to them. Fink and Fink (1981) informally proposed a new arrangement of gonorynchiforms (Fig. 1), with *Chanos* (Chanoidei) as the sister-group of Gonorynchidae and the African freshwater forms. Since this is the only published scheme of gonorynchiform relationships that is supported by characters, it is adopted here. As characters incongruent with their arrangement, Fink and Fink cite the endopterygoid and basibranchial teeth of *Gonorynchus*, whereas all other gonorynchiforms have no teeth anywhere (a condition almost unique amongst teleosts). Fink and Fink assume that endopterygoid and basibranchial teeth were lost independently in chanids and the African freshwater gonorynchiforms.

Fossil Chanidae

There follows an annotated list of fossil taxa recorded as Chanidae.

Oligocene

Chanos brevis (Heckel 1853), *Chanos dezignii* (Heckel 1853). These two species, both from Chiavon, Vicenza, Italy, have not been revised since Bassani's (1888) work.

Neohalecopsis striatus (Weiler 1920), Septarientones, Flörsheim, Rheinhessen, Germany. This species, based on a single specimen, was transferred by Weiler (1928) to the new monotypic genus *Neohalecopsis*, compared with *Halecopsis* (see below).

Eocene

Chanos forcipatus (Heckel 1853), Monte Postale, Bolca, Verona, Italy (Cuisian). According to Blot (1980) this species, untouched since Heckel's work, is now under revision, as is a second undescribed species.

Chanoides macropoma Agassiz, 1835, *Chanoides leptostea* Eastman, 1905. *Chanoides,* represented only by these two species from Monte Bolca, Verona, Italy (Cuisian), is usually placed in the Albulidae by paleontologists (e.g., Blot 1980), following Woodward (1901), and in the Chanidae by neontologists (e.g., Fowler 1974). Acid-prepared material of *C. macropoma* is available, and the fish proves to be an otophysan, with a Weberian apparatus (Patterson in press). *C. leptostea,* unrevised since Eastman's description, may be a chanid.

Halecopsis insignis (Delvaux and Ortlieb 1887), Ypresian, London Clay, southeast England; Argile des Flandres, Hainaut, Belgium and Nord, France; Argile de Hemmoor, northwest Germany. Casier (1946) placed *Halecopsis* and Weiler's Oligocene *Neohalecopsis* in a new family Halecopsidae. Casier (1946, 1966) compared *Halecopsis* with *Gonorynchus,* but not with *Chanos.* The apparent absence of teeth in *Halecopsis,* the parietal penetrated by a sensory canal (middle pit-line of Casier 1966), very long quadrate, and other general similarities are compatible with gonorynchiform relationships. The exoccipitals of *Halecopsis* project posteriorly, in the manner identified by Fink and Fink (1981:313) as characteristic of gonorynchiforms. If, as is suggested below, most fossil Chanidae are fishes with the gestalt of *Chanos,* but of unknown relationships, *Halecopsis* and *Neohalecopsis* are better included here than in a family Halecopsidae which is both uncharacterized and of unknown relationships.

Paleocene

Chanos torosus Danil'chenko 1968, Upper Paleocene, Turkmenia, USSR, compared by Danil'chenko with the Recent *C. chanos,* and with the Oligocene and Eocene species of the genus listed above.

Late Cretaceous

Prochanos rectifrons Bassani 1879, Hvar, Yugoslavia, *Parachanos* sp. d'Erasmo 1952, Komen, Yugoslavia. These fishes are from beds that may range from Cenomanian to Senonian in age (Radovčič 1975).

Early Cretaceous

Chanos leopoldi Costa 1860, ?Aptian, Pietraroia, Benevento, Italy. This species was last revised by d'Erasmo (1915), who regarded *Pro-chanos rectifrons* (the type and only species of that genus) as a possible synonym.

Tharrhias araripis Jordan and Branner 1908, *Tharrhias rochae* Jordan and Branner 1908, *Dastilbe elongatus* Silva Santos 1947. These three species are from the ?Aptian, Santana Formation, Serra do Araripe, Ceara, Brazil (Silva Santo and Valença 1968); *Dastilbe elongatus* also occurs in beds of ?Aptian age at Maranhão, Brazil (Silva Santos 1947).

Dastilbe crandalli Jordan 1910, ?Aptian, Riacho Doce, Alagoas, Brazil, *Dastilbe moraesi* Silva Santos 1955, ?Aptian, Areado Formation, Presidente Olegario, Minas Gerais, Brazil. *Parachanos aethiopicus* (Weiler 1922; see Arambourg and Schneegans 1936; Taverne 1974b, 1981), Neocomian or Aptian, Cocobeach Series, Gabon and Equatorial Guinea. Type and only species of *Parachanos*. According to Taverne (1981:974), *Parachanos* is a synonym of *Dastilbe*, and the species of *Dastilbe* and *Parachanos aethiopicus* may be regarded as populations of a single species, *Dastilbe crandalli*.

Aethalionopsis robustus (Traquair 1911), Neocomian, Bernissart, Belgium (Taverne 1981).

Chanopsis lombardi Casier 1961, Neocomian, Couches de la Loia, Zaïre. Type and only species of *Chanopsis*, known by a single imperfect head.

Otoliths

In addition to the above species, known by more or less complete skeletons, one species is based on otoliths: *Chanos compressus* Stinton 1977, Lower Eocene (Cuisian), Wittering Formation, Hampshire, England.

Discussion

On the face of it, the preceding list is an impressive fossil record, that might be regarded as virtually continuous from the earliest Cretaceous (the only notable gap is Oligocene to Recent). But most of the listed fossil taxa are little more than names: they refer to fossils that have the gestalt of *Chanos*, but few or no anatomical details are known. The only fossil species in which any sort of comprehensive anatomical information is available are those in which acid-prepared specimens exist, the Eocene *Chanoides macropoma* and the Lower Cretaceous *Tharrhias araripis*. The first of these turns out to be an otophysan, whereas the second is the fish that I (Patterson 1975:167) compared with *Chanos* (Fig. 2), writing "The skulls of these two fishes are almost identical, and features such as a suprapreopercular, four infraorbitals, highly specialized jaws and complete absence of teeth leave no doubt that they are closely related." However, as Fig. 2 shows, the caudal skeleton of *Tharrhias* is far more generalized than that of *Chanos*, or any other Recent gonorynchiform or otophysan. It was this dissimilarity that prompted Taverne (1974b, 1974c, 1981) to doubt the relationship between gonorynchiforms and ostariophysans. The caudal skeleton is known in four other Lower Cretaceous "chanids," *Parachanos aethiopicus* (Taverne 1974b), *Dastilbe crandalli* and *D. elongatus* (Silva Santos 1947: Figs. 5, 6), and *Aethalionopsis robustus* (Taverne 1981), and is virtually identical to that of *Tharrhias*. (*Aethalionopsis* has three epurals, not two, and two or three fringing fulcra.)

As for the "unique specializations of the skull" shared by *Tharrhias* and *Chanos*, Fink and Fink's reorganization and characterization of gonorynchiforms (Fig. 1) provides a new and sounder basis for an assessment. Among the ostariophysan characters that are accessible in fossils (nos. 1, 3, 5, 6, 7, 9a–c in Fig. 1), *Tharrhias araripis* shows all except contact between the first neural arch and exoccipital (no. 7, cf. Fig. 3), and fusion between the first pre-ural and first ural centra and the first uroneural (no. 9c, cf. Fig. 2). However, contact between the first neural arch and exoccipital could have been developed in cartilage, since there is no periosteal cover on the opposing margins of the exoccipital and neural arch above the foramen magnum.

Among the gonorynchiform characters accessible in fossils (nos. 10, 11, 13, 14, 15, 16, 17a,b,d,e in Fig. 1), *Tharrhias araripis* exhibits no. 10, though the pterosphenoids are by no means so small as in *Chanos*, nos. 11, 13, 14, 15, 17a, 17b, and 17e (two epurals, Fig. 2). *Tharrhias araripis* lacks characters 16 (form of joint between first neural arch and exoccipitals, Fig. 3) and 17d (there are two postcleithra, but

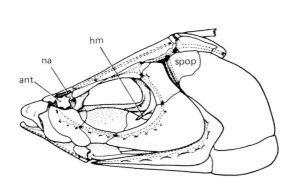

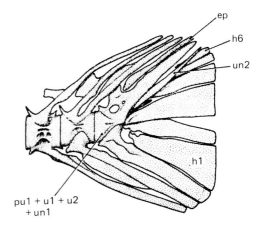

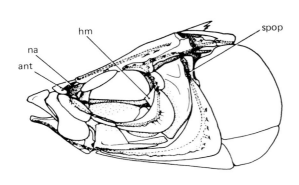

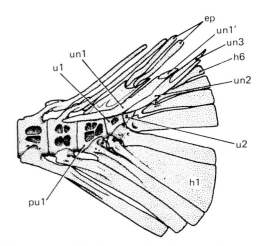

Fig. 2. Above, skull and caudal skeleton of *Chanos chanos,* Recent; below, skull and caudal skeleton of *Tharrhias araripis,* Lower Cretaceous, Santana Formation, Brazil, from Patterson (1975:Figs. 6, 7). ant, Antorbital; ep, epural; h, hypural; hm, hyomandibular; na, nasal; pu, pre-ural centrum; spop, suprapreopercular; u, ural centrum; un, uroneural.

none in Recent gonorynchiforms). *Tharrhias* has none of the characters linking *Gonorynchus* and the African freshwater gonorynchiforms (nos. 18–22 in Fig. 1), and none of the otophysan characters identified by Fink and Fink (1981).

On this evidence, four possibilities are worth summarizing:

1. *Tharrhias* is a member of the Chanidae. This view demands that characters absent in *Tharrhias* but present in *Chanos* and other Recent gonorynchiforms (occipital joint, absence of postcleithra, caudal skeleton fusions) were developed independently in the latter two groups. It also demands charac-

ters uniquely shared by *Tharrhias* and *Chanos.* Since I know no such characters, I reject this hypothesis.

2. *Tharrhias,* and probably other Lower Cretaceous "chanids" (*Dastilbe, Parachanos, Aethalionopsis*), are stem-group gonorynchiforms, unassignable to a Recent subgroup. On this view, Recent gonorynchiforms are united by two characters that *Tharrhias* lacks (no postcleithra, occipital joint), and perhaps by others inaccessible in fossils (nos. 12, 17c,f in Fig. 1). I know of nothing incongruent with this hypothesis.

3. *Tharrhias* and the other Lower Cretaceous "chanids" are stem-group ostariophysans,

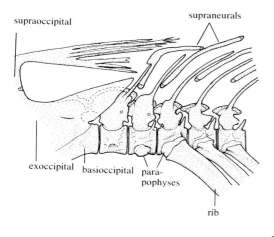

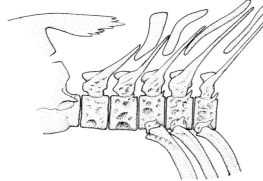

Fig. 3. Occiput and anterior vertebrae in left lateral view. Above, *Chanos chanos*, Recent, after Fink and Fink (1981:Fig. 6); below, *Tharrhias araripis*, Lower Cretaceous, Santana Formation, Brazil, based on BM(NH) specimens. In *Chanos*, cartilage is indicated by heavy stipple, and the broken lines beneath the projecting flange of the exoccipital mark boundaries of bone and cartilage.

unassignable to Gonorynchiformes or Otophysi. On this view, the characters uniquely shared by *Tharrhias* and gonorynchiforms (especially the form of the parietals, suspensorium, and premaxilla, and lack of pharyngeal teeth) were independently acquired, but the caudal skeleton fusions and occipital joint (nos. 7, 9c in Fig. 1) characterizing Recent gonorynchiforms and ostariophysans are synapomorphous. Since similar caudal skeleton fusions are found in clupeoids and other groups of teleosts, and since the form of the occipital joint in *Tharrhias* is not precisely known (cartilage is not preserved), I regard this hypothesis as unparsimonious.

4. *Tharrhias* and other Lower Cretaceous "chanids" are stem-group euteleostean or clupeocephalan (sensu Patterson and Rosen 1977) teleosts, unassignable to subgroup. This demands that the 13 ostariophysan and gonorynchiform characters of *Tharrhias* were independently acquired, and may also be rejected as unparsimonious.

I conclude that *Tharrhias,* and probably *Dastilbe, Parachanos,* and *Aethalionopsis* as well, are best treated as stem-group gonorynchiforms (hypothesis 2, above). As such, they are among the earliest reliably identified euteleosteans. But *Chanos* is not a living fossil, since no fossil taxa have yet been shown to belong in the Chanidae by means of synapomorphies. These conclusions, negative as they are, are another instance of a story that is fairly familiar by now. Every fossil belongs to a Recent taxon of some rank, and our success in interpreting a fossil depends on determining what that taxon is by identifying synapomorphies in the fossil. Ten years ago, it seemed to me that the Lower Cretaceous fossils just discussed were genuine members of the Chanidae. But I was guided largely by gestalt, and by characters whose distribution I had not analyzed thoroughly. The new conclusions reached here were made possible by Fink and Fink's cladistic analysis of Recent gonorynchiforms and ostariophysans, an illustration of the fact that advances in our understanding of fossils generally follow advances in our understanding of the relationships of Recent species. Until Recent groups have been characterized, we cannot assign fossils to them (Patterson 1977:621, 632). Perhaps the most useful general conclusion I can reiterate is that overall similarity is no necessary guide to relationship, particularly when comparing fossil with Recent taxa.

Literature

Arambourg, C., Schneegans, D. 1936. Poissons fossiles du bassin sedimentaire du Gabon. Annls Paléont. 24:139–160.

Bassani, F. 1888. Ricerce sui pesci fossili di Chiavon. Atti Accad. Sci. fis. mat. Napoli (2) 3, 6:1–104.

Blot, J. 1980. La faune ichthyologique des gisements du Monte Bolca (Province de Vérone, Italie). Bull. Mus. Natn. Hist. Nat. Paris (4) 2C:339–386.

Casier, E. 1946. La faune ichthyologique de l'Yprésien de la Belgique. Mém. Mus. R. Hist. Nat. Belg. 104:1–267.

Casier, E. 1961. Matériaux pour la faune ichthyologique Éocretacique du Congo. Annls Mus. R. Afr. Cent. 8° Sci. Géol. 39:xii+1–96.

Casier, E. 1966. Faune ichthyologique du London clay. London: British Museum (Natural History).

Chang, M-M., Chou, C-C. 1976. Discovery of *Plesiolycoptera* in Songhuajiang-Liaoning Basin and origin of Osteoglossomorpha (in Chinese). Vertebr. Palasiat. 14:146–153.

Chang, M-M., Chou, C-C. 1977. On late Mesozoic fossil fishes from Zhejiang Province, China (in Chinese, English summary). Mem. Inst. Vertebr. Palaeont. Palaeoanthrop. Peking 12:1–59.

Danil'chenko, P. G. 1968. Ryby verkhnego paleotsena Turkmenii (Upper Paleocene fishes of Turkmenia), pp. 113–156. In: Obruchev, D. V. (ed.), Ocherki po filogenii i sistematike iskopaemykh ryb i bezcheliustnykh. Moscow: Nauka Press.

d'Erasmo, G. 1915. La faune e l'éta dei calcari a ittioliti di Pietraroia (Prov. di Benevento). Palaeontogr. Ital. 21:1–53.

Fink, S. V., Fink, W. L. 1981. Interrelationships of the ostariophysan fishes (Teleostei). Zool. J. Linn. Soc. Lond. 72:297–353.

Forey, P. L. 1973. A revision of the elopiform fishes, fossil and Recent. Bull. Brit. Mus. Nat. Hist. Geol. Suppl. 10:1–222.

Fowler, H. W. 1974. A catalog of world fishes (XX). Quart. J. Taiwan Mus. 27:479–610.

Gosline, W. A. 1960. Contributions toward a classification of modern isospondylous fishes. Bull. Brit. Mus. Nat. Hist. Zool. 6:325–365.

Gosline, W. A. 1980. The evolution of some structural systems with reference to the interrelationships of modern lower teleostean fish groups. Jap. J. Ichthy. 27:1–28.

Grande, L. 1980. Paleontology of the Green River Formation, with a review of the fish fauna. Bull. Geol. Surv. Wyoming 63:xvii+1–333.

Greenwood, P. H. 1970. On the genus *Lycoptera* and its relationship with the family Hiodontidae (Pisces, Osteoglossomorpha). Bull. Br. Mus. Nat. Hist. Zool. 19:257–285.

Greenwood, P. H. 1973. Interrelationships of osteoglossomorphs, pp. 307–332. In: Greenwood, P. H., Miles, R. S., Patterson, C. (eds.), Interrelationships of fishes. London: Academic.

Greenwood, P. H. 1977. Notes on the anatomy and classification of elopomorph fishes. Bull. Br. Mus. Nat. Hist. Zool. 32:65–102.

Greenwood, P. H., Rosen, D. E., Weitzman, S. H., Myers, G. S. 1966. Phyletic studies of teleostean fishes, with a provisional classification of living forms. Bull. Amer. Mus. Nat. Hist. 131:339–456.

Lenglet, G. 1974. Contribution à l'étude ostéologique des Kneriidae. Annls Soc. R. Zool. Belg. 104:51–103.

Patterson, C. 1975. The distribution of Mesozoic freshwater fishes. Mem. Mus. Nat. Hist. Nat. Paris A88:156–174.

Patterson, C. 1977. The contribution of paleontology to teleostean phylogeny, pp. 579–643. In: Hecht, M. K., Goody, P. C., Hecht, B. M. (eds.), Major patterns in vertebrate evolution. New York: Plenum.

Patterson, C. In press. *Chanoides*, a marine Eocene otophysan fish (Teleostei, Ostariophysi). J. Vert. Paleont.

Patterson, C., Rosen, D. E. 1977. Review of ichthyodectiform and other Mesozoic teleost fishes and the theory and practice of classifying fossils. Bull. Amer. Mus. Nat. Hist. 158:81–172.

Radovčič, J. 1975. Some new Upper Cretaceous teleosts from Yugoslavia with special reference to localities, geology and palaeoenvironment. Palaeont. Jugosl. 17:1–55.

Roberts, T. R. 1973. Interrelationships of ostariophysans, pp. 373–395. In: Greenwood, P. H., Miles, R. S., Patterson, C. (eds.), Interrelationships of fishes. London: Academic.

Rosen, D. E., Greenwood, P. H. 1970. Origin of the Weberian apparatus and the relationships of the ostariophysan and gonorynchiform fishes. Amer. Mus. Novit. 2428:1–25.

Schuster, W. H. 1960. Synopsis of biological data on milkfish *Chanos chanos* (Forskål), 1775. Fish. Biol. Synopses FAO 4.

da Silva Santos, R. 1947. Uma redescrição de *Dastilbe elongatus*, com algumas considerações sobre o genero *Dastilbe*. Notas Prelim. Estud. Div. Geol. Min. Bras. 42:1–7.

da Silva Santos, R. 1955. Descrição dos peixes fósseis. Bolm. Div. Geol. Miner. Bras. 155:17–27.

da Silva Santos, R., Valença, J. G. 1968. A Formação Santana e sua paleoictiofauna. Anais Acad. Bras. Cienc. 40:339–360.

Stinton, F. C. 1977. Fish otoliths from the English Eocene. Part 2. Palaeontogr. Soc. (Monogr.) 1977:57–126.

Taverne, L. 1974a. L'ostéologie d'*Elops* Linné, C., 1766 (Pisces Elopiformes) et son intérêt phylogénétique. Mém. Acad. R. Belg. Cl. Sci. 8° (2) 41, 2:1–96.

Taverne, L. 1974b. *Parachanos* Arambourg et Schneegans (Pisces Gonorhynchiformes) du Crétacé inférieur du Gabon et de Guinée Equatoriale et l'origine des téléostéens Ostariophysi. Rev. Zool. Afr. 88:683–688.

Taverne, L. 1974c. Sur l'origine des téléostéens Gonorhynchiformes. Bull. Soc. Belge Géol. 83:55–60.

Taverne, L. 1979. Ostéologie, phylogénèse et systématique des téléostéens fossiles et actuels du super-ordre des Ostéoglossomorphes. Troisième partie. Mém. Acad. R. Belg. Cl. Sci. 8° (2) 43, 3:1–168.

Taverne, L. 1981. Ostéologie et position systématique d'*Aethalionopsis robustus* (Pisces, Teleostei) de Crétacé inférieur de Bernissart (Belgique) et considérations sur les affinités des Gonorhynchiformes. Bull. Acad. Roy. Belg. Cl. Sci. (5) 67:958–982.

Taverne, L. 1982. Sur *Pattersonella formosa* (Traquair, R. H. 1911) et *Nybelinoides brevis* (Traquair, R. H. 1911), teleostéens salmoniformes argentinoïdes du Wealidien inférieur de Bernissart, Belgique, précédemment attribués au genre *Leptolepis* (Agassiz, L. 1832). Bull. Inst. Roy. Sci. Nat. Belg., Sci. Terre 54, 3:1–27.

Weiler, W. 1928. Beiträge zur Kenntnis der tertiären Fische des Mainzer Beckens II. 3. Teil. Die Fische des Septarientones. Abh. Hess. Geol. Landesanst. 8, 3:1–64.

Woodward, A. S. 1901. Catalogue of the fossil fishes in the British Museum (Natural History). Part 4. London: British Museum (Natural History).

12
Denticeps clupeoides Clausen 1959: The Static Clupeomorph

P. Humphry Greenwood

Department of Zoology, British Museum (Natural History), London SW7 5BD, England

The lineage Denticipitoidei is represented by two monotypic taxa, namely the extant *Denticeps clupeoides* and the Neogene (probably Miocene) fossil *Palaeodenticeps tanganikae* Greenwood (1960). Its known time span is thus about 22 million years, and it is confined to Africa.

The subject organism, *Denticeps clupeoides,* is classified as follows: class: Osteichthyes; subclass: Actinopterygii; infraclass: Actinopteri; subcohort: Clupeomorpha; order: Clupeiformes; suborder: Denticipitoidei. *Denticeps clupeoides* is readily distinguished from all other clupeomorph species (except *Palaeodenticeps*) by the presence of small toothlike bodies (odontodes) on the exposed surfaces of most skull roofing bones (Fig. 1). Odontodes also occur on the opercular and infraorbital bones, the lateral aspects of the maxilla, premaxilla, and dentary; on at least one (but usually two) branchiostegal rays, and on the posttemporal and extrascapular bones. The spacing of these odontodes varies on the different bones; on some it is close enough to impart a furry appearance to the bone. A few odontodes occur on some of the anterior body scales near the lateral line (which, unlike that in other clupeomorphs, is present along the entire length of the flanks).

Denticeps, like *Palaeodenticeps,* also differs from other members of the Clupeomorpha in various anatomical details, particularly in the caudal fin skeleton and in the absence of epi-neural intermuscular bones. *Denticeps* and the clupeoids differ in certain features of their osteology and soft anatomy which, since they cannot be checked in the fossil denticipitid, will not be detailed here (but see Greenwood 1960, 1968a, 1968b).

In several of these features *Denticeps* and, where the characters can be checked, *Palaeodenticeps* appear to be more primitive than members of their sister-group, the Clupeoidei (Greenwood 1968a, 1968b).

The fossil denticipitid material is exceptionally well preserved (Greenwood 1960) and thus allows one to make detailed comparisons of *Palaeodenticeps* and *Denticeps*. In all major osteological features, and in the distribution of odontodes (Fig. 2), the fossil taxon is identical with its living counterpart (see Greenwood 1960 and, especially, 1968a:258–260). Indeed, some of the features first thought to be unique to the fossil taxon (and thus diagnostic of *Palaeodenticeps*) are now known to be present in *Denticeps clupeoides* as well. The relationships of the two taxa, it would seem, could be better expressed if they were treated as species of a single genus; however, until a formal synonymy has been published, two genera will be recognized.

Osteologically, *Denticeps* differs from *Palaeodenticeps* only in having more vertebrae (40 compared to 31–32), more scales in the lateral line series (37–40 compared to 32–33), more

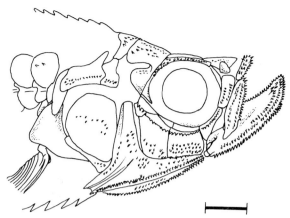

Fig. 1. Denticeps clupeoides; skeleton of head and pectoral girdle to show distribution of odontodes. Modified after Clausen (1959). Scale = 1 mm.

pleural ribs (12 compared to 10 pairs), and in having a slightly longer posterior projection of the preoperculum. In gross morphology the two species are very similar (Figs. 1 and 2), although the dorsal fin is situated a little further posteriorly in *Denticeps*.

The interrelationships of the family Denticipitidae, the sole representative of the suborder Denticipitoidei, are discussed at length by Greenwood (1968a). In brief, the lineage is considered to be the sister-group of the Clupeoidei (i.e., all other clupeomorph fishes). In many features the Denticipitoidei are plesiomorphic relative to the Clupeoidei, but their overall *primitive* status is somewhat obscured by various apomorph features unique to the lineage.

Ecologically, *Denticeps clupeoides* is confined to streams of low ionic content, has a preference for those parts of a stream where the current is strongest, and is a shoaling species (Clausen 1959). It rarely exceeds an adult length of 6 cm.

Virtually no information is available on the feeding habits of the species. Since the gill rakers are few and relatively widely spaced, it seems unlikely that *D. clupeoides* is a microphagous filter-feeder. The few gut analyses available indicate that pupal Diptera, ostracods, plant debris, and sand grains are ingested, suggesting bottom-feeding habits.

From the information available on the biology of *D. clupeoides* it is difficult to reach any conclusions on its ecological status as a generalist or a specialist. Certainly, as compared with many of the clupeoids, the feeding apparatus and habits of *Denticeps* appear to be general-

Fig. 2. Palaeodenticeps tanganikae, head, pectoral girdle and anterior part of vertebral column. From Greenwood (1960), Bull. Brit. Mus. (Nat. Hist.) Geol. 5. Scale in mm.

ized, but its restriction to fresh water would seem to be a specialized condition, although one shared with a few clupeoid families and subfamilies.

There is some suggestion that *Palaeodenticeps tanganikae* may have existed in a lacustrine habitat, and that this lake had a volcanic origin (Greenwood 1960). However, we must take care in drawing conclusions from that information. The crater-lake habitat may represent only the place and conditions where fossilization took place; the species could have lived elsewhere, for example, in streams associated with the lake.

The present distribution of *Denticeps clupeoides* is relatively restricted, confined to streams draining into four small rivers in southwest Nigeria and to certain streams in Dahomey, near the border with Nigeria (Clausen 1959; Greenwood 1965). The small adult size of the species may, however, have allowed it to be overlooked in other areas.

Palaeodenticeps has an even more restricted distribution, being known only from a single locality at Mahenge in the Singida district of Tanzania (Greenwood 1960). The great distance (approximately 3500 km) separating this locality from the area in which *Denticeps* now occurs is noteworthy, and suggests a formerly more widespread distribution of the lineage.

The temporal range of *P. tanganikae* is indeterminable since only one record of the taxon (dated as Miocene or possibly Oligocene) is available.

The small amount of study material for both *Denticeps clupeoides* and *Palaeodenticeps tanganikae* precludes any accurate assessment of intraspecific variation in their morphometric and meristic characters. What few data we have seem to indicate a low level of variability.

As compared with their sister-group (in other words, all other herring-like fishes), the Denticipitoidei have a restricted distribution, few if any ecological specializations, and a low level of intragroup taxonomic diversity. The same conclusions can be drawn if the family Denticipitidae is compared with any family among the Clupeoidei, a possible exception being provided by

the clupeid subfamily Congothrissinae (Poll 1964, 1974).

Like the denticipitids, the congothrissines are an African freshwater taxon of restricted distribution (one region of the Zaire River), and are represented by a single extant taxon of small adult size. *Congothrissa gossei*, like *Denticeps clupeoides*, does not appear to show much intraspecific variability, but again, not much study material is available for review. This comparison is also complicated by the uncertain taxonomic position of the Congothrissinae. It seems most probable that the congothrissines are, in fact, merely a tribe of the taxonomically and ecologically diverse subfamily Pellonulinae, a member of the large and diverse clupeoid family Clupeidae (Whitehead 1973).

Literature

Clausen, H. S. 1959. Denticipitidae, a new family of primitive isospondylous teleosts from west African fresh-water. Vidensk. Medd. Dansk. Naturhist. Foren. 121:141–151.

Greenwood, P. H. 1960. Fossil denticipitid fishes from east Africa. Bull. Brit. Mus. (Nat. Hist.) Geol. 5:1–11.

Greenwood, P. H. 1965. The status of *Acanthothrissa* Gras, 1961 (Pisces, Clupeidae). Ann. Mag. Nat. Hist. (13) 7(1964):337–338.

Greenwood, P. H. 1968a. The osteology and relationships of the Denticipitidae, a family of clupeomorph fishes. Bull. Brit. Mus. (Nat. Hist.) Zool. 16:213–273.

Greenwood, P. H. 1968b. Notes on the visceral anatomy of *Denticeps clupeoides* Clausen, 1959, a west African clupeomorph fish. Rev. Zool. Bot. Afr. 77:1–10.

Poll, M. 1964. Une famille dulcicole nouvelle de poissons africains: les Congothrissidae. Mém. Acad. Roy. Sci. d'Outre-Mer, Bruxelles NS 15, 2:1–40.

Poll, M. 1974. Synopsis et distribution géographique des Clupeidae d'eau douce africains, descriptions de trois especes nouvelles. Bull. Acad. Roy. Belg. Cl. Sci. 60 (5):141–161.

Whitehead, P. J. P. 1973. The clupeoid fishes of the Guianas. Bull. Brit. Mus. (Nat. Hist.) Zool. Supp. 5:1–227.

13
Polypterus and *Erpetoichthys:* Anachronistic Osteichthyans

P. Humphry Greenwood

Department of Zoology, British Museum (Natural History), London SW7 5BD, England

The Polypteridae, sole representative family of the Cladistia, cannot be considered as living fossils on the basis of their inadequate paleontological record. Rather, their claim to that status lies in their having retained a great number of the primitive features lost by other extant members of the Actinopterygii (Rosen et al. 1981). *Polypterus* and *Erpetoichthys* are classified as follows: class: Osteichthyes; subclass: Actinopterygii; infraclass: Cladistia; order: Polypteriformes; family: Polypteridae.

Typical views on the archaic nature of the Cladistia are those of Goodrich (1928:91) ". . . the Polypterini are the survivors of this large and varied group [the Palaeozoic Palaeoniscoidei] hitherto supposed to be extinct," and those of Gardiner (1967:189) "*Polypterus* is merely a much modified chondrostean survivor for which, unfortunately, the connecting links are as yet missing."

Some of the primitive features retained by polypterids are: thick and rhombic ganoid scales articulating by means of a "peg-and-socket" joint, functional spiracles (Magid 1966), a heterocercal tail (externally diphycercal), a quadratojugal, large gular plates, a maxilla firmly articulated with the preoperculum, and an undivided basibranchial copula. (For anatomical details see Daget 1950; Gardiner 1967, 1973; Schaeffer 1973; and further references therein).

The polypterids do, of course, have a number of derived features (some uniquely so), such as for example the form and organization of the separate dorsal finlets, the anatomy of the pectoral fins and, at least when adult, the maxilla fused with part of the infraorbital bone series. The vertebrae have complete and fully ossified centra with a neural canal and ossified but autogenous supraneural and hemal processes, and there is no spiracular canal in the neurocranium (Daget 1950; Schaeffer 1973; Gardiner 1973).

The fossil record for the Polypteridae (Middle Cretaceous to Pleistocene; see Greenwood 1974), like that of the living taxa, is confined to Africa. It is a record remarkably poor in the number of localities and specimens known, and in the variety of skeletal material recovered. The latter is represented mainly by isolated scales, together with a few vertebrae, skull roofing bones, and fragments of jaw bones. Entire fishes are unknown, so we have no idea of the gross morphology of any fossil polypterid, nor do we have any detailed information about their cranial anatomy.

Polypterid interrelationships have long been a matter of debate. For example, Huxley (1861) placed the group in his Crossopterygii (together with the Rhipidistia and Actinistia), whereas previously Müller (1844) had included the polypterids in the actinopterygian order Holostei (as distinct from the Chondrostei of that subclass). Stensiö (1921, 1932) was unable to accept either solution and considered the polypterids to represent a distinct assemblage, intermediate between the Actinopterygii and the Crossopterygii.

Many later workers, unlike Müller, tended to associate polypterids with the chondrostean division of the Actinopterygii (see summary in Daget 1950; also Gardiner 1967). Others, however, still favored Stensiö's approach, and there has been recent support for a possible crossopterygian relationship (Nelson 1970).

When treated as actinopterygian fishes with chondrostean affinities, the polypterids generally were allied with the most primitive group of all actinopterygians, the Palaeonisciformes (Devonian to Triassic; see Gardiner 1967, 1973). But, in any scheme of relationships based on derived characters uniquely shared by the polypterids and Palaeonisciformes, such a relationship cannot be confirmed because the two groups have no known synapomorphic features restricted in that way.

The basic actinopterygian status of the polypterids and their consequent broad relationships have been confirmed, however, by Rosen et al.

(1981) through an extensive reanalysis of osteichthyan apomorph features. Rosen et al. treat the polypterids as a sister-group (the Cladistia) of two other actinopterygian groups combined (i.e., the Chondrostei + Neopterygii); no closer relationship can be determined at present.

On that basis, the Polypteridae may be considered living primitive representatives of a group (Actinopterygii), the majority of whose members (Chondrostei and Neopterygii) have lost or modified the primitive features still retained by the polypterids; in that sense they are "living fossils."

Nine species and seven subspecies of living *Polypterus* are recognized (Poll 1942; Daget 1950); the sister-genus, *Erpetoichthys* (see Swinney and Heppel 1982, regarding nomenclature) is represented by a single species, *E. calabaricus* (Fig. 1).

There is little difference in the gross morphology or, it seems, the anatomy and ecology of any *Polypterus* species. The various taxa are separated principally on the basis of meristic

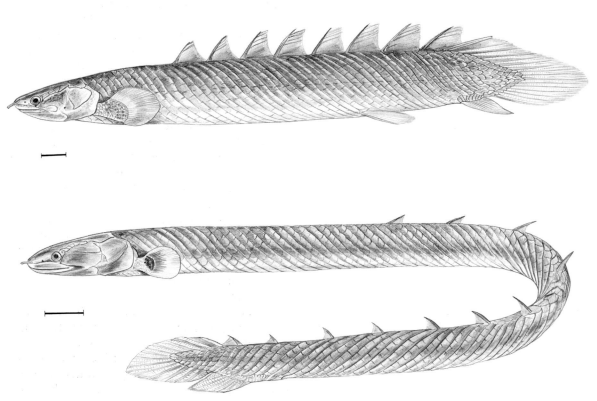

Fig. 1. *Polypterus senegalus* (upper figure) and *Erpetoichthys calabaricus*. Drawn by Gordon Howes. Scale = 1 cm.

and some slight morphometric differences (scale numbers, number of dorsal finlets, and body proportions) as well as on differences in coloration (Daget 1950).

Erpetoichthys calabaricus differs from all *Polypterus* species in its elongate and anguilliform body, absence of pelvic fins, and in some anatomical features as well (see Daget 1950, who suggests that in certain anatomical characters *Erpetoichthys* is pedomorphic, retaining in the adult conditions found in larval *Polypterus*).

Seeing that the fossil record for polypterids is so poor, it is not surprising that no fossil species have been described; scales from a Quaternary deposit were, however, assigned to an extinct subspecies, *Polypterus bichir ornatus* (Arambourg 1947). Most fossil material is referred to the genus *Polypterus* simply because the size of the specimens is commensurate with comparable elements in *Polypterus* rather than in the smaller *Erpetoichthys*. Some of the Eocene scales, however, are larger than those from any extant *Polypterus* species, and are thought to be from fishes approximately 180 cm long. (The maximum length attained by any living species is 80–90 cm.)

All living and apparently all fossil polypterids are freshwater fishes. Surprisingly little ecological information is available for either *Polypterus* or *Erpetoichthys*. In general, *Polypterus* species are confined to the relatively shallow marginal areas of lakes and rivers (but may occur in fast-flowing streams and at some distance offshore in lakes). They are nocturnal, and could be classified as generalized predators feeding on fishes, amphibians, and larger aquatic invertebrates (especially Crustacea and insects). *Erpetoichthys* is apparently more restricted to reedy habitats, and may prey more heavily on invertebrates than do the larger *Polypterus* species.

Because all polypterids use the asymmetrically bilobed and highly vascular swim bladder as a lung, they are able to inhabit relatively deoxygenated swamp areas, regions frequented by very few higher actinopterygian species (usually those with suprabranchial accessory breathing organs). Larval polypterids have a pair of well-developed true external gills (i.e., a pair of gill-like organs originating outside the branchial cavity); for anatomical details see Daget 1950.

From the rather inadequate evidence available, the polypterids would seem to be ecological generalists in both their trophic and habitudinal requirements. Because several primitive Actinopteri (the sister-group of the Cladistia) have a respiratory swim bladder, the ability of polypterids to occupy areas low in dissolved oxygen should probably not be considered an ecological specialization *per se*.

Polypterids are confined to tropical Africa (Fig. 2), their distribution within the continent being associated either with the drainage basins of river systems emptying into the Atlantic, or with the Nile. Polypterids do not occur in any river flowing into the Indian Ocean. *Erpetoichthys* has a restricted distribution along a narrow strip of the west African coastal region.

The majority of fossil remains have been found within the same area as that now occupied by the family (Fig. 2), although some are from localized regions where polypterids are absent nowadays (e.g., Lakes Victoria and Edward). Only one locality, in Tunisia, lies well beyond the present general area of distribution.

Temporally, the rather scanty fossil record extends from the Mid-Cretaceous to the Late Pleistocene (Greenwood 1974). Thus it is short (approximately 100 million years) in comparison with the record for the polypterid's chondrostean sister-group, the Chondrostei (Devonian to Recent, approximately 345 million years), or even that of the palaeonisciform part of the Chondrostei (Mid-Devonian to Jurassic, approximately 200 million years).

Because no restricted sister-group relationship can be determined for the Polypteridae, it is difficult to make detailed and, therefore, meaningful comparisons between the Cladistia and their nearest relatives. In effect, those relatives are all other actinopterygian fishes, living and extinct.

If one treats the polypterids merely as an actinopterygian family endemic to Africa, its present areal distribution is larger than that of some endemic families, similar to others, and smaller than most. As compared with the range of most nonendemic actinopterygian families in Africa, that of the Polypteridae is more restricted because most of these families extend further into subtropical regions and have a broader longitudinal range as well.

Likewise, one cannot really compare levels

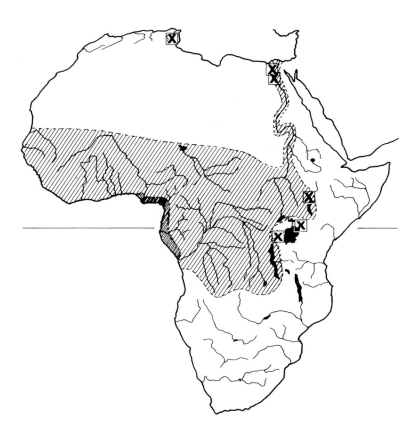

Fig. 2. Distribution map of the Polypteridae, recent and fossil. Spaced hatching shows distribution of *Polypterus* species; the densely hatched area shows that of *Erpetoichthys calabaricus* (which overlaps certain *Polypterus* species). Fossil localities indicated by ⊠.

of intraspecific variation among the polypterids with those of all other Actinopterygii. As a general statement, one could perhaps say that phenotypic variation in polypterids is no greater than in most actinopterygian species; the relatively high number of *Polypterus* subspecies is probably as much a reflection of an individual taxonomist's evaluation of a particular situation as it is a reflection of intraspecific variation. In terms of taxonomic diversity at the intrafamilial level, the Polypteridae would not rate high among the actinopterygians; but there are several families among both advanced and primitive actinopterygian groups with an equally low, or even lower level of generic and specific diversity.

Literature

Arambourg, C. 1947. Contribution à l'étude géologique et paleontologique du bassin du lac Rodolphe et de la basse vallée de l'Omo. Miss. Sci. Omo, 1932–1933. 1(3):469–489. Paris.

Daget, J. 1950. Revision des affinitiés phylogénétiques des polypteridés. Mém. Inst. Fr. Afr. Noire 11:1–178.

Gardiner, B. G. 1967. Further notes on palaeoniscoid fishes, with a classification of the Chondrostei. Bull. Brit. Mus. (Nat. Hist.) Geol. 14:143–206.

Gardiner, B. G. 1973. Interrelationships of teleostomes, pp. 105–135. In: Greenwood, P. H., Miles, R. S., Patterson, C. (eds.), Interrelationships of fishes. London: Academic.

Goodrich, E. S. 1928. *Polypterus* a palaeoniscid? Palaeobiologica 1:87–92.

Greenwood, P. H. 1974. Review of Cenozoic freshwater fish faunas in Africa. Ann. Geol. Surv. Egypt. 4:211–231.

Huxley, T. H. 1861. Preliminary essay upon the systematic arrangement of the fishes of the Devonian epoch. Mem. Geol. Surv., U.K. 1861:1–46.

Magid, A. M. A. 1966. Breathing and function of the spiracles in *Polypterus senegalus*. Anim. Behav. 14:530–533.

Müller, J. 1844. Über den Bau und die Grenzen der Ganoiden und über das natürliche System der Fische. Abh. Deut. Akad. Wiss. Berlin 32:117–216.

Nelson, G. 1970. Subcephalic muscles and intracranial joints of sarcopterygian and other fishes. Copeia 1970:468–71.

Poll, M. 1942. Contribution à l'étude systématique des Polypteridae (Pisc.). Rev. Zool. Bot. Afr. 35:141–179, 269–317.

Rosen, D. E., Forey, P. L., Gardiner, B. G., Patterson, C. 1981. Lungfishes, tetrapods, paleontology and plesiomorphy. Bull. Amer. Mus. Nat. Hist. 167:159–276.

Schaeffer, B. 1973. Interrelationships of chondrosteans, pp. 207–226. In: Greenwood, P. H., Miles, R. S., Patterson, C. (eds.), Interrelationships of fishes. London: Academic.

Stensiö, E. A. 1921. Triassic fishes from Spitzbergen. 1:1–307. Vienna.

Stensiö, E. A. 1932. Triassic fishes from east Greenland collected by the Danish Expeditions in 1929–1931. Medd. Grønland 83(3):1–345.

Swinney, G. N., Heppel, D. 1982. *Erpetoichthys* or *Calamoichthys:* the correct name for the African reed-fish. J. Nat. Hist. 16:95–100.

14
Sturgeons as Living Fossils

Brian G. Gardiner

Biology Department, Queen Elizabeth College, London W8 7AH, England

The sturgeons belong to the family Acipenseridae and to the order Acipenseriformes. The Polyodontidae or paddle fishes are also included in this order.

Order Acipenseriformes Berg 1940

The acipenseriformes are a very distinctive assemblage of fishes. The order is characterized by a palatoquadrate that meets in the midline anteriorly and never articulates with the neurocranium, a maxilla that is firmly connected to the palatoquadrate anteriorly, and a symplectic that unites the latter with the hyomandibula and at the same time articulates with the lower jaw. The parasphenoid reaches back above the carotid arteries and there are no aortic or parabasal canals. The hyomandibula has uniquely expanded, blacklike ends and is devoid of an opercular process. The hyomandibular nerve passes round the hyomandibula instead of through it as in most bony fishes but, in adult *Acipenser guldenstadti* (Holmgren and Stensiö 1936) and *A. fulvescens* (Jollie 1980) it passes through a cartilagenous extension of the bone. There is a pair of cranio-spinal processes and several sclerotomes are incorporated with the occipital region. The post-temporal is expanded with internal processes that extend onto the neurocranial surface. The supraorbital sensory canal passes between the two narial openings and posteriorly anastomoses with the infraorbital canal in the dermopterotic.

The dermal skeleton is devoid of ganoine in the living members, but small pustules of enamel occur on some of the head bones of *Chondrosteus*. The jaws are either toothless or with minute teeth and there is a transverse row of teeth on the first hypobranchial.

The scaling is unique and for the most part reduced to small isolated denticles (as in chondrichthyans) with up to five rows of much larger scales in *Acipenser* and *Scaphirhynchus*. There is no premaxilla and the lower jaw is reduced to a dentary and prearticular.

Phylogenetic Relationships

The acipenseriformes is the sister-group of all extant Actinopteri (Rosen et al. 1981), uniquely sharing with them a segmented basibranchial, a fossa bridgei into which the spiracular canal opens, and a lateral cranial canal (Gardiner 1973).

Family Acipenseridae Bonaparte 1831

The family is characterized by the shape of the palatoquadrate in which the ectopterygoid (palatine?) contacts the middle of the maxilla, and by a quadratojugal that braces the hind end of the palatoquadrate against the postero-dorsal portion of the maxilla. The mouth is small, protractile, and bears a transverse row of barbels anteriorly. Branchiostegal rays are said to be absent, but Jollie (1980) records two in *Acipenser;* the gill rakers are few and a stout spine forms the leading edge of the pectoral fin.

The body has five series of bony scutes along its length; the lateral line passes through the dorsolateral row. Elsewhere the body is naked apart from the skin overlying the postcleithrum and branchiostegals, which is studded with scales (Jollie 1980).

Acipenser Linnaeus 1758

There are 19 species of the genus *Acipenser* (Günther 1870) distributed in northern seas and rivers. The majority of the species occur in Europe and Asia (ten), while four are found in North America, two each in China and Japan, and one is confined to the northern Pacific. Although most live in both fresh and salt water, at least two (*A. fulvescens* and *A. ruthenus*) are usually taken in fresh water. The type species is *Acipenser sturio* L. (Fig. 1).

Huso Brandt 1833

This genus contains two species that are found in the Adriatic, Black, Caspian, and Okhotok Seas and the basin of the Amur River.

Scaphirhynchus Heckel 1835

Like *Acipenser,* this genus contains both North American (two species said to be confined to the Mississippi River Basin and Mexico) and southern Russian representatives (three species, *kaufmanni, rossikowi,* and *fedtschenkoi* in the Aral Sea and the Syr-Daru and Amu Daryu Rivers.

The Russian representatives are often regarded as belonging to a separate genus *Kessleria* Bogdanov 1882 (*Pseudoscaphirhynchus* Nikolski 1900), but since the differences are confined to such trivial characters as the texture of the barbels, the length of the caudal peduncle and its scaling, the shape of the gill rakers, and the size of the air bladder, generic separation seems hardly justified. All five species are shovel nosed, lack open spiracles, and have fan-shaped gill rakers.

Geological Range

Acipenser: Upper Cretaceous–Recent

The earliest undoubted remains of *Acipenser* (*A. albertensis* Lambe) occur in Upper Cretaceous (Campanian) beds (Belly River Series; Edmonton Beds; Oldman Formation) of the Red Deer River, Alberta (Gardiner 1966) and include scutes, pectoral fin spines, and a cleithrum. Scales and spines have also been recorded from the Upper Cretaceous (Lower Ravenscrag) of Eastend, Saskatchewan and from the Upper Cretaceous (Maestrichtian) of Montana (Hell Creek Formation) and Wyoming (Lance Formation). The American form is referred to as *A. eruciferous* Cope.

Throughout the Tertiary, *Acipenser* is recorded from the Eocene of England (*A. toliapicus* Agassiz) and France (*A. lemoinei* [Priem]); the Oligocene of western Siberia, the Isle of Wight, and the Paris Basin (*A. parisiensis* Priem); the Miocene of Virginia (*A. ornatus* Leidy) and of Germany (*A. molassicus* Probst, *A. tuberculosus* Probst); and the Pliocene of Si-

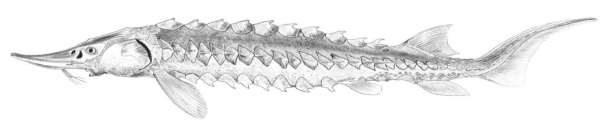

Fig. 1. Acipenser sturio L. BMNH 1860.14.4. Standard length 290 mm.

beria, Rumania, the United States, France, and England.

Huso: Lower Pliocene–Recent

Family Polyodontidae Bonaparte 1838

This family is characterized by jaws and palatoquadrate symphysis with minute teeth, rostrum very elongate, tactile and with two barbels; eyes small and above the anterior end of the upper jaw; a single branchiostegal ray. Minute, isolated scales occur over at least the scapular arch and extend forward the entire length of the isthmus. The preopercular canal joins the infraorbital below the spiracle. The post-temporal and first epibranchial are both enormously expanded and blade like. The extrascapular and part of the temporal series are reduced to tube bones (viz. lamellar portion absent). The cleithrum is separated from supracleithrum above and clavicle below, joined to them by ligaments. The quadratojugal, ectopterygoid, interclavicle, and postcleithrum are all absent.

The family contains two monotypic freshwater genera.

Polyodon Schneider 1801

P. folium Schneider occurs in the rivers and lakes of eastern North America associated with the Mississippi and its tributaries. The rostrum is leaflike and flexible and the gill rakers are fine and very numerous.

Psephurus Günther 1873

P. gladius (Martens 1861) is only found in the Yangtze River, China. This species has a protractile mouth, a swordlike rostrum, and a few large caudal fulcra.

Geological Range

Neither genus is recorded in the fossil record.

Fossil Relatives

Paleopsephurus MacAlpin 1947

This genus is represented by a single species from the Upper Cretaceous (Hell Creek Beds) of Montana and has generally been regarded as a paddle fish (MacAlpin 1947; Lehman 1966; Gardiner 1967; Jollie 1980). However, although it clearly is a member of the Acipenseriformes, sharing with them a blade-shaped hyomandibula, a maxilla firmly fused to the palatoquadrate arch which is joined anteriorly and separate from the overlying neurocranium, an elongate parasphenoid and a post-temporal with an internal process and toothless jaws, it does not possess any derived characters with the Polyodontidae. Instead it shares with *Chondrosteus* and *Acipenser* a similarly shaped palatoquadrate in which the hooklike ectopterygoid sutures with the maxilla and the quadratojugal runs from the dorso-posterior corner of the maxilla to the back end of the palatoquadrate (see MacAlpin 1947:Fig. 1).

It shares with *Acipenser* ventral caudal fulcra, a parasphenoid with notches for the carotids, a pectoral girdle of very similar make-up with identically shaped supracleithra, and a skull roof with a medial nuchal. Thus, *Paleopsephurus* is considered here as the sister-group of the Acipenseridae.

Chondrosteus Egerton 1858

This is a well-known genus occurring in the Lower Lias of Lyme Regis and Barrow-on-Soar, Leicestershire (*C. acipenseroides* Egerton) and in the Upper Lias of Holzmaden, Germany (*C. hindenbergi* Pompeckj). Also from Lyme Regis, a *Chondrosteus* much larger than normal has been referred to as *C. pachyurus* Egerton 1858.

Chondrosteus shares with the Acipenseriformes a large blade-shaped hyomandibula; a palatoquadrate that is joined anteriorly and separated from the neurocranium and has the maxilla attached to it in front. This genus exhibits a large symplectic, an expanded post-temporal, and a pair of cranio-spinal processes.

Chondrosteus shares with the Acipenseridae the hooklike ectopterygoid sutured to the middle of the maxilla, a quadratojugal that braces the maxilla with the palatoquadrate, and the absence of teeth from the jaws. The distribution of these features suggests that *Chondrosteus* is the sister-group of the Acipenseridae plus *Paleopsephurus*.

Gyrosteus Morris 1854

Gyrosteus is confined to the Upper Lias of Whitby, Yorkshire. It is a large fish in which the hyomandibula, maxilla, supracleithrum, and cleithrum are all very similar in shape to those of *Chondrosteus*. The jaws are devoid of teeth and as in *Chondrosteus* there is no trace of body scaling. On this evidence it is possible to regard *Gyrosteus* as the sister-group of *Chondrosteus*.

Crassopholis Cope 1883

C. magnicaudata Cope occurs in Eocene, Green River Shales of Wyoming. When first described it was regarded by Cope (1883, 1885) as a polyodontid, a view with which Woodward (1895), MacAlpin (1947), Lehman (1966), Gardiner (1967), and Grande (1980) have concurred. The features it shares with the Polyodontidae include the much expanded post-temporal and supracleithrum; the enlarged, single, branchiostegal ray related to the interhyal; the maxilla, which is attached to the palatoquadrate both anteriorly and posteriorly; the expanded, bladelike first epibranchial, and the many small stellate bones making up the rostrum or paddle (Grande 1980). These, plus the similar pattern of the skull-roofing bones, strongly indicate that *Crassopholis* is the sister-group of the Polyodontidae as Cope (1883, 1885) originally suggested.

Peipiasteus Liu and Zhoi 1964

Peipiasteus comes from the Upper Jurassic of Tsien-shan-tze-kou, Nanling, Peipiao-Liaoning Province, China and when first described (Liu and Zhou 1964) was regarded as a sturgeon belonging to the new family Peipiasteidae. *Peipiasteus* does not show features of either the Acipenseridae or the Polyodontidae, apart from

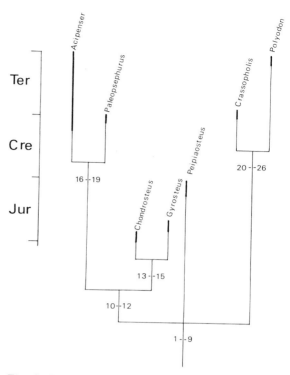

Fig. 2. Tree of Acipenseriformes. 1. Blade-shaped hyomandibula. 2. Palatoquadrate meets in midline anteriorly; separate from the neurocranium. 3. Symplectic. 4. Parasphenoid without parabasal canal. 5. Cranio-spinal process. 6. Supraorbital sensory canal passes between narial openings. 7. Body scaling reduced to isolated denticles. 8. Lower jaw consisting of dentary and prearticular. 9. Ganoine missing from fin rays. 10. Ectopterygoid hook shaped, contacts middle of maxilla. 11. Quadratojugal braces back of maxilla and palatoquadrate. 12. Teeth lost from jaws. 13. Characteristically shaped maxilla. 14. Similar cleithrum and supracleithrum. 15. Body scaling lost. 16. Ventral caudal fulcra. 17. Parasphenoid notched for carotids. 18. Skull roof with medial nuchal. 19. Supracleithrum stout and triangular. 20. Preopercular canal joins infraorbital below spiracle. 21. First epibranchial enormously expanded, bladelike. 22. Post-temporal enormously expanded. 23. Cleithrum separated from supracleithrum and clavicle. 24. Many, small, stellate bones making up the paddle. 25. Enlarged, single branchiostegal ray, related to ceratohyal. 26. Maxilla attached to palatoquadrate anteriorly and posteriorly.

the presence of small, isolated denticles lying over the scapular region. These denticles are similar in shape and size to those in *Polyodon* and in the absence of other characters the most one can say is that *Peipiasteus* forms an unresolved trichotomy with the Acipenseridae and Polyodontidae.

Protoscaphirhynchus Wilimovsky 1956

This genus is found at the same locality as *Paleopsephurus,* that is the Upper Cretaceous, Hell Creek Beds, near Fort Peck, Montana. Wilimovsky (1956) considered it an undoubted member of the Acipenseridae, but the genus is completely armored with ganoine-covered scales and with small post-temporals, frontals, and parietals quite unlike any living or fossil acipenseriform. On this evidence *Protoscaphirhynchus* is certainly not a member of the Acipenseriformes.

Pholidurus Woodward 1889

Known only by a few scales and a partial tail, this genus comes from the Chalk (Senonian) of Gravesend, Kent. It was said by Woodward (1889) to resemble *Psephurus,* but both the fin rays and scales have a superficial covering of ganoine. There is no reason to consider this genus to be a member of the Acipenseriformes.

Conclusions

The information in this entry is summarized in the tree (Fig. 2) on which the time sequence for the origin of the various synapomorphies is indicated.

Literature

Cope, E. D. 1883. A new chondrostean from the Eocene. Amer. Nat. 17:1152–1153.

Cope, E. D. 1885. On two new forms of polyodont and gonorhynchid fishes from the Eocene of the Rocky Mountains. Mem. Nat. Acad. Sci. 3:161–165.

Gardiner, B. G. 1966. A catalogue of Canadian fossil fishes. Roy. Ont. Mus. U. Toronto 68:1–154.

Gardiner, B. G. 1967. Further notes on palaeoniscoid fishes with a classification of the Chondrostei. Bull. Brit. Mus. (Nat. Hist.) Geol. 14(5):143–206.

Gardiner, B. G. 1973. Interrelationships of teleostomes, pp. 105–135. In: Greenwood, P. H., Miles, R. S., Patterson, C. (eds.), Interrelationships of fishes. London: Academic.

Grande, L. 1980. Paleontology of the Green River Formation, with a review of the fish fauna. Bull. Geol. Surv. Wyo. 63:1–333.

Günther, A. 1870. Catalogue of the fishes in the British Museum, Vol. 8. London: Taylor and Francis.

Holmgren, N., Stensiö, E. 1936. Kranium und Visceralskelett der Akranier, Cyclostomen und Fische, pp. 233–500. In: Bolk, L. (ed.), Handuch der Vergleichenden Anatomie, Vol. 4. Berlin: Urban and Schwarzenberg.

Jollie, M. 1980. Development of head and pectoral girdle skeleton and scales in *Acipenser.* Copeia 1980(2):226–249.

Lehman, J. P. 1966. Actinopterygii, pp. 1–242. In: Piveteau, J. (ed.), Traité de paléontologie. 4, 3. Paris: Masson et Cie.

Liu, H. T., Zhou, J. J. 1964. A new sturgeon from the Upper Jurassic of Liaoning, North China. Vertebrata palasiatica 9(3):237–248.

MacAlpin, A. 1947. *Paleopsephurus wilsoni,* a new polyodontid fish from the Upper Cretaceous of Montana, with a discussion of allied fish, living and fossil. Con. Mus. of Paleont. U. of Michigan 6:167–234, 6 pls.

Rosen, D. E., Forey, P. L., Gardiner, B. G., Patterson, C. 1981. Lungfishes, tetrapods, paleontology and plesiomorphy. Bull. Amer. Mus. Nat. Hist. 167(4):159–276.

Wilimovsky, M. J. 1956. *Protoscaphirhynchus squamosus,* a new sturgeon from the Upper Cretaceous of Montana. J. Paleont. 30(5):1205–1208.

Woodward, A. S. 1889. On the palaeontology of sturgeons. Proc. Geol. Assoc. 11:24–44.

Woodward, A. S. 1895. Catalogue of the fossil fishes in the British Museum (Natural History), Vol. 3. London: Brit. Mus. (Nat. Hist.).

15
The Neopterygian *Amia* as a Living Fossil

Hans-Peter Schultze and E. O. Wiley

Museum of Natural History and Department of Systematics and Ecology, University of Kansas, Lawrence, KS 66045

Introduction

The Recent bowfin, *Amia calva,* is a primitive neopterygian fish distinguished by a very long dorsal fin, a posteriorly rounded, hemicercal caudal fin with epaxial fin-rays, "amiid" scales, uniquely ossified centra that are diplospondylous in the caudal region, a double articulation (quadrate and symplectic) with the lower jaw, loss of suborbitals, and other characters. Of the cited characters, the hemicercal caudal fin is primitive for neopterygians; the double articulation of the lower jaw is a derived feature for all Halecomorphi; diplospondylous vertebrae are derived within the Amiidae; and expaxial caudal fin-rays, the loss of suborbitals, and the loss of the endoskeletal basipterygoid process are acquired in parallel by amiids and teleosts. Many additional characters unite amiids with teleosts and certain other neopterygians in the Halecostomi (see Chapter 16, Fig. 2). One obvious feature, a very long dorsal fin, has been picked here to distinguish *Amia* and its closest relative *Kindleia* from all other amiids (Fig. 1).

Today, *Amia calva* is restricted to the eastern and mideastern part of the United States and Canada. They inhabit fresh waters from the Great Lakes (except Lake Superior) south to the Gulf of Mexico (distribution map in Boreske 1974 and Wilson 1982). They prefer sluggish waters rich in vegetation. Pleistocene occur-

rences of the species are known within limits of extant occurrence, while the genus *Amia* had a wide distribution over the whole Northern Hemisphere during the Tertiary. Freshwater environment is generally accepted for all fossil *Amia* and its closest relative *Kindleia.*

Relationships of *Amia* to Other Halecomorphs

Amia calva is the only extant representative of the subdivision Halecomorphi, the sister-group of the teleosts within the Halecostomi (see Chapter 16, Fig. 2). Patterson (1973) discussed the relationships of *Amia* within the families of Halecomorphi (Parasemionotidae, Caturidae, and Amiidae). They are united by a unique double jaw articulation involving the quadrate, symplectic, and articular. The Parasemionotidae are considered the basal grade group, having only this character shared with the other two families. Genera of the grade Caturidae share with Amiidae the derived character of incorporating the dermosphenotic into the skull roof. Other characters, such as the development of ossified centra, membranous bone binding the symplectic to the preopercular, and possible loss of the opisthotic may be parallel developments acquired independently within the Caturidae and Amiidae. The eight haleco-

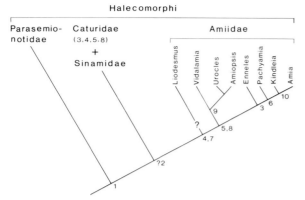

Fig. 1. Relationship of the Halecomorphi after Patterson (1973); 1–8, the halecomorph features after Patterson (1973:262): 1. symplectic articulating with lower jaw; 2. dermosphenotic bound to sphenotic by an anteroventral flange; 3. no pterotic ossification; 4. uppermost branchiostegal enlarged and truncated proximally; 5. solid, perichordally ossified centra, diplospondylous in caudal region; 6. each hypural except the first fused with an ural centrum; 7. only the first one to three ural neural arches ossified; 8. each hypural carrying a single fin-ray; 9. differentiation of diplospondyly into normal and alternating (neural arch on one and hemal arch on the next half-centrum) diplospondyly (Wenz 1977); 10. elongated dorsal fin.

morph features cited by Patterson (1973:262) are present in *Amia* and *Kindleia* only, they are acquired within the Amiidae, and only one or two features relate the other two families to the Amiidae (Fig. 1.)

Patterson (1973) includes two subfamilies, Amiinae and Sinamiinae, within the Amiidae. The two genera of the Sinamiinae, *Sinamia* and *Ikechaoamia,* are characterized by rhombic, ganoin-covered scales and by characters that are occasionally found as variations in *Amia calva* (cf. unpaired parietal, several supraorbitals and supratemporals). But these characters are also found in some genera of Caturidae where they do not vary. Except for two features (dermosphenotic included into skull roof, and each hypural carrying a single fin-ray), all features characteristic for the Amiinae are not known from sinamiines. Except for each hypural carrying a single fin-ray, the features uniting the Amiidae do not occur within the sinamiines. Sinamiines lack diplospondylous

vertebrae in the caudal region and more than only the first one or two neural arches are ossified, the uppermost branchiostegals are not enlarged and truncated proximally (Zhang and Zhang 1980). Thus *Sinamia* and *Ikechaoamia* belong between the Parasemionotidae and Amiidae, a place presently occupied by the grade Caturidae (Brough 1939), and might be considered as a family of their own, the Sinamiidae (Berg 1940; Liu et al. 1963; Su 1973; Zhang and Zhang 1980). Thus, we will restrict the Amiidae to the genera *Liodesmus* (Late Jurassic), *Urocles* (Late Jurassic–?Early Cretaceous), *Vidalamia* (Late Jurassic), *Amiopsis* (?Late Jurassic–Early Cretaceous), *Enneles* (Early Cretaceous), *Pachyamia* (early Late Cretaceous), *Kindleia* (Late Cretaceous–Early Tertiary), and *Amia* (Late Cretaceous–Recent). We consider contrary to Chalifa and Tchernov (1982) that *Pachyamia* is related to *Kindleia* and *Amia* because of character state 6 (hypurals except the first fused with ural centra). *Pachyamia* shows some peculiarities (no supramaxillary but posteriorly divided maxillary, and fused 4 + 5 infraorbital) not known in any other amiid. *Amia* has the longest ''lifespan'' of all Amiidae, about 70 million years.

Some doubts exist about the validity of the genus *Kindleia*. While Wilson (1982) considers *Kindleia* a genus separate from *Amia* with the species *K. fragosa, K. kehreri, K. valenciennesi, K. munieri,* and *K. russeli, Kindleia* is included within *Amia* by Boreske (1974), Estes and Berberian (1969), Janot (1967), and others. Gaudant (1980) proposes *Kindleia* as a subgenus of *Amia*. *Protamia* and *Pappichthys* are only different in size from and thus are synonyms of *Amia*. *Hypamia* is a nomen dubium, and the species *H. elegans* is based on undiagnostic material. *Stylomyleodon* and *Paramiatus* are synonyms of *Kindleia*. *Pseudamia* might be another synonym of *Kindleia* (for more information on the synonymies see Boreske 1974). We are left with the question: Are there two or only one amiid genus in the Late Cretaceous and the Early Tertiary? Wilson (1982) considers the following differences of *Kindleia* from *Amia* as derived: styliform teeth on coronoids, dermopalatines, and vomers; short parietals; short frontals; relatively short, deep body with less separation between the skull and the origin of the dorsal fin. Styliform teeth are a derived character uniting the species of *Kindleia*. The separation between skull and origin of the dorsal fin is significantly less in *Kindleia*; nevertheless, the closeness of the dorsal fin to the caudal fin seems to be a distinct character separating *Kindleia* (about 1% of total length) from *Amia* (about 5%–7% of total length). On the other side, "short" parietals and "short" frontals are very relative characters that overlap with *Amia* species (Boreske 1974:Table 7). A number of vertebrae below 75 (about 12 fewer trunk centra and 8 fewer monospondylous caudal centra), deep orbital notch in the frontals, short vomers, small supramaxilla, narrow maxilla and mandible, and short truncated gular plate also distinguish *Kindleia* from *Amia*, but are primitive features after Wilson (1982). It is not clear if the low number of vertebrae is really primitive; the closest related form, *Pachyamia*, has a number of vertebrae comparable to *Kindleia*; nevertheless, *Urocles* species cited by Wilson (1982) for a low number show wide variation (58–81 vertebrae), thus this character state is uncertain (Gaudant 1980). All these features together justify the separation of the genus *Kindleia* from *Amia*, even though it may sometimes be difficult to determine incomplete fossil specimens as belonging to one or the other genus. These features clearly separate the species *K. fragosa*, *K. kehreri*, *K. ignota* (= *K. munieri*; see Gaudant 1981a), and *K. russeli* from all *Amia* species. The position of *K. valenciennesi* is uncertain. There is complete contradiction in the data on *K. valenciennesi* given by Boreske (1974:46: 68 centra, close approximation of dorsal and caudal fins, fourth infraorbital larger than fifth indicating similarity to *K. kehreri*) and by Gaudant (1978:689–690: 85 vertebrae, of those 30 trunk and 27 monospondylous caudal centra, elongated parietals, and the morphology of the cleithrum indicating similarity to *A. uintaensis*).

Species of both genera are found in the same deposits (for *A. uintaensis* and *K. fragosa* see Grande 1980; for *A. robusta* and *K. russeli* see Janot 1967). They coexisted in the same environment (sympatric?) with adaptation to different diet (*Amia* piscivorous, *Kindleia* mollusc eater; Boreske 1974).

Stratigraphic Occurrence of *Amia* and *Kindleia*

Pleistocene

Amia calva: Chicago, Illinois, Vero Beach, Florida; Itchtucknee River deposits, Columbia Co., Florida (Boreske 1974).

Pliocene

Amia cf. *calva:* Lower Valentine Formation, Nebraska; Ogallala Formation, Kansas (Estes and Tihen 1964).

Miocene

Amia cf. *scutata:* Pawnee Creek Formation, Colorado (Boreske 1974).
Amia sp.: Turtle Butte Formation, South Dakota (Skinner et al. 1968)

Oligocene

Amia scutata: Cypress Hills Formation, Saskatchewan; Chadron Formation, South Dakota; Lower Brule Formation, South Dakota and Nebraska; Florissant Formation, Colorado (Boreske 1974).
Amia uintaensis: Cypress Hill Formation, Saskatchewan (Boreske 1974).
Amia robusta (= *anglica*): Isle of Wight, England (Newton 1899; Janot 1967).
Amia macrocephala: near Bilina, CSSR (Obrhelová 1979).
Amia oligocenica: Sieblos, Germany (Winkler 1880).
Amia longistriata: Sieblos, Germany (Winkler 1880).
Amia sp.: Passamari and Grant Horse Formations, Montana (Boreske 1974); Late Stampanian, Aix-en-Provence, France (Gaudant 1981b); Melanienton, Early Oligocene, Borken, Germany (Weiler 1961).

Eocene

Amia:
Amia uintaensis: Golden Valley Formation, North Dakota; Wind River Formation, Willwood, Wasatch, Green River, Bridger,

Washakie Formations, Wyoming; Uinta Formation, Utah (Boreske 1974).

Amia cf. *A. uintaensis:* Eureka Sound Formation, Ellesmere Island, Northwest Territories, Canada (Estes and Hutchison 1980).

Amia hesperia: Allenby Formation, British Columbia, Canada (Wilson 1982)

Amia mongoliensis: Late Eocene Ulan Shireh beds, regions of Murun, Inner Mongolia (Hussakof 1932).

Amia robusta: Eocene, district of Zaissan, province of Semipalatinsk, West Siberia (Stoyanow 1915).

Amia barroisi: Belgium (Leriche 1902); Suffolk, England (= *A. eocena,* Leriche 1909); southeastern Kazakhstan, USSR (Khisarova 1974).

Amia lemoinei: Belgium (Leriche 1902).

Amia sp.: Clarno Formation, Oregon (Cavender 1968); Allenby Formation, British Columbia, Canada; Klondike Mountain Formation, Washington (Wilson 1982); probably all *A. hesperia* (Wilson 1982:424).

Kindleia:

Kindleia fragosa: Golden Valley Formation, North Dakota; Wind River, Willwood, Wasatch, Green River, Bridger Formations, Wyoming (Boreske 1974); Eureka Sound Formation, Ellesmere Island, Northwest Territories, Canada (Estes and Hutchison 1980).

Kindleia (Pseudamia) heintzi: Eocene, Spitsbergen (Lehman 1951).

Kindleia ignota (= *munieri*): Upper Eocene, Gypsum of Montmartre, France (Gaudant 1981a).

Kindleia kehreri: Middle Eocene, Messel and Geiseltal, Germany (Jerzmanska 1977a).

Paleocene

Amia:

Amia uintaensis: Paskapoo Formation, Alberta, Canada; Fort Union Formation, Montana and Wyoming (Boreske 1974); Tongue River Formation, Montana (Estes 1976).

Amia robusta: Late Paleocene, Cernay, France (Janot 1967).

?Amia valenciennesi: Late Paleocene, Menat (Puy-de-Dôme), France (Gaudant 1978).

Kindleia:

Kindleia fragosa: Paskapoo Formation, Alberta, Canada; Fort Union Formation, Mon-

tana and Wyoming (Boreske 1974); Tongue River Formation, Montana (Estes 1976).

Kindleia russeli: Late Paleocene, Cernay, France (Janot 1966, 1967).

Cretaceous

Maestrichtian:

Amia:

Amia cf. *uintaensis:* Hell Creek Formation, Montana; Lance Formation, Wyoming; Ojo Alamo Formation, New Mexico; Aguja Formation, Texas (Boreske 1974).

Kindleia:

Kindleia fragosa: Edmonton Formation, Alberta, Canada; Hell Creek Formation, Montana; Lance Formation, Wyoming (Boreske 1974; Breithaupt 1982).

Kindleia cf. *fragosa:* St. Mary River Formation, Alberta, Canada (Langston 1976).

Campanian:

Kindleia fragosa: Oldman Formation, Alberta, Canada; Judith River Formation, Montana; "Mesaverde" Formation, Wyoming (Boreske 1974).

The earliest records of both genera are from North America, and both genera occur first in Late Paleocene in Europe. Two routes are discussed for a North Atlantic connection between North America and Europe: the northeastern or DeGeer route over Greenland and Spitsbergen to northern Scandinavia, and the southeastern or Thulean route over Greenland and Iceland to England. The findings of terrestrial tetrapods and amiids in the Canadian Arctic (Ellesmere Island, West and Dawson 1978) is the connecting locality between North America and Greenland. There is no fossil record for the Thulean route, but because of its later subsidence this route is favored by West and Dawson (1978). The amiid remains of the Eureka Formation in Ellesmere Island and *K. (Pseudamia) heintzi* in Spitsbergen are recorded as Eocene. The occurrence of *Amia* and *Kindleia* in the Upper Paleocene of France requires an earlier migration than Eocene over one or the other North Atlantic route.

Two forms identified as west European species of *Amia* reached Kazakhstan and west Siberia during the Eocene; the Late Eocene locality of *A. mongoliensis* in Inner Mongolia is not far east of the former localities. In the Early

Tertiary, the Turgai Straits and the Obik Sea separated Europe from Asia (see McKenna 1975); therefore, Jerzmańska (1977a:42; 1977b) favored, contrary to Boreske (1974), a migration of the Asian forms from the east over the Bering Strait because she considers the marine Turgai Strait between Europe and Asia to be a major barrier for freshwater forms like *Amia*. The Bering Strait migration would be substantiated if a closer relationship between the North American and Asian forms could be demonstrated. On the other side, land bridges over the Turgai Strait are reported (Kurtén 1966); that could support Janot (1967) and Boreske (1974), who argue in favor of a close relationship to the European *A. robusta*.

While *Kindleia* disappeared with the end of the Eocene in Europe and North America, *Amia* survived in Europe into the Oligocene and in North America until today.

Lifespan of *Amia* and *Kindleia* Species

Amia:
A. calva: Pleistocene to Recent, North America.
A. cf. *calva:* Pliocene, North America.
A. cf. *scutata:* Miocene, North America.
A. scutata: Oligocene, North America.
A. macrocephala: Oligocene, Europe.
A. oligocenica: Oligocene, Europe.
A. longistriata: Oligocene, Europe.
The three species *A. macrocephala, oligocenica,* and *longistriata* have not been re-

studied since their first description; we therefore feel uncertain about their validity.
A. mongoliensis: Late Eocene, Asia.
A. hesperia: Middle Eocene, North America.
A. barroisi: Eocene, Europe and Asia.
A. lemoinei: Eocene, Europe.
A. barroisi and *lemoini* have not been included in any recent revision of *Amia* species, so their validity seems to be uncertain. After Boreske's (1974) revision of the North American species of *Amia,* it seems that too many species have been described from Europe.
?*A. valenciennesi:* Late Paleocene, Europe.
A. robusta: Late Paleocene to Oligocene, Europe and Asia.
A. uintaensis: Paleocene to Oligocene, North America.
A. cf. *uintaensis:* Late Cretaceous, North America.

Kindleia:
K. ignota (= *munieri*): Upper Eocene, Europe.
K. kehreri: Middle Eocene, Europe.
K. (Pseudamia) heintzi: Eocene, Spitsbergen.
K. russeli: Late Paleocene, Europe.
K. fragosa: Late Cretaceous to Middle Eocene, North America.

Conclusions

Amia and *Kindleia* are the most advanced, "final" members of the Amiidae, and the only ones that entered fresh water. The marine amiids had a comparable short lifespan, or at least they are recorded within a short geological period only, while the two closest related fresh-

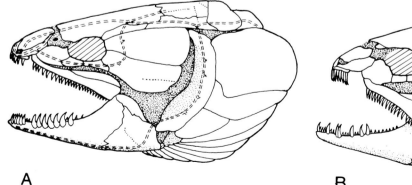
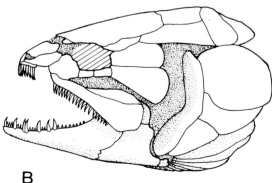

A   **B**

Fig. 2. Comparison of the head of (A) Recent *Amia calva* with (B) Eocene *Amia uintaensis* (after Boreske 1974: Fig. 15D; and Grande 1980: Fig. II20c).

water amiids had the longest lifespan within the family: *Amia* 70 million years and *Kindleia* 30–35 million years. The differences between both genera are small compared with other genera in the family, and those between the species even smaller (Fig. 2). It means that the evolutionary rate of change is very slow in both freshwater amiids; the evolutionary changes are even so minimal that Boreske (1974) used the determinations *A. scutata*, *A.* cf. *scutata*, *A.* cf. *calva*, and *A. calva* in time sequence. It means that only small (presumably gradual) changes occurred between the Oligocene (*A. scutata*) and the Pleistocene to Recent (*A. calva*). Still, three North American and questionably eight Eurasian species are recognized within the timespan Late Cretaceous to Recent. The genus *Amia* has a lifespan of 70 million years, and the species with the longest lifespan is *A. uintaensis* (30–35 million years.)

Literature

Berg, L. S. 1940. Classification of fishes both recent and fossil. Trav. Inst. Zool. Acad. Sci. USSR, 5:197–207. (In Russian.)

Boreske, J. R. 1974. A review of the North American fossil amiid fishes. Bull. Mus. Comp. Zool. Harvard U. 146:1–87.

Breithaupt, B. H. 1982. Paleontology and paleoecology of the Lance Formation (Maastrichtian), east flank of Rock Springs Uplift, Sweetwater County, Wyoming. Contr. Geol. U. Wyoming 21:123–151.

Brough, J. 1939. The Triassic fishes of Besano, Lombardy. London: British Museum (Natural History).

Cavender, T. 1968. Freshwater fish remains from the Clarno Formation Ochoco Mountains of north-central Oregon. Ore Bin 30:125–141.

Chalifa, Y., Tchernov, E. 1982. *Pachyamia latimaxillaris*, new genus and species (Actinopterygii: Amiidae), from the Cenomanian of Jerusalem. J. Vert. Paleont. 2:269–285.

Estes, R. 1976. Middle Paleocene lower vertebrates from the Tongue River Formation, Southeastern Montana. J. Paleont. 50:500–520.

Estes, R., Berberian, P. 1969. *Amia* (=*Kindleia*) *fragosa* (Jordan), a Cretaceous amiid fish, with notes on related European forms. Breviora 329:1–14.

Estes, R., Berberian, P. 1970. Paleoecology of a Late Cretaceous vertebrate community from Montana. Breviora 343:1–35.

Estes, R., Hutchinson, J. H. 1980. Eocene lower vertebrates from Ellesmere Island, Canadian Arctic Archipelago. Palaeogeogr., Palaeoclimat., Palaeoecol. 30:325–347.

Estes, R., Tihen, J. A. 1964. Lower vertebrates from the Valentine Formation of Nebraska. Amer. Midld. Nat. 72:453–472.

Gaudant, J. 1978. Observations sur quelques Amiidae (Pisces) cénozoiques d'Europe occidentale. C.R. Acad. Sci. Paris, 287, Sér. D:689–691.

Gaudant, J. 1980. Sur *Amia kehreri* Andreae (Poisson Amiidae du Lutétien de Messel, Allemagne) et sa signification paléogéographique. C.R. Acad. Sci. Paris, 290, Sér. D:1107–1110.

Gaudant, J. 1981a. Nouvelles recherches sur l'ichthyofaune des gypses et des marnes supragypseuses (Eocène supérieur) des environs de Paris. Bull. B.R.G.M. (Bureau de Recherches Géologiques et Minières), 2 Sér., Sect. IV, No. 1-1980/1981:57–75.

Gaudant, J. 1981b. Mise au point sur l'ichthyofaune oligocène des anciennes plâtrières d'Aix-en-Provence (Bouches-du-Rhone). C.R. Acad. Sci. Paris, 292, Sér. III:1109–1112.

Grande, L. 1980. Paleontology of the Green River Formation, with a review of the fish fauna. Geol. Surv. Wyoming Bull., 63.

Hussakof, L. 1932. The fossil fishes collected by the Central Asiatic expeditions. Amer. Mus. Novit. 553:1–19.

Janot, C. 1966. *Amia russeli* nov. sp., nouvel Amiidé (poisson holostéen) du Thanétien de Berru, près de Reims. C.R. Somm. Séanc. Soc. Geol. France 1966:142.

Janot, C. 1967. A propos des Amiidés actuels et fossiles, pp. 139–153. In: Coll. Internat. C.N.R.S. (Centre National de la Recherche Scientifique), no. 163, Problèmes actuels de paléontologie(évolution des vertébrés).

Jerzmańska, A. 1977a. The freshwater fishes from the Middle Eocene of Geisteltal, pp. 41–65. In: Matthes, H. W. (ed.), Eozäne Wirbeltiere des Geiseltales. Halle-Wittenberg: Tagungsber. Martin-Luther U.

Jerzmańska, A. 1977b. Süsswasserfische des älteren Tertiärs von Europa, pp. 67–78. In: Matthes, H. W. (ed.), Eozäne Wirbeltiere des Geiseltales. Halle-Wittenberg: Tagungsber. Martin-Luther U.

Khisarova, G. D. 1974. Eocene Amiidae from Southeastern Kazakhstan, pp. 16–21. In: *Materials on the history of the fauna and flora of Kazakhstan*, Vol. 6.

Kurtén, B. 1966. Holarctic land connexions in the early Tertiary. Comment. Biol. 29:1–5.

Langston, W. 1976. A Late Cretaceous vertebrate fauna from the St. Mary river Formation in Western Canada, pp. 114–133. In: Athlon. Essays on Palaeontology in honour of L. S. Russell, Royal Ontario Museum. Life Sci. Misc. Publ.

Lehman, J.-P. 1951. Un nouvel Amiidé de l'Eocène du Spitzberg, *Pseudamia heintzi*. Tromsö Mus. Arsh. 70:3–11.

Leriche, M. 1902. Les poissons paleocènes de la Belgique. Mém. Mus. Hist. Natur. Belgique 2:1–48.

Leriche, M. 1909. Note préliminaire sur des poissons nouveaux de l'Oligocène Belge. Bull. Soc. Belge Géol. 22:378–384.

Liu, T.-S., Liu, H.-T., Su, T.-T. 1963. The discovery of *Sinamia zdanskyi* from the Ordos region and its stratigraphical significance. Vertebrata Palasiatica 1963: 1–13 (In Chinese with English summary.)

McKenna, M. C. 1975. Fossil mammals and Early Eocene North Atlantic land continuity. Ann. Missouri Bot. Gard. 62:335–353.

Newton, E. T. 1899. On the remains of *Amia* from the Oligocene strata in the Isle of Wight. Geol. Soc. London Quart. J. 55:1–10.

Obrhelová, N. 1979. Süsswasser-Ichthyofauna im Tertiär der CSSR. Cas. Miner. Geol. 24:135–146.

Patterson, C. 1973. Interrelationships of holosteans, pp. 233–305. In: Greenwood, P. H., Miles R. S., Patterson, C. (eds.), Interrelationships of fishes. London: Academic.

Skinner, M., Skinner, S., Gooris, R. 1968. Cenozoic rocks and faunas of Turtle Butte, south-central South Dakota. Bull. Amer. Mus. Nat. Hist. 138:381–436.

Stoyanow, A. A. 1915. Sur les débris d' *Amia* des dépôts tertiaires du système de l'arête Manrak, district de Zaïssan, province de Sémipalatinsk. Izvest. geol. Kom. USSR 34:487–507. (In Russian with French summary.)

Su, T.-T. 1973. A new *Sinamia* from the Upper Jurassic of Southern Anhui. Vertebrata Palasiatica 11:149–153.

Weiler, W. 1961. Die Fischfauna des unteroligozänen Melanientons und des Rupeltons in der Hessischen Senke. Notizbl. Hess. Landesamt. Bodenforsch 89:44–65.

Wenz, S. 1977. Le squelette axial et l'endosquelette caudal d' *Enneles audax*, poisson Amiidé du Crétacé de Ceara (Brésil). Bull. Mus. Nat. Hist. Nat. 3. Sér., Sci. de la Terre 67:341–348.

West, R. M., Dawson, M. R. 1978. Vertebrate paleontology and the Cenozoic history of the North Atlantic region. Polarforschung 48:103–119.

Wilson, M. V. H. 1982. A new species of the fish *Amia* from the Middle Eocene of British Columbia. Palaeontology 25:413–424.

Winkler, T. C. 1880. Mémoire sur les poissons fossiles des lignites de Sieblos. Arch. Mus. Teyler 5:85–108.

Zhang, M., Zhang, H. 1980. Discovery of *Ikechaoamia* from South China. Vertebrata Palasiatica 18:89–93. (In Chinese with English summary.)

16
Family Lepisosteida (Gars) as Living Fossils

E. O. Wiley and Hans-Peter Schultze

Museum of Natural History and Department of Systematics and Ecology, University of Kansas, Lawrence, KS 66045

Introduction

Gars (Fig. 1). are primitive neopterygian fishes with elongate snout, plicidentine teeth, opisthocoelous vertebrae, heavy dermal bone retaining ganoid ornamentation, ganoid scales, a semiheterocercal tail, and fulcral scales on the median fins. The elongate snout is an ontogenetic product of ethmoid elongation and is correlated with many of the synapomorphies characterizing the family (Wiley 1976). Plicidentine teeth and opistocoelous vertebrae are also synapomorphies (Patterson 1973). The remaining characters are retained plesiomorphies.

Living gars (seven species) are restricted to the Western Hemisphere from Costa Rica to eastern North America and including Cuba and the Isle de Pines. They inhabit rivers, larger streams, and lakes as well as estuarine and coastal marine waters (Suttkus 1963). Gars are lurking predators. Some reach impressive size (3 m) but none are dangerous to humans. Fossil gars (about seven recognizable species) are wider ranging with records from western North America (Cretaceous to Pleistocene), Europe (Cretaceous to Oligocene), Africa (Cretaceous), and India (Cretaceous).

Phylogenetic Affinities

Among Recent fishes, gars have been placed with *Amia* in the Holostei (cf. Nelson 1969; Jes-sen 1972) or as the Recent sister-group of *Amia* and teleosts (the Halecostomi, Patterson 1973; Wiley 1976). Among fossil and extant groups, gars have been allied with semionotids (Westoll 1944; Rayner 1948) or considered a group apart (Patterson 1973). Following Hennig's (1966) principles, Patterson's (1973) conclusions are best supported—gars are a group apart, *Amia* is more closely related to teleosts, and semionotids are a paraphyletic assemblage more derived than gars (see Fig. 2). Although gars have a fossil record extending only to the early Cretaceous, their position on the phylogeny suggests a Permian origin of the clade since *Acentrophorus,* a neopterygian more derived than gars, is Permian. However, no earlier gar relatives have been identified.

Systematics

Gars are placed in the Neopterygii as a division, Ginglymodi, coordinate with their recent sister-group, Halecostomi (Patterson 1973; Wiley 1976; Patterson and Rosen 1977). A single family, Lepisosteidae, is recognized (Suttkus 1963; Wiley 1976). The number of genera is controversial. Wiley (1976) recognized two genera, *Lepisosteus* and *Atractosteus,* whose Recent species correspond to the limits of the subgenera recognized by Suttkus (1963), and Wiley (1976) sets the minimum age of each genus as Late Cretaceous. This division

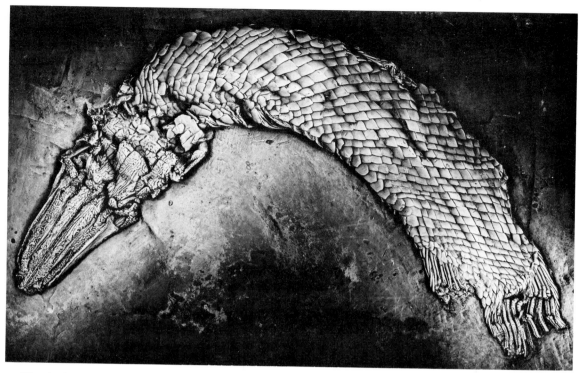

Fig. 1. Atractosteus strausi, an Eocene gar from the "oil shales" of Messel, Germany (Wiley 1976).

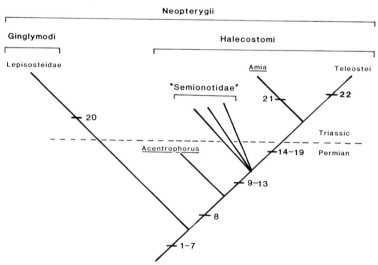

Fig. 2. A phylogeny of neopterygian fishes. Synapomorphies are: 1. basipterygoid process entirely composed of the parasphenoid; 2. a postnarial commissure present between the supra- and infraorbital canals; 3. clavicle lost; 4. uncinate processes on first and second infrapharyngobranchials; 5. infrapharyngobranchials laterally supported; 6. differentiated dorsal gill arch muscles; 7. four basibranchial copulae; 8. interoperculum present; 9. medial neural spines in the caudal region; 10. a posterior myodome; 11. a large post-temporal fossa; 12. a supramaxilla; 13. maxilla with internal articulatory head; 14. dermal intercalar; 15. post-temporal process; 16. loss of internarial commissure; 17. an endochondral rostral; 18. a single supratemporal on each side of the midline; 19. a rectus communis muscle; 20. 27 synapomorphies listed by Wiley (1976); 21. a double lower jaw articulation involving the symplectic and quadrate; 22. a number of synapomorphies discussed by Patterson (1973) and Wiley (1976). *Acentrophorus* may have additional halecostome characters and the "Semionotidae" may share characters now listed as uniting *Amia,* Teleostei, and their fossil relatives.

161

has been followed by Fisher (1978) and Lee et al. (1980) but not by Robins et al. (1980) and Grande (1980).

Stratigraphic Occurrence

Below we list nominal species and the record of doubtful species for different geologic periods. Much gar material is readily identifiable as gar but frequently too fragmentary to be reliably assigned to genus or species. We have listed such material in an appropriate manner and annotated records based on the information we have.

Pleistocene

Lepisosteus sp. indet.: A fair number of records from the United States are cited in Wiley (1976:42).

Lepisosteus platostomus: Schultz (1965), Smith (1964), Kansas.

Lepisosteus osseus: Hay (1923), North Carolina. Schultz (1965), Neff (1975), Kansas.

Lepisosteus platyrhincus: Uyeno and Miller (1963), Florida.

Atractosteus spatula: Hay (1919), Florida. Hay (1926), Uyeno and Miller (1963), Swift (1968), Texas.

Pliocene

Lepisosteidae, gen. and sp. indet.: Smith and Lundberg (1972), Nebraska.

Lepisosteus sp.: Eshelman (1975), Kansas.

Miocene

Lepisosteus ? platostomus: Wilson (1968), Kansas.

Lepisosteus sp.: Smith (1962), Oklahoma.

Atractosteus spatula: Smith (1962).

Lepisosteidae gen. and sp. indet.: *Pneumatosteus nahunticus* Cope (1869:242). North Carolina (specimen not examined).

Lepisosteidae gen. and sp. indet.: *Atractosteus emmonsi* Hay (1929; based on Emmons 1858. North Carolina (doubtful record, specimen not examined).

Oligocene

†*Lepisosteus fimbriatus* Wood 1846:6: Europe (see Eocene).

Lepisosteus sp. indet.: †*Lepidosteus longus* Lambe 1908:13, Saskatchewan, Canada.

Lepisosteus sp.: Weiler (1961), Europe.

Eocene

†*Lepisosteus fimbriatus* (Wood 1846:6): Many Eocene and Oligocene formations of Belgium, France, and England.

†*Lepisosteus cuneatus* (Cope 84:9?): Green River Formation, Utah.

Lepisosteus sp. indet.: †*Clastes cycliferus* Cope 1873:634, various formations in western North America

Lepisosteus sp. indet. (Bjork 1967): Slim Buttes Formation, South Dakota.

Lepisosteus sp. indet. (Estes and Hutchison 1980): Ellesmere Island, Northwest Territories, Canada.

†*Atractosteus strausi* (Kinkelin 1884:244): "Oil shales" of Messel, Germany.

†*Atractosteus simplex* (Leidy 1873:98): Various formations of western North America (see Wiley 1976).

†*Atractosteus atrox* (Leidy 1873:97): Green River and Bridger Formations, western North America.

Lepisosteidae gen. and sp. indet.: †*Naisia apicales* Münster 1846:34, Germany (specimen not examined).

Lepisosteidae gen. and sp. indet.: †*Trichiuridea sagittidens* Winkler 1876:31, Germany (specimen not examined).

Cretaceous

†*Lepisosteus indicus* Woodward 1890:23: Lameta Formation, India.

†*Lepisosteus opertus* Wiley 1976:42: Hell Creek Formation, Montana.

Lepisosteus sp. (Wiley and Stewart 1977): Niobrara Formation, Kansas.

†*Atractosteus africanus* (Arambourg and Joleaud 1943:42): Damerquo Beds, Niger.

†*Atractosteus occidentalis* (Leidy 1856:120): Numerous formations in western North America.

Atractosteus sp. indet.: †*Clastes pastulosus* Sauvage 1897:94, Portugal (specimen not examined).

Lepisosteidae gen. and sp. indet.: †*Paralepidosteus praecursor* Casier 1961:42, Africa.

Lepisosteidae gen. and sp. indet.: †*Lepisosteus knieskerni* Fowler 1911:150, New Jersey (doubtful record, specimen not located).

Classification of Gars

We provide a classification taken from Wiley (1976). The classification follows the plesion convention, the listing convention, the use of informal species group names, and the approximate stratigraphic ranges of the taxa (these conventions and others are reviewed by Wiley 1981).

Family Lepisosteidae Cuvier—Early Cretaceous to Recent

Genus *Lepisosteus* Lacépède, Late Cretaceous to Recent

plesion *L. oppertus* Wiley, Late Cretaceous

plesion *L. cuneatus* (Cope), Eocene

L. platostomus Rafinesque, ? Miocene to Recent

L. osseus species group, Late Cretaceous to Recent

L. osseus (Linnaeus), Pleistocene to Recent

plesion *L. indicus* (Woodward), Late Cretaceous

L. oculatus species group, ? Paleocene to Recent

plesion *L. fimbriatus* (Wood), ? Paleocene to Oligocene

L. oculatus Winchell, Recent

L. platyrhincus De Kay, Pleistocene to Recent

Genus *Atractosteus* Rafinesque, Late Cretaceous to Recent

plesion *A. strausi* (Kinkelin), Eocene

A. tropicus Gill, Recent

plesion *A. simplex* (Leidy), Eocene

A. spatula species group, Cretaceous to Recent

incertae sedis: †*A. africanus* (Arambourg and Joleaud),

Cretaceous:

†*A. occidentalis* (Leidy), Late Cretaceous

plesion *A. atrox* (Leidy), Eocene

A. spatula (Lacépède), Miocene to Recent

A. tristoechus (Bloch and Schneider), Recent

Discussion

Cretaceous and Eocene gars look like Recent gars. Sinch the Cretaceous, evolution within the family has been restricted to changes in snout proportions, relative sizes and numbers of some bones, reduction trends in ganoin ornamentation, and presumably changes in gill raker morphology and reductions in gill arch tooth plates (but we can't check the latter two trends on the fossils). If hardly changing in morphology and low taxonomic diversity count, gars are prime living fossils. However, no one species has an impressive longevity, at least as measured by known specimens. Several Recent species have Pleistocene records, and at least *A. spatula* is known from the Miocene. Several others may be very old as judged by relationships with fossils. For example, *Atractosteus tropicus* is a primitive member of its genus and may be Cretaceous in origin. And if *Lepisosteus osseus* is the closest relative of *L. indicus*, then it and the phylogenetically older *L. platostomus* may be ancient indeed.

Acknowledgment: We thank J. D. Stewart (University of Kansas) for helping us sort out Miocene and Pliocene records and for his help with the literature of the Late Cenozoic.

Literature

Arambourg, C., Joleaud, L. 1943. Vertébrés fossiles du bassin du Niger. Bull. Dir. des Nîmes A.O.F. 7:1–74.

Bjork, P. R. 1967. Latest Eocene vertebrates from Northwestern South Dakota. J. Paleont. 41:227–236.

Casier, E. 1961. Matériaux pour la faune ichthyologique éocrétacique du Congo. Ann. Mus. roy. Afr. Centr. 8 Vol. Sci. Geol. 39:1–96.

Cope, E. D. 1869. Second edition to the history of fishes of the Cretaceous of the United States. Proc. Amer. Philos. Soc. 9:240–244.

Cope, E. D. 1873. On the extinct *Vertebrata* of the Eocene of Wyoming, observed by the expedition of 1872, with notes on the geology. U.S. Geol. Surv. Terr. 1872: 565–649.

Cope, E. D. 1884. The Vertebrata of the Tertiary formations of the West. U.S. Geol. and Geog. Surv. Terr., III: 1–1009.

Emmons, E. 1858. Agriculture of the eastern counties, together with descriptions of the fossils of the Marl beds. Report of the North Carolina Geological Survey, Raleigh, NC.

Eshelman, R. E. 1975. Geology and paleontology of the Early Pleistocene (Late Blancan) White Rock fauna from north-central Kansas. U. Michigan Paper Paleont. No. 13:1–60.

Estes, R., Hutchison, J. H. 1980. Eocene lower vertebrates from Ellesmere Island, Canadian Artic Archipelago. Paleogeo., Paleoclim., Paleoecol. 30:325–347.

Fisher, W. (ed.). 1978. FAO species identification sheets for fishery purposes. Western Central Atlantic (Fishing Area 31). Food and Agriculture Organization of the United Nations, Rome. VI volumes, looseleaf.

Fowler, H. W. 1911. A description of the fossil fish remains of the Cretaceous, Eocene and Miocene Formations of New Jersey. Bull. Geol. Surv. N.J. 4:22–182.

Grande, L. 1980. Paleontology of the Green River Formation, with a review of the fish fauna. Geol. Surv. Wyoming Bull. 63:1–331.

Hay O. P. 1919. Descriptions of some mammalian and fish remains from Florida of probably Pleistocene age. Proc. U.S. Nat. Mus. 56:103–112.

Hay, O. P. 1923. The Pleistocene of North America and its vertebrate animals from the states east of the Mississippi River and from the Canadian Provinces east of longitude 95°. Carnegie Inst. Washington. Publ. No. 322:1–499.

Hay, O. P. 1926. A collection of Pleistocene vertebrates from Southwestern Texas. Proc. U.S. Nat. Mus. 68:1–18.

Hay, O. P. 1929. Second bibliography and catalogue of the fossil vertebrates of North America, Vol. I. Carnegie Inst. Washington.

Hennig, W. 1966. Phylogenetic systematics. Urbana: U. Illinois Press.

Jessen, H. 1972. Schultergürtel und Pectoralflosse bei Actinopterygiern. Fossils and Strata 1:1–101.

Kinkelin, G. F. 1884. Die Schleusenkammer von Frankfurt-Niederrad und ihre Fauna. Ber. Senckenberg. Naturforsch. Ges. 1884:219–257.

Lambe, L. M. 1908. The vertebrates of the Oligo-

cene of the Cypress Hills, Saskatchewan. Contrib. Canad. Paleont. 3:1–65.

Lee, D. S., C. R. Gilbert, C. H. Hocutt, R. E. Jenkins, D. E. McAllister, and J. R. Stauffer, Jr. 1980. Atlas of North American freshwater fishes. North Carolina State Mus. Nat. Hist., Raleigh, NC.

Leidy, J. 1856. Notice of remains of extinct reptiles and fishes, discovered by Dr. F. V. Hayden in the Bad Lands of the Judith River, Nebraska Territory. Amer. J. Sci. 2:118–120.

Leidy, J. 1873. Notice of remains of fishes in the Bridger Tertiary formation of Wyoming. Proc. Acad. Nat. Sci. Philadelphia 1873:97–99.

Münster, G. von, 1846. Ueber einen in den tertiären Ablagerungen der Gegend von Magdeburg gefundenen neuen Fischzahn. Beitr. Petrefactenkunde 7:34–35.

Neff, N. A. 1975. Fishes of the Kanopolis local fauna (Pleistocene) of Ellsworth County, Kansas. U. Michigan Pap. Paleont. No. 12:39–48.

Nelson, G. J. 1969. Gill arches and the phylogeny of fishes with notes on the classification of vertebrates. Bull. Amer. Mus. Nat. Hist. 141:475–552.

Patterson, C. 1973. Interrelationships of holosteans. In: Greenwood, P. H., Miles, R., Patterson, C. (eds.), Interrelationships of fishes. Zool. J. Linnean Soc. 53, supp. 1:233–305.

Patterson, C., Rosen, D. E. 1977. Review of ichthyodectiform and other Mesozoic teleost fishes and the theory and practice of classifying fossils. Bull. Amer. Mus. Nat. Hist. 158(2):81–172.

Rayner, D. H. 1948. The structure of certain Jurassic holostean fishes with special reference to their neurocrania. Phil. Trans. Roy. Soc. London 223B:287–345.

Robins, C. R., R. M. Bailey, C. E. Bond, J. R. Brooker, E. A. Lachner, R. N. Lea, and W. B. Scott. 1980. A list of common and scientific names of fishes from the United States and Canada. (4th ed.) Amer. Fish. Soc. Spec. Publ. No. 12:1–74.

Sauvage, H. E. 1897. Note sûr les lépidosteides du terrain garumnien du Portugal. Bull. Soc. Geol. France 25:92–96.

Schultz, G. E. 1965. Pleistocene vertebrates from the Butler Springs local fauna, Meade County, Kansas. Pap. Michigan Acad. Sci. 50:235–265.

Smith, C. L. 1962. Some Pliocene fishes from Kansas, Oklahoma, and Nebraska. Copeia 1962:505–520.

Smith, G. R. 1964. A late Illinoian fish fauna from southwestern Kansas and its climatic significance. Copeia 1964:278–285.

Smith, G. R., and J. G. Lundberg. 1972. The Sand Draw fish fauna. Bull. Amer. Mus. Nat. Hist. 148:40–54.

Suttkus, R. D. 1963. Order Lepisostei. Mem. Sears Found. Marine Res., 1:61–88.

Swift, C. 1968. Pleistocene freshwater fishes from Ingleside Pit, San Patricio County, Texas. Copeia 1968:62–69.

Uyeno, T., Miller, R. R. 1963. Summary of Late Cenozoic freshwater fish records of North America. Occ. Pap. Mus. Zool. U. Michigan 631:1–34.

Weiler, W. 1961. Die Fischfauna des unteroligozänen Melanientons und des Rupeltons in der Hessischen Senke. Notizbl. Hess. Landesamt, Bodenforsch.

Westoll, T. S. 1944. The Haplolepidae, a new family of Late Carboniferous bony fishes. Bull. Amer. Mus. Nat. Hist. 85:1–122.

Wiley, E. O. 1976. The phylogeny and biogeography of fossil and Recent gars (Actinopterygii:Lepisosteidae). U. of Kansas Mus. Nat. Hist. Misc. Publ. No. 64:1–111.

Wiley, E. O. 1981. Phylogenetics: the theory and practice of phylogenetic systematics. New York: Wiley.

Wiley, E. O., Stewart, J. D. 1977. A gar (*Lepisosteus* sp.) from the marine Cretaceous Niobrara Formation of Western Kansas. Copeia 1977:761–762.

Wilson, R. L. 1968. Systematics and faunal analysis of a Lower Pliocene vertebrate assemblage from Trego county, Kansas. Cont. Mus. Paleont. U. Michigan 22:75–126.

Winkler, T. C. 1876. Deuxième mémoire sur les dents des poissons fossiles du terrain bruxellian. Arch. Mus. Teyler 4:16–48.

Wood, S. V. 1846. On the discovery of an alligator and of several new Mammalia in the Hordwell Cliff, with observations upon the geological phenomena of that locality. Lond. Geol. J. 1:1–17, 117–122.

Woodward, A. S. 1890. Description of a fish skull (from the Lameta beds, Nagpur, India). Rec.-Geol. Surv. India 23:23–24.

17
The Coelacanth as a Living Fossil

Peter Forey

Department of Palaeontology, British Museum (Natural History), London SW7 5BD, England

The living coelacanth, *Latimeria chalumnae* Smith, occupies a rare position in the history of the study of vertebrates, being a member of a group recognized on the basis of fossils and only subsequently found to be extant. In this sense then, the self-contradictory term "living fossil" carries a slightly more literal meaning than most other examples in this book. Lungfishes (Dipnoi) were also known as fossils before a Recent example (*Lepidosiren*) was recognized (Fitzinger 1837), but, in this case, a mere 9 years separated the identification of fossil and Recent representatives. For coelacanths, nearly one-hundred years separates Agassiz's (1839) naming of the Permian *Coelacanthus* and the landing of the first specimen of *Latimeria* late in 1938. During that interval, a large number of fossil coelacanths had been described, with *Macropoma* (Albian–Turonian) as the youngest and presumed last member of the group. It was understandable that *Latimeria* should be tagged a living fossil.

The term living fossil is, however, somewhat perplexing. Darwin is accredited (Stanley 1979:126) with coining the term. He considered them to be "remnants of a once preponderant order" and as forms "which like fossils, connect to a certain extent orders at present widely sundered in the natural scale" (Darwin 1859:107). The latter idea is contained in Huxley's "intercalary types" (1908); that is, forms that are morphologically but not temporally in-termediate. The difference between Huxley and Darwin is that the former regarded intercalary types as exact morphological intermediates and the latter as approximate.

Three separate ideas are embodied in the term living fossil: (1) that it is a living species known to have a long fossil record; (2) that it represents a "missing link" and is "a type of animal that has survived long beyond its appropriate era" (White 1953:114); and (3) that it is a Recent member of natural group known to have been more diverse in the past. *Latimeria* fails to comply with (1) and (2) and can only comply with (3) if the group is specified.

Latimeria is unknown in the fossil record, and it cannot be regarded as a missing link, although it has been so dubbed, for the characters of *Latimeria* preclude it from being regarded as an intercalary type. *Latimeria chalumnae** is the sole living representative of the Infraclass Actinistia (coelacanths), recognized by a rostral organ, extracleithrum, and a tandem jaw articulation in which a quadrate articulation is accompanied by an articulation between the symplectic and the articular. The sister-group of the Ac-

---

*The second discovered specimen of a living coelacanth was named *Malania anjouanae* by Smith because it differed from the first in the absence of a first dorsal fin and middle lobe of the tail, scale ornament, and (supposedly) in details of jaw articulation. Subsequent examination has led to the view that *Malania* represents a badly mutilated specimen of *Latimeria chalumnae* (see White 1953).

tinistia is the infraclass Choanata (lungfishes and tetrapods) (Rosen et al. 1981). There seems little point in trying to compare *Latimeria* with choanates so, for present purposes, I will consider the group Actinistia with respect to parameter (3) mentioned above.

Coelacanths are known by approximately 70 species distributed among 28 genera. As usual the number of species recognized varies according to authority and to whether one is prepared to accept geographically and stratigraphically based species as well as those based on morphology. The figure given here is conservative. Fossil coelacanths are known from the Upper Devonian and show an irregular distribution throughout the stratigraphic column to the Upper Cretaceous. The numbers of species recognized for each of the geological periods is given as a histogram (Fig. 1). The greatest taxic diversity occurs in the Lower Triassic.

Attempts to answer questions about rates of change, be they morphological or taxonomic changes, requires an explicit hypothesis of interrelationships among coelacanths. In other words, a classification is required that can be justified independent of convention. Previous classifications such as those of Berg (1940), Ro-

mer (1945), or Vorobyeva and Obruchev (1967) are essentially gradal and as such are of limited use.

Here, I present a different classification. The tree (Fig. 2). is derived from a cladogram. It must be regarded as preliminary because it only incorporates 9 (25 species) of the better known genera. Many genera are, as yet, too incompletely known to be inserted. Against each bifurcation the synapomorphies are listed and explained in the legend. The interesting fact for our present task is the number of synapomorphies (evolutionary novelties, Eldredge 1979) at each successive dichotomy. It should be noted that loss characters are not considered (they do not specify groups); nor are the charac-

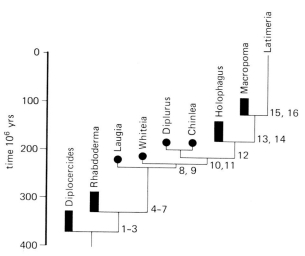

Fig. 2. Tree of the better known coelacanth genera showing acquisition of synapomorphies through time. The thick vertical lines represent geological range with circles denoting those known only from very limited time ranges. Characters are numbered 1–16 and are as follows: 1. rostral organ; 2. tandem jaw articulation including a symplectic; 3. extracleithrum; 4. descending process of supratemporal; 5. one-to-one relation between caudal fin rays and internal support; 6. process from frontal bracing basisphenoid; 7. metapterygoid with single point of braincase articulation; 8. medial branch of otic sensory canal; 9. medial position of parietal ossification center; 10. posterior lamina outgrowth of prootic; 11. descending process of intertemporal; 12. jugal meeting tectal series; 13. dorsal laminae at anterior end of parasphenoid; 14. jugal canal running at ventral edge of squamosal; 15. ascending process on prootic; 16. anterior branches of supratemporal commissure.

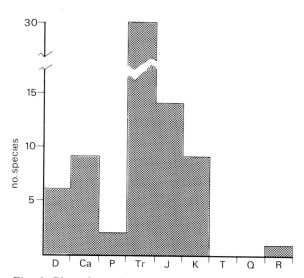

Fig. 1. Plot of species diversity through time from Devonian (D) to Recent (R). The plot does not include the Devonian *Dictyonosteus* because its coelacanth affinities are in doubt. Several undescribed species from the Lower Triassic of Greenland have been omitted since these remains have been simply identified as coelacanth.

ters of each genus listed. This admittedly crude plot suggests that the rate of acquisition of characters has remained approximately the same throughout coelacanth history.

This is rather different from that plot produced by Schaeffer (1952a, 1952b), in which it was suggested that there was a marked slowing down in the rate of character acquisition between Upper Devonian and Upper Carboniferous followed by near stasis. In other words, the graph produced by Schaeffer shows a marked asymmetry with one showing taxonomic diversity (Fig. 1). Since "evolutionary rates" of coelacanths are sometimes cited in literature, a brief comment is in order. It is not my intention to try to reconcile the two patterns, chiefly because they were produced by different means that influence the results. However, it needs to be said that the Schaeffer pattern, which at first sight is agreeable to a punctuationist view, is based on a complex procedure developed by Westoll (1949) for lungfishes. That procedure suffers because it is not based on any scheme of relationship of the organisms concerned but relies on an intuitively derived sequence of structural grades and allows reversals.

Despite the fact that the patterns of character acquisition given here and by Schaeffer may differ, it must be emphasized that both are based on relatively few characters. Both show that there has been an overall low rate of character acquisition. In Fig. 2 only 16 characters are listed between *Diplocercides* and the *Macropoma/Latimeria* dichotomy, and this represents a time span of some 260 million years. The inclusion of more genera would add more characters (particularly those of Triassic genera, largely omitted here) to this list. But, as far as *Latimeria* is concerned, this would only be of importance if these extra genera were inserted so as to intersect the right-hand axis of the tree.

The sister-group of *Latimeria* is *Macropoma*, with three species that together are known from the Albian-Turonian of southeast England and the Planerkalk of Bohemia. There are only minor points of difference between *Latimeria* and *Macropoma*: (1) The skull roofing bones of *Latimeria* are sunk below the skin surface, except for small raised areas on the posterior frontals and the parietals, contrasted with ornamented (presumably superficial) bones in *Macropoma*; (2) the cheek bones of *Latimeria*

are well separated from one another, whereas those of *Macropoma* fit more closely together; (3) the snout of *Latimeria* is covered by a tessellated pattern of small ossicles whereas that of *Macropoma* is covered with a large hemispherical ossification; (4) the swim bladder of *Latimeria* is soft-walled whereas that of *Macropoma*, like that of many coelacanths, has a mineralized wall; and (5) there are slight proportional and meristic differences and minor differences in scale ornament both between species of *Macropoma* and between *Macropoma* and *Latimeria*. It is said (Berg 1940) that the dorsal, anal, and paired fins are more pedunculate in *Latimeria* than that in other coelacanths, but preservation of *Macropoma* does not really confirm or deny this claim.

Our knowledge of the biology of *Latimeria* is largely inferential since only dead or moribund specimens have been studied (for reviews see Lockett 1980: McCosker 1979). Apart from the first specimen, reported to have been captured near East London, *Latimeria* has only been found off the islands of Grande Comore and Anjouan in the Comoran chain. *Latimeria* has been hooked at depths between 70 and 600 m, but the majority have been recovered from 150 to 300 m. McCosker (1979) notes that these depths straddle a deep water (150–200 m) thermocline across which the temperature drops from 23°–26°C to 15°C. The implication is that *Latimeria* is tolerant of, at least, a 10°C range in temperature. *Latimeria* appears to be a truly marine fish despite claims (Forster 1974) that it may seek out submarine aquifers. The osmoregulatory mechanism (for review see Griffith and Pang 1979) included retention of high urea content and the presence of rectal salt-glands. Stomach contents include several species of fish (some deep-water species), and entire cuttlefish have also been reported. The jaw structure and short gut-length, together with deductions about the jaw mechanism (Lauder 1980) suggest that this carnivore adopts a suction inhalation mechanism. It is reasonable to presume that *Latimeria* is near the top of its local food chain. There are, to my knowledge, no reports of coelacanths as stomach contents of other species.

Comparison of these points with the sister-group *Macropoma* is difficult, all the more because there is considerable controversy about the environmental conditions of the chalk sea in the Anglo-Paris basin. That *Macropoma* was a marine fish is not in dispute. Hancock (1976) cautiously proposed a scenario of a sea of normal salinity (35‰), a depth of 100–600 m but generally not less than 200 m, warm water, and a generally quiet bottom. Osteological features of the jaw structure and evidence of the gut are similar to those of *Latimeria*. The jaw structure of other coelacanths is also similar to that of *Macropoma,* although there is some slight variation in the form of the parasphenoid and palatal dentition.

The vast majority of other coelacanths are also known from marine rocks. Only six species (one Permian and five Triassic) can be said to have been confined to fresh water, and at least four more may have been euryhaline. Information about coelacanth distribution in space and time is not very helpful for the purposes of this book. Most are known only from single localities and horizons, and this reflects a taxonomy that includes purely stratigraphic/geographic species. *Rhabdoderma elegans* (Newberry) is the longest lived and most widespread species known. It persists for about 30 million years (Namurain–Stephanian) and is known from Ohio to the Donetz basin, USSR.

It is difficult to draw conclusions from the disparate information given in the preceding paragraphs. *Latimeria* is very similar to its sister-group by any criteria of assessing similarity, and there is relatively little morphological difference between *Latimeria* and the Devonian *Diplocercides*. The stratigraphic plot shows maximum taxonomic diversity in the Triassic, but there is no clear evidence of substantial changes in the rate of morphological change. When it becomes possible to insert more species into the cladogram, a clearer pattern may emerge, but I would predict that when this is done the morphological rate would reflect taxonomic rate, meaning an acceleration in the Triassic, not at the time of origin of the group.

Literature

Agassiz, J. L. R. 1839. Recherches sur les poissons fossiles. Neuchatel 2: pl. 42.

Berg, L. S. 1940. Classification of fishes, both recent and fossil. Trudy Zoologicheskogo Instituta. Akademiya Nauk SSSR. Leningrad 5:87–517.

Darwin, C. R. 1859. On the origin of species. (1960 reprint.) London: John Murray.

Eldredge, N. 1979. Cladism and common sense, pp. 165–198. In: Cracraft, J., Eldredge, N. (eds.), Phylogenetic analysis and paleontology. New York: Columbia U. Press.

Fitzinger, L. J. F. J. 1837. Vorläufiger Bericht über eine höchst interessante Entdeckung Dr. Natterers in Brasil. Isis, Jena:379–380.

Forster, G. R. 1974. The ecology of *Latimeria chalumnae* Smith: results of field studies from Grande Comore. Proc. Roy. Soc. London, Ser. B 186:291–296.

Griffith, R. W., Pang, P. K. T. 1979. Mechanisms of osmoregulation in the coelacanth: evolutionary implications. Occ. Papers Calif. Acad. Sci. 134:79–93.

Hancock, J. 1976. The petrology of the Chalk. Proc. Geol. Assoc. 86:499–535.

Huxley, T. H. 1908. Discourses: biological and geological. London: Macmillan.

Lauder, G. V. 1980. The role of the hyoid apparatus in the feeding mechanism of the coelacanth *Latimeria chalumna*. Copeia 1980:1–9.

Lockett, N. A. 1980. Review lecture: some advances in coelacanth biology. Proc. Roy. Soc. London, Ser. B 208:265–307.

McCosker, J. E. 1979. Inferred natural history of the living coelacanth. Occ. Papers Calif. Acad. Sci. 134:17–24.

Romer, A. S. 1945. Vertebrate paleontology. Chicago: U. of Chicago Press.

Rosen, D. E., Forey, P. L., Gardiner, B. G., Patterson, C. 1981. Lungfishes, tetrapods, paleontology, and plesiomorphy. Bull. Amer. Mus. Nat. Hist. 167:159–276.

Schaeffer, B. 1952a. Rates of evolution in the coelacanth and dipnoan fishes. Evolution 6:101–111.

Schaeffer, B. 1952b. The Triassic fish *Diplurus* with observations on the evolution of the Coelacanthini. Bull. Amer. Mus. Nat. Hist. 99:25–78.

Stanley, S. M. 1979. Macroevolution. San Francisco: Freeman.

Vorobyeva, E. I., Obruchev, D. V. 1967. Subclass *Sarcopterygii,* pp. 480–498. In: Obruchev, D. V. (ed.), Fundamentals of palaeontology, 11. Jerusalem: Israel Program for Scientific Translations.

Westoll, T. S. 1949. On the evolution of the Dipnoi, pp. 121–184. In: Jepsen, G. L., Simpson, G. G., Mayr, E. (eds.), Genetics palaeontology and evolution. Princeton: Princeton U. Press.

White, E. I. 1953. The coelacanth fishes. Discovery 14:113–116.

18
"Notidanus"

John G. Maisey and Katherine E. Wolfram

Department of Vertebrate Paleontology, American Museum of Natural History, New York, NY 10024

Introduction

The old lithographic limestone quarries around Solnhofen, Bavaria, are perhaps best known for the beautiful *Archaeopteryx* fossils that have been found there. For us, however, the Solnhofen Limestone provides an intriguing glimpse of a Late Jurassic shark fauna that contains early representatives of several extant families, including horn-sharks (Heterodontidae), monk-fish (Squatinidae), carpet sharks (Orectolobiformes), dogfishes (Scyliorhinidae), primitive rays (batoids), and also hexanchoids (cow-sharks, "notidanids"). Any of these groups would be worth consideration here, since all of them include fossil species that are "anatomically very similar (bordering on identity)" to living species, as requested for this paper. Apart from the heterodontids, however, only the hexanchoids have been credited with the venerable and patriarchal status of "living fossil." For example, it is clear from Woodward's (1886a, 1886b) works that he regarded living hexanchoids as relics of the archaic hybodont sharks from the Paleozoic and Mesozoic, a view reiterated by Romer (1966:42) and Schaeffer (1967:22). Thenius (1973:167, Fig. 87) actually includes *Hexanchus* as one of "the most important 'living fossils' among the vertebrates" although it is not discussed further in the text. Since the hexanchoid fossil record extends back to the Early Jurassic, we will examine the evidence for regarding hexanchoids as living fossils under their old, all-embracing generic name of *"Notidanus."* All living hexanchoids were placed in this genus by Cuvier (1817) and Günther (1870), and in the early literature all fossil hexanchoids were also referred to this genus. We do not regard *"Notidanus"* as a valid taxon, only as a convenient title for this article.

Modern Hexanchoids

Three living genera of hexanchoid sharks are recognized here: *Hexanchus* (the "six-gill shark") with two species, *H. griseus* and *H. vitulus; Heptranchias* (the "sharp-nosed seven-gill shark") with one species, *H. perlo;* and *Notorynchus* (the "blunt-nosed seven-gill shark"), also with one species, *N. cepedianus*. These generic distinctions are based not only on the number of gill clefts but also on head shape, tooth morphology, and dental variation. A number of endoskeletal differences are also recognizable (see below and Fig. 1.).

Among living sharks and rays, hexanchoids are generally thought to be very primitive (e.g., Huxley 1876; Gegenbaur 1872; Goodrich 1909; White 1937; Compagno 1973, 1977), and also closely related to *Chlamydoselachus* (another candidate for the title of living fossil whose biggest problem is its lack of a fossil record, apart

Fig. 1. Living "notidanids" (hexanchoids): (A–F) *Hexanchus griseus;* (G–L) *Heptranchias perlo;* (M–R) *Notorynchus cepedianus.* (A) *Hexanchus* braincase, dorsal view; (B) head, ventral view; (C) lower lateral tooth; (D) some upper tooth outlines to show variation; (E) lower symphyseal tooth; (F) body form; (G) *Heptranchias* braincase, dorsal and (G′) ventral view; (H) head, ventral view; (I) some upper teeth; (J) lower symphyseal and (K) some lower lateral teeth; (L) body form; (M) *Notorynchus* braincase, dorsal and (M′) ventral view; (N) head, ventral view; (O) some upper teeth; (P) lower lateral tooth; (Q) lower symphyseal tooth; (R) body form.

from a couple of Pliocene and late Tertiary teeth), e.g., Garman (1885), Gudger and Smith (1933), and Compagno (1977). We wish to avoid the controversy over the supposed affinities of *Chlamydoselachus* to certain Paleozoic sharks, but an excellent (if dated) review of that issue will be found in Gudger and Smith (1933). For the sake of presenting the evidence as lucidly as possible, we will regard *Chlamydoselachus* as the sister-taxon of hexanchoids.

Chlamydoselachus and living hexanchoids share the following characteristics (regarded here as synapomorphies).

1. general body configuration: the single dorsal, the pelvics, and the anal fin are all located posteriorly, close to the elongate caudal fin; dorsal fin lying between pelvic and anal; elongated trunk region between pelvics and pectorals; short snout region.

2. more than five gill clefts (a condition otherwise known only in a genus of sawshark, *Pliotrema,* and a trygonid ray, *Hexatrygon*).

Chlamydoselachus and hexanchoids share some apparently primitive characteristics also, including:

(a) shoulder-girdle consisting of two separate scapulocoracoids, not fused at the midline.
(b) occipital part of the braincase extending somewhat behind the auditory capsules (except *Heptranchias;* see below).

In *Notorynchus* the scapulocoracoids are reportedly separated by a median "sternal" cartilage (Haswell 1885; Parker 1891). In *Hexanchus*, *Heptranchias*, and *Chlamydoselachus*, the scapulocoracoids are generally connected by fibrous tissue in the midline, although a "sternal" cartilage was reported in some *Hexanchus griseus* by White (1895). In all other living sharks and rays, the scapulocoracoids are fused ventrally to form a continuous girdle in the adult. Unfused and separate scapulocoracoids are found in many fossil sharks including "*Notidanus*" *muensteri* (see below), *Hybodus*, *Xenacanthus*, *Synechodus*, *Palaeospinax*, *Ctenacanthus*, *Tristychius*, and *Cobelodus*. In some sharks (e.g., *Xenacanthus*, *Cobelodus*, edestids) there are paired ventral cartilages in the "sternal" region. The occiput of most living sharks and rays is located level with the posterior margin of the auditory capsules (Schaeffer 1981), but extends farther posteriorly in many fossil sharks, including *Hybodus*, *Xenacanthus*, *Tristychius*, and *Cobelodus*.

Living hexanchoids are united by various dental characters:

3. teeth compressed labio-lingually; lateral teeth bladelike but with several cusps in a rectilinear series along the cutting edge.

4. upper and lower teeth distinctly different, the lowers generally being longer and having more cusps.

5. posteriormost upper and lower teeth are small and button-like, unserrated and lacking cusps.

Additionally hexanchoids share the following supposedly primitive character that does not occur in other living sharks and rays (including *Chlamydoselachus*):

(c) palatoquadrate expanded posteriorly, articulating with the postorbital process of the braincase ("amphistylic" condition).

A postorbital articulation was present in many fossil sharks, e.g., *Synechodus*, *Ctenacanthus*, *Goodrichthys*, *Xenacanthus*, *Tamiobatis*, "*Cladodus*" *wildungensis*, *Cobelodus*, *Cladoselache*, and *Hopleacanthus*. Its absence in most modern sharks and rays is probably apomorphic. *Hybodus* appears to have lost the postorbital articulation independently of modern sharks and rays (Maisey 1982, 1983). An opposing view was put forward by Edgeworth (1935), who regarded the hexanchoid postorbital articulation as secondary,

based on the lateral position of the postorbital process relative to the dorsal constrictor muscle of the palatoquadrate. This line of reasoning does not seem particularly constructive, since many fossil sharks unrelated to hexanchoids also probably had their dorsal constrictors mesial to the postorbital process (e.g., *Xenacanthus*, *Tamiobatis*, *Cobelodus*, *Cladoselache*). Furthermore, the muscle is ventro-mesial to the process in *Squatina* (Luther 1908:plate II, Fig. 18). Nonetheless, Edgeworth's notion regarding hexanchoids is of interest in the light of the phylogenetic hypothesis we shall advance here (see below).

The Fossil Record

Fossil hexanchoids are mainly represented by isolated teeth. Nonetheless, fairly complete hexanchoid fossils are known from the Late Cretaceous of Sahel Alma (Davis 1887; Cappetta 1980) and Late Jurassic of Solnhofen and Eichstadt (Beyrich 1849; Wagner 1861; Hasse 1882; Schweizer 1964). Although de Beaumont (1960a) referred another articulated fossil species from the Eocene of Monte Bolca to "*Notidanus*," it actually seems to be a galeomorph (Eastman 1904; Blot 1980) and is therefore ignored here. Apart from the fin positions, tooth morphology, and vertebral structure, little can be determined in the Cretaceous form. Therefore most of the present comparison is based on the Kimmeridgian material that is referred to "*Notidanus*" *muensteri*.

"*Notidanus*" *muensteri* agrees with living hexanchoids in characters 1, 3, 4, and 5, as well as (a), (b), and (c). We therefore concur that, as far as can be established, "*N.*" *muensteri* is a hexanchoid. We do not know how many gill arches and clefts it possessed (character 2).

Living hexanchoids are probably more derived than "*Notidanus*" *muensteri* in the following respects:

6. Precaudal vertebral centra are weakly calcified (*Heptranchias*) or uncalcified (*Hexanchus*, *Notorynchus*); strongly calcified tectospondylous vertebrae in "*Notidanus*" *muensteri*.

7. Teeth are strongly flattened labio-lingually, and the tooth root is almost parallel to the crown in cross section, with its basal surface

enlarged to form the "lingual" side of the root; tooth less flattened, with root at a decided angle to the crown, in *"Notidanus" muensteri*.

Ridewood (1921) and Compagno (1977) contend that the "notochordal" condition of living hexanchoids is secondary; the notochord is divided up by transverse septa that constrict the notochord precaudally, as in other living sharks, and vertebral centra are present caudally (in *Chlamydoselachus* the notochord is partly constricted but not septate precaudally). Not only was the notochordal sheath of *"Notidanus" muensteri* heavily calcified, but the notochord was strongly constricted within each vertebral centrum.

The vertebral column is also strongly calcified in the Cretaceous *"Notidanus" gracilis*. In both articulated fossil species, the vertebrae have radial calcifications of tectospondylous type (Hasse 1882; Schweizer 1964; Cappetta 1980).

Schweizer's (1964) detailed study of *"N." muensteri* tooth morphology and dental variation indicates that the peculiar semisynchronized replacement pattern of modern hexanchoid lateral teeth was lacking. Instead, all the teeth were gradually replaced as in many other sharks. The upper anterior teeth of living hexanchoids are still replaced gradually, with two or three functional teeth in each replacement series. These teeth have fewer cusps and are less bladelike than the synchronously replaced ones. There are also fewer rows of lower lateral teeth in living hexanchoids (five per side in *Heptranchias perlo* and *Hexanchus vitulus;* six in *Hexanchus griseus* and *Notorynchus cepedianus*) than in *"Notidanus" muensteri* (in which there may be eight or nine, based on Beyrich 1849:plate 17). The number of tooth rows in *"N." gracilis* has not been determined. The absence of small cusplets or serrations anterior to the primary (tallest) cusp of *"N." muensteri* could represent a primitive condition. Serrations are also absent from *"N." arzöensis* (de Beaumont 1960b). Nonetheless, anterior serrations are present in *"N." serratus* teeth from Solnhofen (Fraas 1855; Quenstedt 1858; Woodward 1886b, 1889; Schweizer 1964). These teeth seem closest to those of living *Notorynchus* in the form of their anterior serra-

tions and their low cusp count (see below), although these similarities may simply be primitive.

A suite of characters having a disjunct distribution among living hexanchoids and *"Notidanus"* include:

8. high cusp count (eight to ten) in *Hexanchus* and *Heptranchias* lower lateral teeth (four or five in *Notorynchus*, three or four in *"Notidanus";* teeth totally different in *Chlamydoselachus;* Figs. 1C, 1K, 1P; Figs. 2E, 2F).

9. (a) vertical median cusp present in the lower symphyseal tooth of *Heptranchias;* (b) almost vertical in *Hexanchus* (strongly inclined in *Notorynchus* and *"Notidanus"*).

10. short postorbital process in *Heptranchias* (elongate in *Hexanchus griseus, Notorynchus, Chlamydoselachus,* and *"Notidanus";* Figs. 1A, 1G, 1M; Fig. 2B).

11. otico-occipital region only half the length of the orbitotemporal–ethmoid region in *Heptranchias* (regions of almost equal length in *Hexanchus, Notorynchus, Chlamydoselachus,* and *"Notidanus"*).

12. very slender hyomandibula and ceratohyal in *Heptranchias* (fairly slender in *Hexanchus vitulus,* broad and well developed in *Hexanchus griseus, Notorynchus, Chlamydoselachus,* and *"Notidanus"*).

13. very strong articular surface between the postorbital process and palatoquadrate in *Heptranchias* (ill defined on the braincase of *Hexanchus, Notorynchus,* and *"Notidanus";* absent in *Chlamydoselachus;* Gegenbaur 1872; Holmgren 1941).

14. five lateral teeth in *Heptranchias* and *Hexanchus vitulus* (more in other hexanchoids and *Chlamydoselachus*).

15. six gill slits in *Hexanchus* (seven in *Chlamydoselachus, Heptranchias,* and *Notorynchus;* unknown in *"Notidanus"*).

16. adductor muscle of jaws with anterior and posterior divisions in *Heptranchias* (undivided in other living hexanchoids and *Chlamydoselachus;* unknown in *"Notidanus"*).

In the hypothesis of relationships presented here (Fig. 3), *"Notidanus"* is the sister-taxon to living hexanchoids, with which it is united by

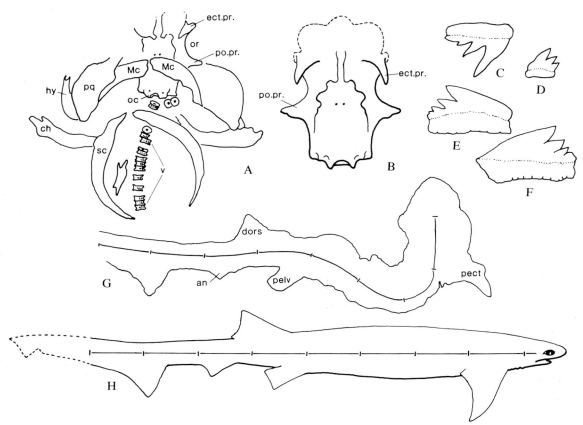

Fig. 2. The Late Jurassic *"Notidanus" muensteri*. (A) outline of cranial elements in ventral view as preserved in Geologisch-Paläontologisches Institut Tübingen; ch., ceratohyal; ect. pr., ectethmoid process; hy, hyomandibula; Mc, Meckel's cartilages (lower jaw); oc, occiput, or, orbit; po. pr., postorbital process; pq. palatoquadrate; sc, scapulocoracoid; v, vertebral centra; (B) outline of braincase, ventral view (restored); (C) upper lateral tooth; (D) lower symphyseal (?) tooth; (E,F) lower lateral teeth; (G) outline of specimen described by Beyrich. Vertebral axis divided into arbitrary units for purposes of measurements; an, anal fin; dors., dorsal fin; pect., pectoral fin; pelv, pelvic fin. (H) restoration of body form based on (G). Length between pectorals and pelvis is much greater than in living hexanchoids, and is reminiscent of *Chlamydoselachus*.

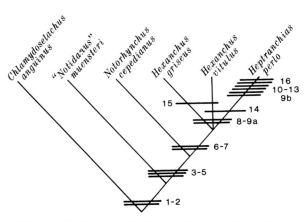

Fig. 3. Cladogram of relationships based on characters in the text.

characters 3, 4, and 5. The most primitive living hexanchoid seems to be *Notorynchus*, which shares characters 6 and 7 with *Hexanchus* and *Heptranchias* and is distinguished (?primitively) by having few cusps on the anterior margin of its lateral teeth (Fig. 1P). *Hexanchus* and *Heptranchias* are united by character 8.

Hexanchus griseus and *H. vitulus* are united by having six gill clefts (character 15) and several fine cusps on the anterior margin of the lateral teeth. *Heptranchias* is distinguished from other hexanchoids by several characters (9b, 10–13, 16). Character 14 (number of lower lateral tooth rows) conflicts with 15, producing an unresolved trichotomy between *Heptranchias perlo* and the two *Hexanchus* species.

Species-Level Diversity

There are three living genera, of which two *(Notorynchus, Heptranchias)* are regarded as monotypic and one *(Hexanchus)* probably contains two species. Somewhat larger numbers of fossil species have been described, but most are known only from isolated teeth and are of dubious taxonomic validity. Paleontologists have tended to refer fossil hexanchoid teeth to *Notidanus* on the grounds that the number of gill clefts is unknown. At the same time, however, it has been admitted "that in the living Notidanidae there are very decided differences in the teeth of the various species" (Woodward 1886b:207). The long-standing view that taxonomic characters of modern hexanchoid teeth could not be applied to the fossils (see also Woodward 1889; Arambourg 1952; de Beaumont 1960a, 1960b; Pledge 1967; Herman 1975) has been challenged (e.g., Jordan 1907; Fowler 1911; Applegate 1965; Applegate and Uyeno 1968; Waldman 1971; Cappetta 1975; Kemp 1978; Ward 1979; Case 1980). It is certainly possible to assign fossil hexanchoid teeth of Cenozoic and Cretaceous age to the three living genera. A review of fossil species-level diversity based on isolated teeth is impractical, however, since tooth morphology is variable among living species, and dental characters of the kind used by paleomammalogists are unavailable in sharks. Many fossil hexanchoid species seem founded only on stratigraphic grounds.

Distribution in Space and Time

Living hexanchoids are fairly widespread. *Hexanchus griseus* occurs almost worldwide in temperate and tropical oceans (e.g., Mediterranean, Madagascar, Australia, Gulf of Mexico) and may also range into cooler waters of the north Atlantic and off the coast of Argentina (Bigelow and Schroeder 1948; Springer and Waller 1969). *Hexanchus vitulus* is more restricted, being recorded in tropical waters off Florida, in the Gulf of Mexico, and the West Indies, as well as in the Philippines and south-

west Indian Ocean (Springer and Waller 1969; Forster et al. 1970; Bass et al. 1975). In more temperate waters it is apparently replaced by juvenile *H. griseus*. *Heptranchias perlo* occurs in the western and eastern Atlantic, the Mediterranean, southwestern Indian Ocean, off New Zealand and Japan, and possibly in the eastern Pacific (Bigelow and Schroeder 1948; Bass et al. 1975).

Quaternary records of hexanchoid teeth include northwestern Europe and the Mediterranean. Mostly these records can be referred to *Hexanchus* and *Notorynchus* but not to *Heptranchias*, almost the opposite of their present-day distribution. Among the "species" we would refer to *Hexanchus* are *"Notidanus" anomalus, "N." targionii,* and *"N." gigas* (see Lawley 1875; Woodward 1886b, 1889). Kemp (1978, 1982) records *Notorynchus cepedianus* teeth from the Late Quaternary of Australia. *"Notidanus" primigenius* Agassiz 1843 has also been referred to *Notorynchus* by Kemp (1978, 1982), who notes its earliest Australian occurrence in the Miocene. Other Tertiary records include eastern and western North America as far north as British Columbia (*Hexanchus;* Waldman 1971), which corresponds to the northerly limit of living *H. griseus* (Herald and Ripley 1951). *Notorynchus* and *Heptranchias* teeth are recorded from the Miocene (and Cretaceous) of New Jersey (Welton 1974; Case 1980). *Heptranchias* has also been recorded from the Oligocene of Japan (Applegate and Uyeno 1968). *Hexanchus* and *Heptranchias* occur in the Late Eocene of Australia; this is the latest record of fossil *Hexanchus* in that region (Kemp 1982). In the European Tertiary, we would refer *"Notidanus" loozi* to *Heptranchias* and *"N." serratissimus* in part to *Hexanchus* (see Agassiz 1843; Vincent 1876; Woodward 1886b; Cappetta 1976; Kemp 1978, 1982; but cf. Pledge 1967). *"Notidanus" microdon* occurs in the Early Tertiary and Late Cretaceous of Europe and North Africa (Agassiz 1843; Arambourg 1952); its teeth are characteristic of *Hexanchus*. Cappetta (1975) recognized *Notorynchus* from the Aptian (Lower Cretaceous) of France. He also erected a new genus, *Notidanodon,* for *"Notidanus" pectinatus* (Agassiz 1843:221, plate 36, Figs. 3, 3A) in which there are up to four extremely well-developed cusps anterior to the principal one; we regard this tooth morphology as falling within the range of *Heptranchias*. Other *Hexanchus*- and *Heptranchias*- like teeth occur in the "Greensand" of New Zealand, e.g., *"Notidanus" marginalis, "N." dentatus* (see Woodward 1886b; Davis 1888). The Senonian Lebanese species *"N." gracilis,* known from articulated remains, has been referred to *Hexanchus* (Cappetta 1980). In the Jurassic at least one modern hexanchoid tooth pattern *(Notorynchus)* can be recognized in *"Notidanus" serratus* (Kimmeridgian; Fraas 1855; Quenstedt 1858; Schweizer 1964), *"N." daviesii* (Oxfordian; Woodward 1886b), and *"Hexanchus" wiedenrothi* (Lower Lias; Thies 1983). The fossil records are unfortunately strongly biased toward western nations, and it may well turn out that elsewhere all three living hexanchoid tooth patterns will eventually be traced back to the Jurassic, and they certainly seem to be established by the Cretaceous.

In addition to these patterns, however, there is a fourth, represented by the articulated Jurassic species *"Notidanus" muensteri*. In this pattern there are no serrations or cusps anterior to the primary one (Figs. 2C–2F). Other species with teeth of this type include the Liassic *"N." arzöensis* (de Beaumont 1960b) and the lower Eocene *"N." ancistrodon* (Arambourg 1952).

Ecological Considerations

Although hexanchoids are commonly thought to be deep-water sharks, they in fact occur over a wide range of ocean depths. *Hexanchus griseus* is demersal and is usually caught by trawls or long lines at considerable depth (200–1100 m) and within a few meters of the bottom (Forster et al. 1970; Bass et al. 1975). Its stomach contents include teleosts, other sharks, and squid (Backus 1957; Springer and Waller 1969). *Hexanchus vitulus* is also considered a deep-water demersal species, but it may take excursions to the surface, especially at night (Fourmanoir 1961; Forster et al. 1970). Teleosts are known to be taken by this species. *Heptranchias perlo* is most common on or near the margins of the continental shelf, usually between 50 and 100 m, but it is commonly found at the surface off the coast of New Zealand (Bass et al. 1975), perhaps at an upwelling cold current. *Notorynchus cepedianus* is often caught in shallow

waters by shore anglers, for example in San Francisco Bay (Herald and Ripley 1951), off the Cape coast of South Africa (Bass et al. 1975), and in Ralph's Bay, south of Hobart, Tasmania, where it has been caught in less than 3 m of water (Kemp 1982). The young of this species are apparently born in shallow waters, moving into deeper waters only when they are larger. Thus the presence of fossil hexanchoids in shallow marine epicontinental sediments is consistent with the habitats of living representatives. The Solnhofen limestones were probably deposited in a subtidal environment at depths of up to 60 m. (Janicke 1969; Barthel 1970, 1972).

References to immature hexanchoids in the Solnhofen fauna (e.g., Wagner 1861; Woodward 1886b) are unfounded, the specimens subsequently having been identified as scyliorhinids (*Pristiurus;* see Woodward 1889:158, 344). Thus we have no evidence to support the happy scenario of *"N." muensteri* pupping in the shallow tropical lagoons of southern Germany. In fact, small fishes probably did not fare at all well, since the specimen of *"N." muensteri* described by Beyrich (1849) is over nine feet long (Figs. 2G, 2H).

The chalky Senonian limestone of Sahel-Alma that have yielded *"Notidanus" gracilis* were also probably deposited in a shallow shelf environment, on the southern margin of the Tethys ocean, but little work has been done on this depositional environment. Some of the osteichthyan fossils in the fauna appear to be allied to species that today only frequent deeper ocean habitats (Donn Rosen, personal communication).

The dentition of living hexanchoids is strongly specialized. The teeth on their "fixed" upper jaw have elongate cusps, adapted for gripping prey rather than cutting (Figs. 1D, 1I, 1O). The lower dentition could provide a cutting ability, since semisynchronized tooth replacement, an elongated cutting edge, and series of serrations provide a long and almost continuous sawlike dentition from symphysis to near the back of the jaw (its effectiveness on people is amply illustrated by Herald 1968). Nonetheless, X-ray plates in our possession show entire teleosts in the stomach of *Heptranchias perlo* and *Hexanchus griseus*. Hexanchoids are unusual, although not unique, in having awl-shaped "grasping" teeth in the upper jaw and bladelike "cutting" lower teeth (a comparable arrangement occurs in some squaloids). Carcharhinids and lamnoids (in which the jaws are extremely protrusible; Moss 1972) have virtually the reverse arrangement. The dentition of *"Notidanus" muensteri* suggests a less extreme feeding method than in living hexanchoids (Schweizer 1964).

If Cappetta (1980) is correct in assigning the Cretaceous *"Notidanus" gracilis* to *Hexanchus,* we may surmise that reduced calcification of the vertebral centra in this genus has occurred independently from the other genera (*Heptranchias* and *Notorynchus*). The fossil record seems to lend support to the suggestion that vertebral calcification is secondarily reduced in modern hexanchoids (Ridewood 1921; Compagno 1977). Note, however, that in the cladogram (Fig. 3) we have left reduction of vertebral calcification as a single character (character 6), because we have not examined the Senonian species and consider its affinities uncertain.

Is *"Notidanus"* a Living Fossil?

Simply expressing this question in a Victorian taxonomic fashion (i.e., all hexanchoids are "sub-genera" of *"Notidanus"*; Woodward 1886b), *"Notidanus"* clearly ranges from the Jurassic to the present day, spanning some 180 million years. As far as the fossil record from Solnhofen and Eichstädt is concerned, however, *Squatina*, *"Scyllium,"* *Heterodontus,* and *Rhinobatus* may equally be considered living fossils.

It is clear from the data and cladogram presented earlier that "significant" (i.e., recognizable) changes have occurred to hexanchoids since the Late Jurassic. Were these changes progressive or saltatory? From the fossil record, all we can say is that the different patterns of tooth morphology seen in living hexanchoids were all developed by the Late Cretaceous, and that at least one of these (the *Notorynchus* pattern) appeared in the Jurassic.

Development of these dental peculiarities suggests increasing ecological specialization related to the feeding mechanism. Referring again to our cladogram (Fig. 3), it is interesting that the most derived member *(Heptranchias)* has

the "best" postorbital articulations between the braincase and palatoquadrate, and the weakest hyoid support (Gegenbaur 1872; Holmgren 1941), whereas in *Hexanchus, Notorynchus,* and *"Notidanus" meunsteri* the articular facet on the postorbital process is ill defined but the hyoid arch is strong. Furthermore, the postorbital articulation is absent in their sister-taxon (*Chlamydoselachus*), although the palatoquadrate and postorbital process may contact each other during some phases of jaw operation (Compagno 1977). Taken in conjunction with Edgeworth's (1935) suggestion that the hexanchoid postorbital articulation is "secondary," we cannot help wondering whether hexanchoids have reacquired a postorbital articulation and are really much less primitive than our Victorian forebears thought—which brings us to the final twist in this tale. Why did Woodward and subsequent writers pick on hexanchoids as a relict group of hybodont sharks? The presence of a postorbital articulation was certainly a key factor, but we must relate this to what was known of other sharks at that time, especially the fossils.

The first significant discovery of Mesozoic *Hybodus* was of a *H. basanus* head (Egerton 1845) from the Lower Cretaceous. Jaws and teeth of another, later Cretaceous shark were subsequently described as *Hybodus dubrisiensis* by Mackie (1863) and Woodward (1886a). Although Woodward considered that the jaws of *H. dubrisiensis* differed significantly from those of older (then undescribed) *Hybodus* species in having a postorbital articulation, he nevertheless regarded *H. dubrisiensis* as an "advanced hybodont" intermediate between earlier hybodonts and "notidanids" (Woodward 1886b). Although this fossil was subsequently separated into a new genus, *Synechodus* (Woodward 1888), it was left as a "hybodont," although the only feature it shares with *Hybodus* is multicuspid, pointed teeth. Not surprisingly, therefore, other Mesozoic hybodonts were interpreted after *Synechodus* and hexanchoids, e.g., Brown (1900), Jaekel (1906), and Koken (1907). Only Woodward's (1916) account of *Hybodus basanus* gave a radically different interpretation. In a subsequent review of *Heterodontus* and hybodonts, Smith (1942) declared "there must be a flaw in the data somewhere," but came down on the side of Woodward's (1886b) earlier hypothesis of a *Hybodus–Synechodus*–hexanchoid relationship. This hypothesis would clearly be refuted if it could be shown either that *Synechodus* and *Hybodus* are not closely related, or that hexanchoids are not allied to *Synechodus*. In fact, both these relationships can be refuted. It is now known that the cranial morphology of Mesozoic hybodonts is highly specialized and that a postorbital articulation is absent (Maisey 1980, 1982, 1983). Moreover, the braincase of *Synechodus* is profoundly different from those of *Hybodus* and living hexanchoids (Maisey, in preparation), suggesting that neither *Hybodus* nor *Synechodus* are closely related to hexanchoids. *Synechodus* may be closer to hexanchoids than *Hybodus,* inasmuch as calcified vertebrae are present as in almost all living sharks, but no synapomorphies of *Synechodus* and hexanchoids have been established. We conclude that there is no evidence to support a relationship between modern hexanchoids and Mesozoic or Late Paleozoic hybodonts. If hexanchoids are a group of living fossils, it is only because they have a record extending back to the Early Jurassic, not because of their putative relationship with some other fossil sharks. Furthermore, some morphological changes have taken place, albeit to a limited extent, since hexanchoids first appeared on the scene.

Literature

Agassiz, L. 1833–1844. Récherches sur les poissons fossiles. Neuchatel 5:1420.

Applegate, S. P. 1965. Tooth terminology and variation in sharks with special reference to the sand shark, *Carcharias taurus* Rafinesque. Contrib. Sci. Los Angeles Co. Mus. 86:3–18.

Applegate, S., Uyeno, T. 1968. The first discovery of a fossil tooth belonging to the shark genus *Heptranchias,* with a new *Pristiophorus* spine, both from the Oligocene of Japan. Bull. Nat. Sci. Mus. 11:195–200.

Arambourg, C. 1952. Les vertébrés fossiles des gisements des phosphates (Maroc, Algérie, Tunisie). Serv. Geol. Maroc, Notes et Mém. 92:1–372.

Backus, R. H. 1957. Notes on western north Atlantic sharks. Copeia (3):246–248.

Barthel, K. W. 1970. On the deposition of the Solnhofen lithographic limestone. Neues Jb. Geol. Paläont. Abh. 135:1–18.

Barthel, K. W. 1972. The genesis of the Solnhofen lithographic limestone (Lower Tithonian): further data and comments. Neues Jb. Geol. Paläont. Mh. 133–145.

Bass, A. J., d'Aubrey, J. D., Kistnasamy, N. 1975. Sharks of the east coast of southern Africa. v. The families Hexanchidae, Chlamydoselachidae, Heterodontidae, Pristiophoridae and Squatinidae. Oceanogr. Res. Inst. Investig. Rept. 43:1–50.

de Beaumont, G. 1960a. Un *Notidanus* de l'Éocène du Mont Bolca. Compte Rendu de la Societé Paléontologique Suisse, Assemblée annuelle. Eclog. Geol. Helv. 53:308–314.

de Beaumont, G. 1960b. Contribution a l'Étude des Genres *Orthacodus* Woodw. et *Notidanus* Cuv. (Selachii). Mém. Suisses Paleont. 77:1–46.

Beyrich, H. E. 1849. Reise nach Kelheim, Ingolstadt, Eichstädt, Solnhofen und Pappenheim. Zeitschr. Deut. Geol. Gesell. 1:423–447.

Bigelow, H. B., Schroeder, W. C. 1948. Fishes of the Western North Atlantic: no. 1 sharks. Mem. Sears Found. Mar. Res. 17:576.

Blot, J. 1980. La faune ichthyologique des gisements du Monte Bolca (Province de Verone, Italie). Catalogue systematique presentant l'état actuel des récherches concernant cette faune. Bull. Mus. Natl. Hist. Nat. 2:339–396.

Brown, C. 1900. Über das Genus *Hybodus* und seine systematische Stellung. Palaeontographica 46:149–174.

Cappetta, H. 1975. Sélaciens et Holocephale du Gargasien de la region de Gargas (Vaucluse). Geol. Méditer. 2:115–134.

Cappetta, H. 1976. Sélaciens nouveaux di London Clay de l'Essex Ypresien du Bassin de Londres. Geobios 5:551–575.

Cappetta, H. 1980. Les sélaciens du Crétacé superieur du Liban. Palaeontographica Abt. A. 168:69–148.

Case, G. R. 1980. Selachian Fauna from the Trent Formation, Lower Miocene (Aquitanian) of Eastern North Carolina. Palaeontogr. Abt. A. Palaeozool.-Stratigr. 171:75–103.

Compagno, L. J. V. 1973. Interrelationships of living elasmobranchs. In: Greenwood, P. H., Miles, R. S., Patterson, C. (eds.), Interrelationships of fishes. Zool. J. Linnean Soc. 53:15–61.

Compagno, L. J. V. 1977. Phyletic relationships of living sharks and rays. Amer. Zool. 17:303–322.

Cuvier, G. 1817. Le règne animal distribué d'après son organisation, pour servir de base à l'histoire naturelle des animaux et d'introduction a l'anatomie comparée. Paris 2:532.

Davis, J. W. 1887. The fossil fishes of the Chalk of Mount Lebanon in Syria. Sci. Trans. Roy. Dubl. Soc. 2:457–636.

Davis, J. W. 1888. On fossil fish-remains from the Tertiary and Cretaceo-Tertiary formations of New Zealand. Sci. Trans. Roy. Dubl. Soc. 4:1–48.

Eastman, C. R. 1904. Descriptions of Bolca Fishes. Bull. Mus. Comp. Zool. 46:1–36.

Edgeworth, F. H. 1935. The cranial muscles of vertebrates. Cambridge: University Press.

Egerton, P. M. G. 1845. Description of *Hybodus* found by Mr. Boscawen Ibbetson in the Isle of Wight. Geol. Soc. Lond. Quart. J. 1:197–199.

Forster, G. R., Badcock, J. R., Longbottom, M. R., Merrett, N. R., Thomson, K. S. 1970. Results of the Royal Society Indian Ocean Deep Slope Fishing Expedition, 1969. Roy. Soc. Lond. Proc. Ser. B 175:367–404.

Fourmanoir, P. 1961. Requins de la cote ouest de Madagascar. Mém. Inst. Sci. Madag. 4:1–81.

Fowler, H. W. 1911. A description of the fossil fish remains of the Cretaceous, Eocene and Miocene formations of New Jersey. Geol. Surv. New Jersey Bull. 4:192.

Fraas, O. 1855. Beiträge zum obersten weissen Jura in Schwaben. Jahresh. Vereins Vaterl. Naturk. Württemberg, Stuttgart 11:77–107.

Garman, S. W. 1885. *Chlamydoselachus anguineus* Garm., a living species of cladodont shark. Bull. Mus. Comp. Zool. 12:1–35.

Gegenbaur, C. 1872. Untersuchungen zur vergleichenden Anatomie der Wirbelthiere. III. Das Kopfskelet der Selachier, ein Beitrag zur Erkenntniss der Genese des Kopfskeletes der Wirbelthiere. Leipzig: Wilhelm Englemann.

Goodrich, E. S. 1909. Cyclostomes and fishes, p. 519. In: Lankester, E. R. (ed.), A treatise on zoology, Vol. 9, Vertebrata craniata.

Gudger, E. W., Smith, B. G. 1933. The natural history of the frilled shark *Chlamydoselachus anguineus*. Amer. Mus. Nat. Hist. 5:245–319.

Günther, A. 1870. Catalogue of the fishes in the British Museum, Vol. VIII. Brit. Mus. Publ. 25:549.

Hasse, C. 1882. Das Natürliche System der Elasmobranchier-Besonderer Theil; I. Lieferung Jena: Gustav Fischer 1–94.

Haswell, W. A. 1885. Studies on the elasmobranch skeleton. Proc. Linnean Soc. New South Wales 9:71–119.

Herald, E. S. 1968. Size and aggressiveness of the sevengill shark *(Notorynchus maculatus)*. Copeia (2):412–414.

Herald, E. S., Ripley, W. E. 1951. The relative abundance of sharks and bat stingrays in San Francisco Bay. Calif. Fish and Game 37:315–329.

Herman, J. 1975. Les Sélachiens des terrains néocretacés et paléocènes de Belgique et des contrées limitrophes. Éléments d'une biostratigraphie intercontinentale. Mémoires pour servir a l'explication des Cartes geologiques et minières de la Belgique. Mem. 15:1–450.

Holmgren, N. 1941. Studies on the head in fishes. Part II. Comparative anatomy of the adult selachian skull with remarks on the dorsal fins in sharks. Acta Zool. 22:1–100.

Huxley, T. H. 1876. On *Ceratodus forsteri*, with observations on the classification of fishes. Proc. Zool. Soc. 1876 24–59.

Jaekel, O. M. J. 1906. Neue Rekonstruktionen von *Pleuracanthus sessilis* und von *Polyacrodus (Hybodus) hauffianus*. Sitz. Ber. Gesell. Naturf. Freunde Ber. 1906 155–159.

Janicke, V. 1969. Untersuchungen über den Biotop der Solnhofener Plattenkalke. Mitt. Bayer. Staatsamml. Paläont. Hist. Geol. 9:117–181.

Jordan, D. S. 1907. The fossil fishes of California with supplementary notes on other species of extinct fishes. U. Calif. Pub. Bull. Dep. Geol. 5:95–144.

Kemp, N. R. 1978. Detailed comparisons of the dentitions of extant hexanchoid sharks and Tertiary hexanchid teeth from South Australia and Victoria, Australia (Selachii: Hexanchidae). Mem. Nat. Mus. Vict. 39:61–83.

Kemp, N. R. 1982. Chondrichthyans in the Tertiary of Australia, pp. 87–118. In: Rich, P. V., Thompson, E. M. (eds.), The fossil vertebrate record of Australia. Clayton: Monash U.

Koken, E. 1907. Ueber *Hybodus*. Geol. Palaeont. Abh. 5:261–276.

Lawley, R. 1875. Monografia del genere *Notidanus* rinvenuri allo stato fossile del Pliocene subappennino Toscano. Firenze.

Luther, A. 1908. Untersuchungen über die vom N. trigeminus innervierte Muskulatur der Selachier (Haie und Rochen). Acta Soc. Sci. Fenn. 36:1–168.

Mackie, S. J. 1863. On a new species of *Hybodus* from the lower chalk. Geologist. 6:332–347.

Maisey, J. G. 1980. An evaluation of jaw suspension in sharks. Amer. Mus. Nat. Hist. Novit. 2706: 1–17.

Maisey, J. G. 1982. The anatomy and interrelationships of Mesozoic hybodont sharks. Amer. Mus. Nat. Hist. Novit. 2724:1–48.

Maisey, J. G. 1983. Cranial anatomy of *Hybodus basanus* Egerton from the Lower Cretaceous of England. Amer. Mus. of Nat. Hist. Novit. 2758: 1–64.

Moss, S. A. 1972. The feeding mechanisms of sharks of the Family Carcharhinidae. J. Zool. 167:423–436.

Parker, T. J. 1891. On the presence of a sternum in *Notidanus indicus*. Nature 43:142, 516.

Pledge, N. S. 1967. Fossil elasmobranch teeth of

South Australia and their stratigraphic distribution. Trans. Roy. Soc. S. Aust. 91:135–160.

Quenstedt, F. A. 1858. Der Jura. Tübingen: Laupp' sche Buchhandlung.

Ridewood, W. G. 1921. On the calcification of the vertebral centra in sharks and rays. Trans. Roy. Soc. Lond. 210:311–407.

Romer, A. S. 1966. Vertebrate paleontology. Chicago and London: Univ. of Chicago Press.

Schaeffer, B. 1967. Comments on elasmobranch evolution. 3–35 In: Gilbert, P. W., Mathewson, R. F., Rall, D. P. (eds.), Sharks, skates and rays. Baltimore: John Hopkins Press.

Schaeffer, B. 1981. The xenacanth shark neurocranium, with comments on elasmobranch monophyly. Bull. Amer. Mus. Nat. Hist. 169: 3–66.

Schweizer, R. 1964. Die Elasmobranchier und Holocephalen aus den Nusplinger Plattenkalken. Palaeontographica Abt. 123:58–110.

Smith, B. G. 1942. The heterodontid sharks: their natural history, and the external development of *Heterodontus japonicus* based on notes and drawings by Bashford Dean. Amer. Mus. Nat. Hist., Bashford Dean Memorial volume: Archaic fishes, part II. 8:649–784.

Springer, S., Waller, R. A. 1969. *Hexanchus vitulus,* a new sixgill shark from the Bahamas. Bull. Mar. Sci. 19:159–174.

Thenius, E. 1973. Fossils and the life of the past. London: English Universities Press Ltd.

Thies, D. 1983. Jurazeitliche Neoselachier aus Deutschland und S. England. Courier Forschungsinstitut Senckenberg 58:1–116.

Vincent, G. 1876. Description de la faune de l'étage Landénien inférieur de Belgique. Ann. Soc. Malacol. Belg. 1:111–160.

Wagner, J. A. 1861. Monographie der fossilen Fische aus den Lithographischen Schiefern Bayerns: Plakoiden und Pyknodonten. Mun. Abh. Bay. Akad. Wiss. 9:138

Waldman, M. 1971. Hexanchid and Orthacondontid shark teeth from the Lower Tertiary of Vancouver Island, British Columbia. Canad. Jour. Earth Sci. 8:166–170.

Ward, D. J. 1979. Additions to the fish fauna of the English Palaeogene. 3. A review of the Hexanchid sharks with a description of four new species. Tert. Res. 2:111–129.

Welton, B. J. 1974. *Heptranchias howellii* (Reed, 1946), (Selachii-Hexanchidae) in the Eocene of the United States and British Columbia. Paleobioscience 117:1–15.

White, E. G. 1937. Interrelationships of the elasmobranchs with a key to the order Galea. Bull. Amer. Mus. Nat. Hist. 74:25–138.

White, P. J. 1895. A sternum in *Hexanchus griseus.* Anat. Anz. 11:222–224.

Woodward, A. S. 1886a. On the relations of the mandibular and hyoid arches in a Cretaceous shark (*Hybodus dubrisiensis* Mackie). Proc. Zool. Soc. Lond. 1886 218–224.

Woodward, A. S. 1886b. On the palaeontology of the Selachian Genus *Notidanus,* Cuvier. Geol. Mag., N. S. 3:205–217, 253–259.

Woodward, A. S. 1888. On the Cretaceous Selachian Genus *Synechodus.* Geol. Mag. 3:496–499.

Woodward, A. S. 1889. Catalogue of the fossil fishes in the British Museum (Natural History). Part I. Brit. Mus. (Nat. Hist.) Publ. Lond. 47:474.

Woodward, A. S. 1916. The fossil fishes of the English Wealden and Purbeck Formations. Part I. Mon. Palaeont. Soc. Lond. 1916 1–48.

19

Cephalocarida: Living Fossil Without a Fossil Record

Robert R. Hessler

Scripps Institution of Oceanography, A-002, La Jolla, CA 92093

General Background

The Cephalocarida (Sanders 1955) is a small class of crustaceans that has had major impact on phylogenetic studies of the subphylum. Among crustaceans, its morphology has been regarded as most primitive and thus most like that of the hypothetical crustacean ancestor (Sanders 1957, 1959a, 1959b, 1963a, 1963b; Hessler 1964, 1969; Cisne 1974, 1982; Hessler and Newman 1975; Manton 1977). As such, it has been especially influential in extrapolations on the origin of crustaceans and relationships with other taxa (Sanders 1957; Cisne 1974; Hessler and Newman 1975). Four aspects of the Cephalocarida loom large in these interpretations.

Cephalocarids display a greater degree of serial homeomorphy than any other crustacean. Considering the limbs, where this issue is particularly important, all the thoracopods are alike except for some sexual modification of the sixth and seventh, and reduction or loss of the eighth (Hessler and Sanders 1973). Most significantly, the second maxilla is nearly identical to the thoracopods. Similarities with the thoracic limb series occur in larvae on the first maxilla and, to decreasing degrees, on the mandible and second antenna (Sanders 1963a; Hessler 1964).

The postoral limbs are sufficiently generalized (both morphologically and functionally) that limbs of other crustaceans could be derived from their basic form (Sanders 1963a). The discovery of the cephalocarid mixopodium (Sanders 1963a) largely resolved the dispute over the phyllopodium versus stenopodium as the primitive morphology (Borradaile 1926; Siewing 1960; Hessler 1982).

Much of the cephalocarid functional plan is based on food acquisition by the thoracic limbs, followed by transport to the posteriorly directed mouth using protopodal endites (Sanders 1963a). The plesiomorphic nature of this life style is supported by its presence in branchiopods and leptostracan malacostracans (Sanders 1963a; Hessler and Newman 1975; Hessler 1982). This functional system is the most important piece of evidence linking crustaceans to trilobitomorphs and indirectly to chelicerates (Cisne 1974; Hessler and Newman 1975).

Finally, cephalocarids have the most gradual anamophic development known in the Crustacea (Sanders 1963a).

These interpretations on the phylogenetic position of the cephalocarids are not universally accepted (Schram 1982).

Fossil Record

There are no known fossil cephalocarids. The Pennsylvanian *Tesnusocaris goldichi* has been considered one by Brooks (1955), as has the Upper Cambrian *Dela peilertae* by Müller

(1981). In neither case are there sufficient similarities to justify this alignment (Hessler 1969; Schram 1982). Thus, cephalocarids may represent a primitive grade of evolution, but cannot be called "living fossils" in the strict sense. Even granting their primitive features, several cephalocarid attributes are specializations that might not apply to extinct representatives of the class. The lack of eyes and abdominal appendages are in this category. The fossil record of Crustacea extends back into the Cambrian (Schram 1982), and by current interpretations, it would not be unreasonable to find a cephalocarid-like form even in the earliest parts of the record.

Relation to Other Taxa

To date, there are nine crustacean classes: Cephalocarida, Branchiopoda, Malacostraca, Remipedia (Yager 1981), Copepoda, Branchiura, Mystacocarida, Cirripedia, and Ostracoda. Many attempts have been made to group them into larger phylogenetic clusters, but none have been satisfactory (Hessler 1982).

Cephalocarids have been related to branchiopods, either by submerging them in that taxon (Tiegs and Manton 1958) or clumping the two within the Gnathostraca (Dahl 1956; Siewing 1960). These interpretations were based on misidentification of the second maxilla (Sanders 1963a). Hessler and Newman (1975) suggested cephalocarids were most closely allied to branchiopods and malacostracans because of their multisegmented trunk and similar body plan and feeding mode. However, the resulting Thoracopoda must be rejected because it is based entirely on plesiomorphies.

No other special affinity of Cephalocarida with other classes has been proposed. Nor is there reason to believe that the other classes should be clumped in a group coordinate to the cephalocarids. Because of this uncertainty about the interrelationships of the crustacean classes, the sister-group of the cephalocarids is yet to be identified.

Taxonomic Composition

The nine known species (Table 1) are currently divided among four genera: *Hutchinso-niella* Sanders 1955, *Lightiella* Jones 1961a, *Sandersiella* Shiino 1965, and *Chiltonella* Knox and Fenwick 1977.

Cephalocarida is a conservative class; in general habitus all cephalocarids look alike. Body segment and podomere counts are constant. In general, even setal counts vary little. Character variation is sufficiently modest that only a single family is recognized. The biggest variation, one that distinguishes *Lightiella,* is presence or absence of the reduced eighth thoracopod. Other variations that differentiate genera include body slenderness, the size of pleura, presence and size of posterior abdominal combs, presence of an antennal knob, proportions of the penultimate endopodal segment of the second maxilla, endopodal claw number, setal count on exites of the second maxilla and anterior thoracopods, degree of modification of the shape of the sixth thoracopodal exites, and size of the caudal rami (Hessler and Sanders 1973; Knox and Fenwick 1977).

Within *Lightiella,* differences between species center on body size, shape of the labrum, presence of pleura on thoracomere 8, presence of dorsal spines on the telson, segment proportions in the antennae, dentition of the mandibular incisor process, limb setal counts, presence of a notch on thoracic limb exopods, and length of the caudal rami (Gooding 1963; Cals and Delamare Deboutteville 1970; McLaughlin 1976). The three species of *Sandersiella* have been differentiated on the basis of body size, presence of marginal hooks on the cephalic shield, size of posterior thoracic pleura, shape of the telsonic comb, pattern of thoracopodal claws, setal counts on the second antenna, maxillae and thoracopods, morphology of exites of thoracopod 6, shape of the pseudepipod of thoracopod 7, and size and shape of the distal exopodal segment of the last thoracopod.

Only a few species have been caught in sufficient numbers to reveal intraspecific variation. *Hutchinsoniella macracantha* has been taken frequently at Woods Hole, Massachusetts, but has never been subjected to careful variation analysis. However, it also has been collected from 300 m (Hessler and Sanders 1964) on the continental slope and as far away as Brazil (Wakabara 1970). The 300-m specimens had some slightly lower setal counts than those from Woods Hole. The variation reported by Wakabara is difficult to evaluate.

Table 1. Known distribution of the Cephalocarida.

Hutchinsoniella macracantha	Sanders 1955	NW Atlantic from Cape Cod to Chesapeake Bay (3, 6, 7, 14, 15)[a]; SW Atlantic from southern Brazil (19)[a]	1–300 m	Mud–muddy sands
Lightiella serendipita	Jones 1961	NE Pacific at San Francisco Bay (7, 8)[a]	1.0–1.4 m	Muddy sand
incisa	Gooding 1963	Eastern equatorial Atlantic (from Florida?) to W. Indies (5, 7, 16)[a]	Intertidal–2.5 m	*Thallassia* beds, muddy silt–sands
floridana	McLaughlin 1976	Gulf of Mexico at Florida (11, 13, 18)[a]	0.75–2.25 m	Macrophyte beds with fine sands–hard shelly sands
moniotae	Cals and Delamare Deboutteville 1970	South Pacific at New Caledonia (2, 7)[a]	4 m	Coraline sand
Sandersiella acuminata	Shiino 1965	NW Pacific at S. Japan (7, 9, 17)[a]	2–32 m	Mud–muddy sand
calmani	Hessler and Sanders 1973	Eastern equatorial Pacific at Peru (7)[a]	85 m	Mud
bathyalis	Hessler and Sanders 1973	SE Atlantic at Namibia (4, 7)[a]; SW Atlantic at S. Brazil (20)[a]	15 m	Mud
Chiltonella elongata	Knox and Fenwick 1977	SW Pacific at New Zealand (10)[a]	13–16 m	Mud
Genus undetermined Species A		NE Pacific at Anaheim Bay (12)[a]	Intertidal	*Spartina* beds, mud; gut contents of the goby *Clevelandia*
Species B		SW Atlantic off Rio de la Plata (1)[a]	256–293 m	Mud

[a] 1. Bastida 1973, 2. Cals and Delamare Deboutteville 1970, 3. Daugherty and Van Engel 1969, 4. Dinet 1973, 5. Gooding 1963, 6. Hessler and Sanders 1964, 7. Hessler and Sanders, 1973, 8. Jones 1961a,b, 9. Kikuchi 1969, 10, Knox and Fenwick 1977, 11. McLaughlin 1976, 12. Reish et al. 1975, 13. Saloman 1978, 14. Sanders 1955, 15. Sanders 1960, 16. Sanders and Hessler 1964, 17. Shiino 1965, 18. Stoner 1981, 19. Wakabara 1970, 20. Wakabara and Mizoguchi 1976

Limited setal variation is suggested in the descriptions of other species. In *L. incisa* and *L. serendipita,* the exopod of the second antenna has incomplete annulation that probably varies (Gooding 1963; Jones 1961a). The size of the caudal rami varies between two populations of *L. floridana* (Saloman 1978).

Wakabara and Mizoguchi (1976) collected a single specimen of *S. bathyalis* from 15 m off Brazil. Otherwise, the species has only been collected from bathyal depths on the other side of the Atlantic, off Namibia (Hessler and Sanders 1973; Dinet 1973). The setal differences and differences in morphology of the exopod of thoracopod 6 and endopod of thoracopod 7 are minimized by Wakabara and Mizoguchi, but in view of the great depth difference and distance between the two localities, the Brazilian speci-

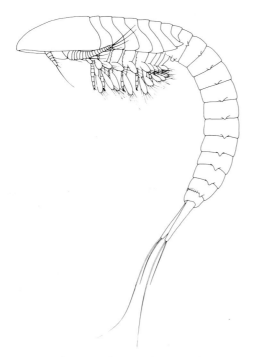

Fig. 1. The cephalocarid *Hutchinsoniella macracantha* (from Hessler and Newman 1975).

men probably belongs to a different species. Similarly, the slight differences in *H. macracantha* along its range could signify distinct species, as is known to be the case in mystacocarids (Hessler 1972).

Geographic Distribution

Several species have been collected in more than one locality (Table 1), but only two display a broad range. *Hutchinsoniella macrancantha* is known from Cape Cod (41.5°N) to southern Brazil (23°S). *Sandersiella bathyalis* is reported from both sides of the South Atlantic. As already mentioned, the breadth of these ranges may result from lumping species. *Lightiella incisa* is reported from the West Indies and Florida (Hessler and Sanders 1973), but the single larva from the Florida locality could belong to *L. floridana* as well.

Cephalocarids, being small and transparent, are difficult to detect. The recentness of their discovery and the steady subsequent discovery of new species and localities suggest that many future sightings are likely. The worldwide distribution of localities (Hessler and Sanders 1973; Abele 1982), even including the deep sea (Table 1), makes it reasonable to suspect their distribution is patchily cosmopolitan. Few areas, such as Europe, have been so well studied that it is safe to predict they will not be found there.

Habitat

Cephalocarids have only been collected from sedimentary substrates. They are regarded as being limited to the surficial zone, although they have been taken rarely from thalassinid burrows (Gooding 1963). Most localities are subtidal, at depths less than 100 m, but specimens have been taken intertidally (Gooding 1963) and bathyally, at depths of 300 m (Hessler and Sanders 1964) and 1200–1600 m (Hessler and Sanders 1973; Dinet 1973).

Most localities have had sediment types ranging from muds to silty sands (Cals and Delamare Deboutteville 1970; Gooding 1963; Hessler and Sanders 1973; Jones 1961a,b; Kikuchi 1968, 1969; Knox and Fenwick 1977; McLaughlin 1976; Reish et al. 1975; Sanders 1955, 1960; Sanders and Hessler 1964; Wakabara 1970; Wakabara and Mizoguchi 1976). In some cases, marine grasses grew on the surface (Gooding 1963; McLaughlin 1976; Reish et al. 1975; Sanders and Hessler 1964; Stoner 1981). In general, the organic content of these sediments has been high, and Sanders (1963a; Hessler and Sanders 1973) has postulated that the life style of cephalocarids restricts them to environments with fine-grained, loosely settled, organic detritus on the sediment surface (his "flocculent zone"). Saloman (1978) collected cephalocarids from a "hard, firm gray shelly sand," one with a high percent of organics, but without the flocculent layer. Assuming his analysis of the environment is correct, our understanding of the basic environmental limitations on cephalocarids is still incomplete. A careful study of living cephalocarids under natural circumstances or a microdistributional study has yet to be made.

Cephalocarids are not found in environmentally extreme circumstances, and where known, the associated fauna is rich in species and numbers (for example, Sanders 1960), including potential competitors and predators. Cephalocarid fecundity is very low compared to most marine invertebrates (belying Schram's [1982] suggestion that they are *r*-selected species). At Woods Hole, an individual (they are synchronous hermaphrodites [Hessler et al. 1970]) probably produces only about six young each year (Sanders 1957). Life span is unknown. One must conclude that while they are primitive and generalized, they are quite well adapted to survive in a dynamic marine habitat, not needing the protection of geographic or ecological refuge conditions.

Literature

Abele, L. G. 1982. Crustacean biogeography, pp. 241–304. In: Abele, L. G. (ed.), Biology of the Crustacea, Vol. 1. New York: Academic.

Bastida, R. 1973. On the finding of Crustacea Cephalocarida off the Argentine Uruguayan coast. Physis, sect. A 32(84):220.

Borradaile, L. A. 1926. Notes upon crustacean limbs. Ann. Mag. Nat. Hist., Ser. 9 17:193–213.

Brooks, H. K. 1955. A crustacean from the Tesnus Formation (Pennsylvanian) of Texas. J. Paleont. 29:852–856.

Cals, P., Delamare Deboutteville, C. 1970. Une nouvelle espèce de Crustacé Cephalocaridé de l'Hemisphere austral. C. R. Acad. Sci. Paris 270:2444–2447.

Cisne, J. L. 1974. Trilobites and the origin of arthropods. Science 186:13–18.

Cisne, J. L. 1982. Origin of the Crustacea, pp. 65–92. In: Abele, L. G. (ed.), Biology of the Crustacea, Vol. 1. New York: Academic.

Dahl, E. 1956. Some crustacean relationships, pp. 138–147. In: Bertil Hanstrom, Zoological papers in honour of his sixty-fifth birthday, 20 November 1956.

Daugherty, S. J., Van Engel, W. A. 1969. Record of *Hutchinsoniella macracantha* Sanders, 1955 (Cephalocarida) in Virginia. Crustaceana 16(1):107, 108.

Dinet, A. 1973. Distribution quantitative du meiobenthos profond dans la region de la dorsale de Walvis (Sud-Ouest Africain). Mar. Biol. 20(1):20–26.

Gooding, R. U. 1963. *Lightiella incisa,* sp. nov. (Cephalocarida) from the West Indies. Crustaceana 5:293–314.

Hessler, A. Y., Hessler, R. R., Sanders, H. L. 1970. Reproductive system of *Hutchinsoniella macracantha.* Science 168:1464.

Hessler, R. R. 1964. The Cephalocarida: comparative skeletomusculature. Mem. Conn. Acad. Arts Sci. 16:1–97.

Hessler, R. R. 1969. Cephalocarida, pp. 120–128. In:

Moore R. C. (ed.), Treatise on invertebrate paleontology. Part R (Arthropoda 4). Geol. Soc. Amer. and U. Kansas Press, Lawrence, KS.

Hessler, R. R. 1972. New species of Mystacocarida from Africa. Crustaceana 22:259–273.

Hessler, R. R. 1982. Evolution within the Crustacea, pp. 149–185. In: Abele, L. G. (ed.), Biology of the Crustacea, Vol. 1. New York: Academic.

Hessler, R. R., Newman, W. A. 1975. A trilobitomorph origin for the Crustacea. Fossils and Strata 4:437–459.

Hessler, R. R., Sanders, H. L. 1964. The discovery of Cephalocarida at a depth of 300 meters. Crustaceana 7:77–78.

Hessler, R. R., Sanders, H. L. 1973. Two new species of *Sandersiella* (Cephalocarida), including one from the deep sea. Crustaceana 24:181–196.

Jones, M. L. 1961a. *Lightiella serendipita* gen. nov., sp. nov., a cephalocarid from San Francisco Bay, California. Crustaceana 3:31–46.

Jones, M. L. 1961b. A quantitative evaluation of the benthic fauna off Point Richmond, California. U. Calif. Publ. Zool. 67:219–320.

Kikuchi, T. 1968. The zoological environment of the Amakusa Marine Biological Laboratory. Publ. Amakusa Mar. Biol. Lab., Kyushu U. 1(2):117–127.

Kikuchi, T. 1969. New locality of *Sandersiella acuminata* Shiino (Crustacea, Cephalocarida). Publ. Amakusa Mar. Biol. Lab., Kyushu U. 2(1):33–36.

Knox, G. A., Fenwick, G. D. 1977. *Chiltoniella elongata* n. gen. et sp. (Crustacea: Cephalocarida) from New Zealand. J. Roy. Soc. N.Z. 7:425–432.

Manton, S. M. 1977. The Arthropoda. Clarendon (Oxford), xx, 527 pp.

McLaughlin, P. A. 1976. A new species of *Lightiella* (Crustacea; Cephalocarida) from the west coast of Florida. Bull. Mar. Sci. 26:593–599.

Müller, K. J. 1981. Arthropods with phosphatized soft parts from the Upper Cambrian "Orstein" of Sweden, pp. 147–151. In: Taylor, J. M. (ed.), Second International Symposium on the Cambrian System. U.S. Geol. Surv., Open File Rept. #81-743.

Reish, D. J., Kauwling, T. J., Schreiber, T. C. 1975. Annotated checklist of the marine invertebrates of Anaheim Bay. Calif. Dept. Fish and Game, Fish Bull. 165:41–51.

Saloman, C. H. 1978. Occurrence of *Lightiella floridana* (Crustacea: Cephalocarida) from the west coast of Florida. Bull. Mar. Sci. 28:210–212.

Sanders, H. L. 1955. The Cephalocarida, a new subclass of Crustacea from Long Island Sound. Proc. Nat. Acad. Sci. 41:61–66.

Sanders, H. L. 1957. The Cephalocarida and crustacean phylogeny. Syst. Zool. 6(3):112–129.

Sanders, H. L. 1959a. The significance of the Cephalocarida in crustacean phylogeny. XV Int. Congr. Zool. Lond. Proc. 1958:337–340.

Sanders, H. L. 1959a. New light on the crustaceans. Nat. Hist. Mag. 68:86–91.

Sanders, H. L. 1960. Benthic studies in Buzzards Bay. III. The structure of the soft-bottom community. Limnol. Oceanogr. 5:138–153.

Sanders, H. L. 1963a. The Cephalocarida: functional mophology, larval development, and comparative external anatomy. Mem. Conn. Acad. Arts Sci. 15:1–80.

Sanders, H. L. 1963b. Significance of the Cephalocarida, pp. 163–176. In: Whittington, H. B., Rolfe, W. D. I. (eds.), Phylogeny and evolution of Crustacea. Spec. Publ. Mus. Comp. Zool. Harv. Cambridge: Harvard U. Press.

Sanders, H. L., Hessler, R. R. 1964. The larval development of *Lightiella incisa* Gooding (Cephalocarida). Crustaceana 7:81–97.

Schram, F. R. 1982. The fossil record and evolution of the Crustacea, pp. 93–147. In: Abele, L. G. (ed.), Biology of the Crustacea, Vol. 1. New York: Academic.

Shiino, S. M. 1965. *Sandersiella acuminata* gen. et sp. nov., a cephalocarid from Japanese waters. Crustaceana 9:181–191.

Siewing, R. 1960. Neuere Ergebnisse der Verwandschaftsforschung bei den Crustaceen. Wiss. Zeitschr. U. Rostock, Math.-Nat. Reihe 9:343–358.

Stoner, A. W. 1981. Occurrence of the cephalocarid crustacean *Lightiella floridana* in the northern Gulf of Mexico with notes on its habitat. N.E. Gulf Sci. 4(2):105–107.

Tiegs, O. W., Manton, S. M. 1958. The evolution of the Arthropoda. Biol. Rev. 33:255–337.

Wakabara, Y. 1970. *Hutchinsoniella macracantha* Sanders, 1955 (Cephalocarida) from Brazil. Crustaceana 19:102–103.

Wakabara, Y., Mizoguchi, S. M. 1976. Record of *Sandersiella bathyalis* Hessler and Sanders, 1973 (Cephalocarida) from Brazil. Crustaceana 30:220–221.

Yager, J. 1981. A new class of Crustacea from a marine cave in the Bahamas. J. Crust. Biol. 1:328–333.

20
Leptostraca as Living Fossils

Robert R. Hessler and Frederick R. Schram

Scripps Institution of Oceanography, A-002, La Jolla, CA 92093
Natural History Museum, San Diego, CA 92112

General Background

The Leptostraca is an order of malacostracan crustaceans that has always been prominent in discussions of crustacean phylogeny. Their importance stems from two issues. Some aspects of leptostracan morphology bridge the gap between the vast class Malacostraca and other crustaceans, particularly primitive ones (see below). They are the only living representatives of the Phyllocarida, a group whose fossil record extends back to the Early Paleozoic. *Nebalia* is the best known genus, but the order shows so little variability that Leptostraca as a whole is the proper unit for consideration in the context of this book.

The Malacostraca was traditionally divided into two subclasses, Phyllocarida and Eumalacostraca (Calman 1909; Hessler 1983). There is, however, an increasing tendency to separate the stomatopods out of the Eumalacostraca as a separate subclass Hoplocarida (Schram 1969; Bowman and Abele 1982); this issue need not be aired in the present context. Viewed from the perspective of the eumalacostracans *sensu lato,* whose diversity and ecological importance overwhelmingly dominate within the class, the gulf that separates malacostracans from the other ten classes has seemed so wide that one long-surviving classification combined all the non-malacostracan classes into the Entomostraca (Latreille 1806), a taxon that was considered equivalent in status to the Malacostraca.

The Leptostraca occupies an intermediate position, as is highlighted by the fact that it was once considered a branchiopod rather than a malacostracan (for example, Sars 1896). The important characters with which leptostracans span the "entomostracan"–malacostracan gap are: (1) presence of well-developed caudal rami; (2) foliaceous thoracopods similar to the cephalocarid mixopodium (Sanders 1963); (3) mode of feeding fundamentally like that of cephalocarids and primitive branchiopods (Sanders 1963; Hessler 1982); (4) details of internal anatomy (Siewing 1960; Hessler 1964); and (5) presence of one more abdominal somite than in other malacostracans. Considerable controversy attends the possible primitiveness of the leptostracan carapace (Dahl 1976; Lauterbach 1974; Schram 1982; Hessler 1983). Some of these characters bespeak a grade of evolution close to the base of the Crustacea (Hessler and Newman 1975).

Leptostracans are, however, disappointingly uninformative about the origins of eumalacostracan synapomorphies. Leptostracans do not herald the thoracic stenopodium, the caridoid abdomen, or antennal scale (Calman 1909). Further, they display several clear autapomorphies: no thorocopodal locomotion; scale on first antenna; uniramous second antenna; vestigal fifth and sixth pleopods; direction of water flow under the carapace; and epimorphic development. These features underline the conclusion that leptostracans are highly derived in their own right and not on the stem line leading to the higher malacostracans.

Affinities and Fossil Record

The subclass Phyllocarida is currently divided into five orders (Rolfe 1981): Leptostraca Claus 1880, Archaeostraca Claus 1888, Hymenostraca Rolfe 1969, Hoplostraca Schram 1973, and Canadaspidida Briggs 1978. Unity of the subclass is based on possession (where known) of a caudal furca, bivalved carapace, and abdomen with seven somites in addition to the telson. Prior to elucidation and erection of the Hoplostraca and Canadaspidida, the subclass was also characterized by a movable articulated rostral plate (Rolfe 1969). However, canadaspids may not actually be malacostracans (Hessler 1983).

The taxonomic unity of the Phyllocarida is not above suspicion. The caudal rami and carapace are probably plesiomorphic features in the Malacostraca as a whole. The movable rostrum may also be a malacostracan plesiomorphy, but this is also found in hoplocarids, causing some to believe that it is an apomorphy that links the Phyllocarida and Hoplocarida. The only character remaining to unify the subclass is abdominal somite count. However, even this is not known with certainty in the Hymenostraca and many archaeostracans. Therefore, if leptostracans are to be thought of as "living fossils," it should be remembered that the group of which they are the living example may not be a real taxon, or at least one that can be defined with a series of unambiguous derived characters. Even if the Phyllocarida finally proves to be a unified taxon, the relationship of the Leptostraca to the other orders remains obscure.

The only fossil ascribed to the Leptostraca is *Rhabdouraea bentzi* (Malzahn 1962), from the Upper Permian of Germany (Schram and Malzahn, 1984). The species is based on the posterior portion of one individual, showing a corner of the carapace, the abdominal somites, pleopods, telson, and distinctive, long, rod-like caudal rami. Rolfe (1969) relegated a supposed carapace of this species to the cumaceans. The relationship of *Rhabdouraea* to other leptostracans is obscure, and the genus occupies its own family.

Taxonomic Composition of Living Forms

To date, four genera have been recognized: *Nebalia* Leach 1814, *Paranebalia* Claus 1880, *Nebaliopsis* G. O. Sars 1887, and *Nebaliella* Thiele 1904. These have been included in one family, Nebaliidae, but Hessler (in press) will place *Nebaliopsis* in its own family, based on numerous differences that result from its holopelagic habits. Recent exploration of unusual habitats has yielded two additional genera, one from deep-sea hydrothermal vents (Hessler, in press), the other from marine caves (Bowman, personal communication).

Genera are sharply differentiated from each other on the basis of characters of the eyestalk, rostral keel, blade on the distal article of the antennular peduncle, antennal peduncle segmentation, endopod segmentation on the maxilla, and size and shape of thoracopodal endopod, exopod, and epipod (Thiele 1904; Cannon 1960). Other minor differences exist, and many special features characterize *Nebaliopsis*.

At the species level leptostracan taxonomy is chaotic for several important reasons. They have not been collected systematically, and therefore known distributions are patchy and based on few individuals. Leptostracan morphology is conservative, and analyses of individual variation have not been made. For example, until recently, sexual dimorphism was not fully appreciated (Wägele 1983). Developmental changes are rarely examined, and variation within an instar has not been studied. Thus, no one truly understands what set of characters will allow species differentiation.

Currently, one species is recognized in *Nebaliopsis* (*N. typica* G. O. Sars 1887), two in *Paranebalia* [*P. longipes* (Willemoës-Suhm 1875); *P. fortunata* Wakabara 1976], three in *Nebaliella* (*N. antarctica* Thiele 1904; *N. extrema* Thiele 1905; *N. caboti* Clark 1932), and nine in *Nebalia* [*N. bipes* (Fabricius 1780); *N. geoffroyi* Milne Edwards 1828; *N. typhlops* G. O. Sars 1869; *N. longicornis* Thomson 1879;

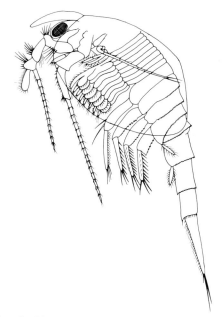

Fig. 1. The leptostracan *Nebalia bipes* drawn as though the carapace were transparent (modified from Calman 1909).

N. japanensis Claus 1888, Jankowski 1973; *N. capensis* Barnard 1914; *N. pugettensis* (Clark 1932); *N. ilheoensis* Kensley 1976; *N. marerubri* Wägele 1983]. Several subspecies of *Nebalia* have been suggested, but are rarely used. Even customarily recognized species are not on firm footing. Some of the older species of *Nebalia* have been at times called junior synonyms of others. Specimens from new localities are frequently misidentified. For example, *Nebalia bipes* from Scandinavia is more than one species (Dahl, personal communication), a problem that applies to many other species as well. Finally, many new localities harbor undescribed species. As often stated before, leptostracan species taxonomy badly needs reworking.

Distribution

Making allowances for inadequate sampling in many areas, the Leptostraca as a whole is cosmopolitan. Thiele (1905), Brattegard (1970), and Abele (1982) give summaries. *Nebaliopsis*

typica is bathy- and abysso-pelagic and is patchily widespread in the world ocean (Linder 1943). *Nebalia typhlops* (Vader 1973) and *Nebaliella* spp., all blind species, are found primarily in the deep sea, but may occur more shallowly where the permanent thermocline is poorly developed, as at high latitudes or in the Mediterranean.

The other species are all from shallow, coastal environments. *Paranebalia longipes* is reported from Bermuda, the Caribbean, Japan, and the Malay Archipelago. This unlikely species distribution highlights the inadequacy of leptostracan taxonomy. Indeed, additional specimens from the latter two areas prove to be a different species (Dahl, in preparation). One would not expect highly disjunct species distributions among shallow, benthic leptostracans, for they display direct development and leave the mother's brood chamber as juveniles adapted for a benthic existence; they have no planktonic dispersal stage. *Paranebalia fortunata* is known from New Zealand (Wakabara 1976). Another undescribed species comes from Chile.

Shallow-living species of *Nebalia* reported from more than one locality tend to be broadly distributed, but display latitudinal zonation. *Nebalia bipes* appears in Arctic and boreal Atlantic waters. Less likely records include the Red Sea, Ceylon, and Japan. *Nebalia pugettensis* borders the North Pacific. *Nebalia geoffroyi* is northern tropical and subtropical, including Puerto Rico, Madeira, the Mediterranean, and the Red Sea. *Nebalia longicornis* is a circumglobal, south temperate to antarctic species, but has also been reported from India and Cuba. Species known from restricted localities are *N. capensis* from South Africa, *N. ilheoensis* from Namibia, and *N. marerubri* from the Red Sea. Unidentified *Nebalia* are available from numerous new localities.

Except for *Nebaliopsis*, leptostracans are all benthic and demersal. They are general detritovores or scavengers and are known to be effective large carrion feeders (Nishimura and Hamebe 1964). They are frequently found where oxygen tensions are low, such as in rotting seaweed.

In summary, the portrayal of leptostracans as living fossils has serious difficulties. The taxonomic status of the fossil subclass they represent is uncertain. In many ways they are morphologically, physiologically, and developmentally specialized. However, they are taxonomically isolated, and thus represent the earliest branching of which living malacostracans give knowledge.

Literature

Abele, L. G. 1982. Biogeography, pp. 241–304. In: Abele, L. G. (ed.), The biology of the Crustacea, Vol. 1. New York: Academic.

Barnard, K. H. 1914. Contributions to the crustacean fauna of South Africa. 4. A new species of *Nebalia*. Ann. S. Afr. Mus. 10:443–446.

Bowman, T. E., Abele, L. G. 1982. Classification of the Recent Crustacea, pp. 1–27. In: Abele, L. G. (ed.), The Biology of the Crustacea, Vol. 1. New York: Academic.

Brattegard, T. 1970. Marine biological investigations in the Bahamas. 13. Leptostraca from shallow water in the Bahamas and southern Florida. Sarsia 44:1–8.

Briggs, D. E. G. 1978. The morphology, mode of life, and affinities of *Canadaspis perfecta* (Crustacea, Phyllocarida), Middle Cambrian, Burgess Shale, British Columbia. Philos. Trans. Roy. Soc. London, Ser. B. 281:439–487.

Calman, W. T. 1909. Crustacea. In: Lankaster, R. (ed.), A treatise on zoology, Part 7. London: A. and C. Black.

Cannon, H. G. 1960. Leptostraca. Bronn's Klassen und Ordnungen des Tierreichs 5. Bd., 1. Abt., 4. Buch, I Teil:1–81.

Clark, A. E. 1932. *Nebaliella caboti* n. sp., with observations on other Nebaliacea. Trans. Roy. Soc. Canada 26:217–235.

Claus, C. 1880. Grundzüge der Zoologie, 4th ed. 2:576.

Claus, C. 1888. Ueber den Organismus der Nebaliden und die systematische Stellung der Leptostraken. Arb. Zool. Inst. Wien 8:1–149.

Dahl, E. 1976. Structural plans as functional models exemplified by the Crustacea Malacostraca. Zool. Scripta 5:163–166.

Fabricius, O. 1780. Fauna Groenlandica. Leipzig.

Hessler, R. R. 1964. The Cephalocarida: comparative skeletomusculature. Mem. Conn. Acad. Arts Sci. 16:1–97.

Hessler, R. R. 1983. A defense of the caridoid facies, wherein the early evolution of the Eumalacostraca is discussed, pp. 145–164. In: Schram, F. R. (ed.), Crustacean phylogeny, Rotterdam: A. A. Balkema.

Hessler, R. R. in press. *Dahlella caldariensis* n. gen., n. sp.: leptostracan (Crustacea, Malacostraca) from deep-sea hydrothermal vents. Jour. Crustacean Biol.

Hessler, R. R., Newman, W. A. 1975. A trilobitomorph origin for the Crustacea. Fossils and Strata 4:437–459.

Jankowski, A. W. 1973. Recent phyllocarids (Leptostraca). 1. *Nebalia japanensis*, a forgotten species. Zool. Zhur. 52:598–601.

Kensley, B. 1976. The genus *Nebalia* in South and South West Africa (Crustacea, Leptostraca). Cimbebasia Ser. A 4(8):155–162.

Latreille, P. A. 1806. Genera Crustaceorum et Insectorum secundum ordinem naturalem in familias disposita, iconibus exemplisque plurimis explicata. König: Parisii and Argentorati.

Lauterbach, K.-E. 1974. Über die Herkunft des Carapax der Crustaceen. Zool. Beitr. N.S. 20:273–327.

Linder, F. 1943. Über *Nebaliopsis typica* G. O. Sars nebst einigen allgemeinen Bemerkungen über die Leptostracen. Dana Rept. 25:1–38.

Malzahn, E. 1962. Beschreibung der Arten, Teil 1. In: Glaessner, M. R., Malzahn, E. (eds.), Neue Crustaceen an dem niederrheinischen Zechstein. Forstschr. Geol. Rheinld. u. Westfal. 6:245–264.

Milne Edwards, H. 1828. Mémoire sur quelques Crustacés nouveaux. Ann. Sci. Nat. 13:299–300.

Nishimura, S., and Hamebe, M. 1964. A case of economical damage done by *Nebalia*. Publ. Seto Mar. Biol. Lab. 12(14): 173–175.

Rolfe, W. D. I. 1969. Phyllocarida, pp. R296–R331. In: Moore, R. C. (ed.), Treatise on invertebrate paleontology, Part R. Arthropoda 4: Crustacea (except Ostracoda), Myriapoda, Hexapoda 1. Geol. Soc. Amer. and U. Kansas Press, Lawrence, Ks.

Rolfe, W. D. I. 1981. Phyllocarida and the origin of the Malacostraca. Geobios no. 14, fasc. 1:17–27.

Sars, G. O. 1869. Nye Dybvandscrustaceear fra Lofoten. Forhandlinger i Videnskabs-Selskabet i Christiana, Aar 1869:147–174.

Sars, G. O. 1887. Report on the Phyllocarida collected by the H.M.S. Challenger during the years 1873–76. Challenger Repts. 19.

Sars, G. O. 1896. Phyllocarida and Phyllopoda. Fauna Norvegiae 1.

Sanders, H. C. 1963. The Cephalocarida: functional morphology, larval development, comparative external anatomy. Mem. Conn. Acad. Arts Sci. 15:1–80.

Schram, F. R. 1969. Some Middle Pennsylvanian Hoplocarida (Crustacea) and their phylogenetic significance. Fieldiana, Geol. 12:235–289.

Schram, F. R. 1973. On some phyllocarids and the origin of the Hoplocarida. Fieldiana, Geol. 26:77–94.

Schram, F. R. 1982. The fossil record and evolution of Crustacea, pp. 93–147. In: Abele, L. G. (ed.), The biology of the Crustacea, Vol. 1, New York: Academic.

Schram, F. R., Malzahn, E. 1984. The fossil leptostracan *Rhabdouraea bentzi* (Malzahn), 1958. Trans. San Diego Soc. Nat. Hist. 20:95–98.

Siewing, R. 1960. Neuere Ergebnisse der Verwandtschaftsforschung bei den Crustaceen. Wiss. Zeitschr. U. Rostock, Math.-Naturwiss. Reihe 9:343–358.

Thiele, J. 1904. Die Leptostraken. Wiss. Ergebnisse der Deutschen Tiefsee-Exped. auf dem Dampfer "Valdivia" 1898–1899 8.

Thiele, J. 1905. Über die Leptostraken des Deutschen Südpolar-Expedition 1901–1903. Deutsche Südpolar-Exped. 1901–1903, Vol. 9. Berlin: Zool. 1:59–68.

Thomson, G. W. 1879. On two new isopods and a new *Nebalia* from New Zealand. Ann. Mag. Nat. Hist. (5)4:415–419.

Vader, W. 1973. *Nebalia typhlops* in western Norway (Crustacea, Leptostraca). Sarsia 53:25–28.

Wägele, J. W. 1983. *Nebalia marerubi,* sp. nov. aus dem Roten Meer (Crustacea: Phyllocarida: Leptostraca). J. Nat. Hist. 17:127–138.

Wakabara, Y. 1976. *Paranebalia fortunata* n. sp. from New Zealand (Crustacea, Leptostraca, Nebaliacea). J. Roy. Soc. N.Z. 6(3):297–300.

Willemoës-Suhm, R. v. 1875. On some Atlantic Crustacea from the "Challenger Expedition." Trans. Linn. Soc. London, ser. 2. Zool. 1:23–59.

21
Anaspidid Syncarida

Frederick R. Schram and Robert R. Hessler

Natural History Museum, San Diego, CA 92112
Scripps Institution of Oceanography, La Jolla CA 92093

The Syncarida are a moderately diverse array of living eumalacostracan crustaceans that originated in the Early Carboniferous. There are several distinct groups recognized within the syncarids. The Palaeocaridacea were a Permo-Carboniferous radiation. The Bathynellacea are small groundwater forms of rather aberrant morphology, worldwide in distribution but with no fossil record. The Anaspidacea are relatively large animals that bear a very close resemblance to some of the palaeocaridaceans, occupy surface water, and are found only in South America, New Zealand, Australia, and Tasmania.

Schminke (1975) and Knott and Lake (1980) provide the most recent survey of phylogeny and taxonomy of the syncarids (Fig. 1). Within the Anaspidacea, the Anaspididae are closely related to another family, the Koonungidae, in the Anaspidinea characterized by appendages on all segments including annulate pleopods, a maxillary palp, females with a prominent seminal receptacle, a short second endopod of the male petasma, and a tailfan. These in turn are a sister-group to two other families, the Psammaspididae and Stygocarididae, which comprise the Stygocaridinea characterized by a reduction of abdominal appendages, no maxillary palp, a small female seminal receptacle, a long second endopod of the male petasma, and no tailfan. The Anaspidacea are in turn considered a sister-group to the extinct Palaeocaridacea,

the latter in contrast to the former having a free first thoracomere, a reduced first thoracopod, the eighth thoracopod oriented similarly to the other thoracopods, and apparently lacking a petasma and seminal receptacle.

The morphological differences between the anaspidines (Fig. 2A) and the divergent stygocaridines (Fig. 2B), which lack a pleopod series and tailfan, are greater than those that exist between the anaspidines and the extinct palaeocaridaceans (Fig. 2C), which both have complete appendage series and tailfans. If the fusion of the first thoracopod into the head were complete in a form like *Praeanaspides praecursor,* it would make a fairly good anaspidid (Schram 1979); whereas, stygocaridines appear to be the result of extensive paedomorphosis of the basic anaspidacean body plan (Schminke 1978).

There are four living families of Anaspidacea, but only the Anaspididae are of interest in the context of "living fossils" since they bear the closest resemblance to the extinct Palaeocaridacea. Only the anaspidids have a fossil record, albeit meager: *Anaspidites antiquus* occurs in the Triassic of New South Wales, and what little is known of that species is barely distinguishable from living forms like *Anaspides tasmaniae.* No other fossil syncarids are known at present to occur between the Triassic and Recent. The four living species in three genera *(Anaspides tasmaniae, Paranaspides lacustris,*

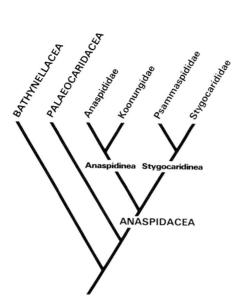

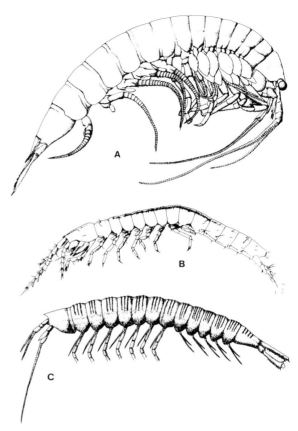

Fig. 1. Cladistic relationships of syncarids (derived from Schminke[1975] and Knott and Lake [1980]).

Fig. 2. Some representative syncarid types. (A) *Anaspides tasmaniae,* an anaspidid anaspidacean. (B) *Stygocarella pleotelson,* a stygocarid anaspidacean (A and B from Schminke [1978]). (C) *Praeanaspides praecursor,* a squillitid palaeocaridacean (from Schram [1979]).

Allanaspides helonomus, and *A. hickmani)* all conform to a single basic plan and occur only in Tasmania.

The nature of the syncarid radiation varies depending on viewpoint. Modern syncarids as a whole are physiological specialists to fresh water. This specialization appears to be a feature generally characteristic of the group, since the Paleozoic and Triassic taxa are restricted to situations that represent freshwater, or at best, brackish water habitats. In addition, most anaspidaceans are ecological specialists; species of three of the four living families occupy either quasi-groundwater or limited transitory surface pools. However, the fourth family, the Anaspididae, are ecological generalists that can occupy surface pools *(Allanaspides),* lakes *(Paranaspides* and *Anaspides),* and streams and caves *(Anaspides).* One species, *Anaspides tasmaniae,* occupies a variety of habitats and exhibits considerable morphological diversity within subspecies types. This versatility is perhaps especially relevant in the context of living fossils, since the diversity and generalism is exhibited by the one species that is most closely aligned in morphology with the Triassic and Paleozoic fossils. However, *A. tasmaniae* is the only species to date in which syncarid phenotypic variability has been studied.

An analysis of syncarid biogeographic history (Schram 1977) reveals that the nonbathynellaceans had a particularly diverse radiation on the Late Paleozoic tropical island continent of Laurentia. Subsequently, with the formation of the supercontinent Pangaea in the Permian, the syncarids were able to disperse to other regions of the world and radiate further. The evolution of the anaspidaceans has been subsequently restricted to their Gondwanan refugium.

Syncarids have been remarkably effective in their radiation where they have been able to maintain themselves. The great constraining factor in their evolution is that they have come to occupy ecologic and geographic refugia. The aberrant bathynellaceans (which generally do not enter into discussions of living fossils) are ubiquitous but restricted to groundwater habitats. Though the anaspidaceans have a very limited distribution, they can be found within many different habitats within their geographic limits. Modern anaspidaceans are more diverse than the Paleozoic palaeocaridaceans only because we are able to sample more habitats containing modern syncarids than we can hope for in the fossil record. For example, we do not have fossils that occur in situations that represent small, surface, upland pools, nor do we have fossils from situations that represent true groundwater habitats.

Contrary to explanations sometimes advanced for refugial existence, Anaspididacea are not restricted because of inability to compete with better adapted forms. The general diversity exhibited within their range, coexisting as they do with a diverse array of caridean and astacidean decapods, isopods, and amphipods in Australian and Tasmanian waters, seem to disprove any such theories of competition. Despite potential competition from these other eumalacostracan groups, the anaspidaceans, and the anaspidids in particular, have held their own.

Rather, the anaspidids seem to have been unable to adapt to predation pressures. In those parts of Tasmania where European settlers have introduced trout, anaspidids have declined or are absent altogether. The anaspidids occur today on Tasmania in drainage systems in the south and west of the island that have been effectively isolated since Permo-Triassic time and have few or no introduced predators from European settlement. Unlike most other eumalacostracans, the anaspidids do not execute a sustained caridoid escape reaction, i.e., they do not flick their tails under the body repeatedly to achieve a quick, posteriorly directed retreat. Anaspidids, when startled or probed, will flick the abdomen once to project themselves effectively into the water column. Once there, however, they become easy targets for predators. *Paranaspides lacustris* rights itself in the water and then slowly swims back to the bottom. *Anaspides tasmaniae* is even more vulnerable, because once in the water column it drifts helplessly until gravity returns it to the bottom. Anaspidids for this reason are easy to collect in the field, by simply sweeping the water with a net; and likewise, a potential predator has no trouble gorging itself on syncarids. The other anaspidacean families (as well as the Bathynellacea) have reduced body size and moved into

transitory surface water or groundwater habitats where any lack of a sustained escape reaction is no disadvantage.

Anaspidid syncarids appear to offer only limited pedagogical value as living fossils as defined by Stanley (1979). Though they can be related directly to Paleozoic forms and contain ecological, if not physiological, generalists, anaspidids are not taxonomically isolated as are many bradytelic lines. Anaspidids do have closely allied sister-groups that exhibit considerable morphological divergence from the basic type. In these terms, the Anaspididae proper only approximate a living fossil condition to the extent that they have a meager fossil record and are morphologically generalized. In regard to the latter, they are of interest to the phylogeneticist because of such generalized eumalacostracan features as an antennal scaphocerite, a pediform first thoracopod (not modified as a maxillipede), thoracopods with a full complement of rami (endopods, expopods, as well as epipodites), a full set of locomotory pleopods, and a caridoid abdomen. In addition, Schram feels the lack of a carapace may be important in this regard. Therefore, the phylogenetic significance of anaspidids is as morphologically generalized forms rather than as fortuitous survivors within the confines of certain restricted parameters of geography, ecology, and past history.

Anaspidid syncarids are an example of a general case of which living fossils in Stanley's sense are only an exception. They are important because they have survived with little morphological change. That anaspidids are closely related to a suite of diverse and extant derived taxa does not detract from this central point. Alternatively, long-lived lineages, if they do not go extinct, might only persist as restricted bradytelic lines that have never had an opportunity to radiate in their refugium, i.e., classic living fossils. The whole issue of living fossils in the strict sense may then be an artifact of the taxonomic level of the groups in question, and of distortions (because of the limits of the fossil record) in our perception of secondary or tertiary radiations of on-going lines.

Acknowledgments: Many of the observations recorded above are derived from field and laboratory observations made of living and fossil forms under research grant DEB 79-03602 (FRS) and DEB 77-24614 (RRH).

Literature

Knott, B., Lake, P. S. 1980. *Eucrenonaspides oinotheke* gen. et sp. n. (Psammaspidae) from Tasmania, and a new taxonomic scheme for Anaspidacea (Crustacea: Syncarida). Zool. Scripta 9: 25–33.

Schminke, H. K. 1975. Phylogenie und Verbreitungsgeschichte der Syncardia (Crustacea: Malacostraca). Verk. Dtsch. Zool. Gesell. 1974:384–388.

Schminke, H. K. 1978. Die phylogenetische Stellung der Stygocarididae—unter besonderer Berücksichtigung morphologischer Ähnlichkeiten mit Larvenformen der Eucarida. Zeit. Zool. System. Evolut.-Forsch. 16:225–239.

Schram, F. R. 1977. Paleozoogeography of Late Paleozoic and Triassic Malacostraca. Syst. Zool. 26:376–379.

Schram, F. R. 1979. British Carboniferous Malacostraca. Fieldiana Geol. 40:1–129.

Stanley, S. M. 1979. Macroevolution: pattern and process. San Francisco: Freeman.

22
The Xiphosurida: Archetypes of Bradytely?

Daniel C. Fisher

Museum of Paleontology, University of Michigan, Ann Arbor, MI 48109

Introduction

The Xiphosurida, or horseshoe crabs, are often cited as a classic example of arrested evolution. They have been so consistently associated with this concept, in both professional and popular literature, that their reputation for extreme conservatism in form and behavior is probably more widely known than any other single aspect of their biology or history. Ironically, however, there have been no published measurements of evolutionary rates (either morphologic or taxonomic) for horseshoe crabs, let alone any rigorous comparisons of evolutionary rates between horseshoe crabs and groups that supposedly evolve more rapidly. Rather, their status as an archetypal bradytelic group has been founded primarily on the judgment and authority of specialists who have perceived only minor morphologic differences between the living forms and certain fossil relatives (Fig. 1A–D). More quantitative studies of the evolutionary history of horseshoe crabs are badly needed, but are beyond the scope of this contribution. I intend instead to accept provisionally the overall appraisal of bradytely (with certain reservations noted below) and to concentrate on some of the attributes thought to be important in explaining it. There may remain some question as to whether bradytely is a real evolutionary phenomenon or only a function of the way we perceive evolutionary history, but we can still evaluate the extent to which it is associated with particular biological and historical patterns.

The term bradytely has been applied in two potentially distinguishable ways. The sense in which it is most commonly used and in which it has been used for selecting case histories for this compilation is, appropriately, the less restrictive of the two and refers to unusually slow evolution within a *lineage* (broadly construed, i.e., not necessarily restricted to a single, direct, species-level, ancestor–descendant sequence). Since lineages may be circumscribed so as to exclude at least some of their members' descendants (i.e., lineages may be paraphyletic), this version of bradytely is not compromised if some of those descendants diverge and take up more horotelic, or even tachytelic, ways. The second, more restrictive formulation of bradytely would limit its application to evolution within a whole *clade* (i.e., a strictly monophyletic group; Eldredge and Cracraft 1980). Each of its component subclades need not be equally bradytelic, but all would have to be considered when the whole clade is characterized. In the case of horseshoe crabs, most previous interpretations of the phylogenetic structure of the whole clade so closely approximated a single, if somewhat diffuse, lineage that the difference between the two versions of bradytely would have been irrelevant. However, in the interpretation of horseshoe crab phylogeny adopted here (Fisher 1975a, 1981, 1982, in prep-

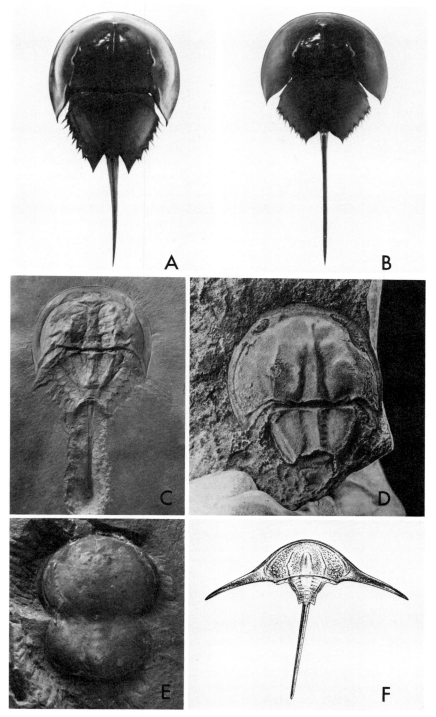

Fig. 1. Representative recent and fossil horseshoe crabs. (A) *Limulus polyphemus* (juvenile; ca. 0.5x natural size). (B) *Carcinoscorpius rotundicauda* (juvenile; ca. 0.3x nat. size). (C) *Mesolimulus walchi* (juvenile; ca. 0.5x nat. size) Jurassic, Germany: Jura-Museum 753 (original specimen for Van der Hoeven (1838, Plate 7, Fig. 2); access to specimen generously allowed by Dr. Günter Viohl, Jura-Museum, Eichstätt). (D) *Limulus vicensis* (ca. nat. size); Triassic, France; reproduced from Bleicher (1897, Plate 1, Fig. 1); taxonomic consistency would now require placing this species in a new genus. (E) *Pringlia birtwelli* (probable adult; ca. 2.0x nat. size); Carboniferous, Great Britain; British Museum (Natural History) I 13882 (loan of specimen generously allowed by Dr. Richard Fortey, BM(NH), London). (F) *Austrolimulus fletcheri* (ca. 0.3x nat. size); Triassic, Australia; reproduced from Riek (1968, Fig. 1).

aration), the difference is substantial. I will endeavor to discuss bradytely in such a way as to accommodate both definitions.

Morphologic Comparisons and Temporal Scale

One strategy for assessing the validity of the conventional portrayal of horseshoe crabs as bradytelic is to compare the living species and determine their times of divergence. This approach provides access to only a fraction of the evolutionary history of the group, but it does permit more extensive comparisons than are typically available when dealing with fossil material alone. Since the analysis depends on a phylogenetic framework of known temporal scale, I will begin by considering this aspect of the problem.

Cladistic analysis of the relationships among the three extant Indo-Pacific species (Fisher 1975a, 1982) suggests that *Tachypleus gigas* and *T. tridentatus* are more closely related to each other than either is to *Carcinoscorpius rotundicauda* (Fig. 2). All of these are then more closely related to each other than to the North American *Limulus polyphemus*. These results are generally consistent with phenetic evaluations such as are derived from serological studies (Shuster 1955, 1962) and hybridization experiments (Sekiguchi and Sugita 1980). They are also compatible with the current classification (Pocock 1902). At present, however, there is no satisfactory control on the age of the most recent common ancestor of any of the three Indo-Pacific species. Our only information concerns the split between the lineages leading to these and to *L. polyphemus*. The fossil species *Tachypleus decheni* and *Limulus coffini* can be incorporated into the pattern of relationships of extant species as shown in Fig. 2, and the Maestrichtian *Casterolimulus kletti* (not shown in Fig. 2) may also be associated with the lineage leading to the extant Indo-Pacific species (Holland et al. 1975). Given these interpretations, the Upper Campanian occurrence of *L. coffini* (Reeside and Harris 1952) gives a minimum age of about 75 million years for the most recent common ancestor of *L. polyphemus* and the extant Indo-Pacific species.

A different approach to dating this same divergence event would be to consider it a result of the opening of the North Atlantic and the consequent isolation of North American and Eurasian (Tethyan) limulid lineages. Alternatively, the occurrence of *Casterolimulus kletti* in North America and the restriction of the subfamily Limulinae to North America could be interpreted as indicating that the divergence between the Limulinae and Tachypleinae occurred subsequent to the opening of the North Atlantic. Although other scenarios could be entertained as well, it is at least plausible to suggest that the beginning of the opening of the North Atlantic, dated to approximately 90–95 million years (Kristoffersen 1978), might constitute a lower bracket for the time of divergence of the Limulinae and Tachypleinae.

Morphologic comparisons between *L. polyphemus* and the extant Indo-Pacific horseshoe crabs have followed one of two patterns. Where the intent is taxonomic, they have focused on a small number of easily recognizable characters that suffice to diagnose the various taxa. In more general discussions, comparisons have been largely subjective assessments emphasizing what is considered to be a high degree of similarity. Differences in the shape of the prosoma and opisthosoma, the placement and relative length of certain spines, and the structure of certain appendages are noted, but are treated as minor variations on a broadly consistent theme. It is not clear, however, to what extent this judgment may have been influenced by the conspicuous *differences* between horseshoe crabs and other arthropods as well as by the obvious *similarities* among horseshoe crabs.

More quantitative comparisons between *L. polyphemus* and the extant Indo-Pacific species include Shuster's (1955, 1962) serological studies, which he interprets as indicative of differences typical for a suite of congeneric species, rather than for members of three different genera. Hybridization studies (Sekiguchi and Sugita 1980) have had limited usefulness for measuring rate of divergence since successful fertilization and development have occurred only for hybrids involving the Indo-Pacific species, whose times of divergence cannot yet be specified. A. Riggs (personal communication) and others have recently begun amino acid sequencing of the blood pigment (hemocyanin) of

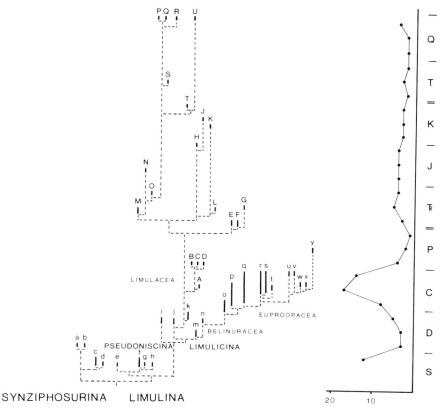

Fig. 2. Phylogenetic relationships and diversity of representative Xiphosurida (following analysis and compilations in Fisher 1975a, 1981, 1982). The phylogenetic tree includes all valid xiphosuran taxa whose morphology is sufficiently well known to enter into the phylogenetic analysis. If a taxon is entirely plesiomorphic relative to its sister, it is shown as continuing their parent lineage. If both sisters show autapomorphies, they are both displaced from the parent lineage, the more plesiomorphous one to the left. The diversity profile on the right is explained in the text. a, *Legrandella lombardii;* b, *Weinbergina opitzi;* c, *Limuloides* sp.; d, *Bunodes* sp.; e, *Bunaia woodwardi;* f, *Pseudoniscus aculeatus;* g, *Cyamocephalus loganensis;* h, *Neolimulus falcatus;* i, *Elleria morani;* j, *Neobelinuropsis rossicus;* k, *Paleolimulus randalli;* m, *Belinurus carteri;* n, *B. alleghenyensis;* o, *B. reginae;* p, *B. bellulus;* q. *B.* koenigianus; r, *Euproops danae;* s, *E. rotundatus;* t, *E. anthrax;* u, *Pringlia demaistrei;* v, *P. fritschi;* w, *P. birtwelli;* x, *Liomesapis laevis;* y, *Prolimulus woodwardi;* A, *Paleolimulus* sp.; B, *P. avitus;* C, *P. signata;* D, *Anacontium carpenteri;* E, *Austrolimulus fletcheri;* F, *Psammolimulus gottingensis* G, *Limulitella bronni;* H, *Mesolimulus walchi;* J, *M. syriacus;* K, *Victalimulus mcqueeni;* L, *Heterolimulus gadeai;* M, *Limulus priscus;* N, *L. woodwardi;* O, *L. vicensis;* P, *Tachypleus gigas;* Q, *T. tridentatus;* R, *Carcinoscorpius rotundicauda;* S, *Tachypleus decheni;* T, *Limulus coffini;* U, *L. polyphemus.* Xiphosurans are classified in the suborders Synziphosurina (a–d) and Limulina, the latter of which is composed of the infraorders Pseudoniscina (e–h) and Limulicina. Limulicina includes the superfamilies Belinuracea (i–q), Euproopacea (r–y), and Limulacea (A–U).

horseshoe crabs, but interspecific comparisons are not yet available.

A more commonly cited comparison in support of xiphosuran bradytely is between the extant species (collectively) and *Mesolimulus walchi* (Desmarest 1822), known from the uppermost Jurassic Solnhofen limestone. *Mesolimulus walchi* is indeed quite similar in overall

form to its extant relatives (Fig. 1), but most accounts of its morphology ignore or gloss over even easily observable differences, such as a much flatter profile to both prosoma and opisthosoma, a distinctive body outline in dorsoventral aspect, a set of ridges on the opisthosoma not found on any living species, or details of the size and position of spines. Much less obvious differences include: more gracile appendages; a smaller, more lightly constructed proventriculus; and different patterns of distribution of setae (Fisher 1975a, 1975b). Functional analysis of *M. walchi,* based on experimental studies of morphology, analysis of trace fossils, taphonomic considerations, and the distribution of epibionts (Fisher 1975a, 1975b, in preparation), reveals an organism broadly similar to living horseshoe crabs, but much more highly specialized for swimming and feeding on relatively soft-bodied prey. It was not as proficient a burrower or bivalve-predator as is *L. polyphemus.*

The role of *M. walchi* in initiating and sustaining the xiphosuran reputation for conservatism deserves both emphasis and clarification. As one of the first fossil horseshoe crabs to be described, and as a species represented by hundreds of well-preserved specimens, it became widely known and originally bore almost the entire burden of comparison with the extant taxa. Ironically, some of the first discussions of its morphology do note certain differences from the living forms (e.g., van der Hoeven 1838), but later workers tend to mention only the impressive similarities. Even though subsequent discoveries of other species of fossil horseshoe crabs turned up greater morphologic diversity, frequent references to *M. walchi* illustrate the inertial effects of first impressions:

Few classes offer so remarkable an instance of longevity as the Crustacea [in which horseshoe crabs were then placed], and few orders can be compared to the Xiphosura for persistency. The Jurassic forms appear to differ little, if at all, from those of our own day; and even those of the Carboniferous epoch were at once recognized as belonging to the same family. (Woodward 1867:35)

Another factor that has drawn particular attention to *M. walchi* is the remarkable association of some specimens with trace fossils formed just prior to the death of their maker. These trace fossils are actually of two distinct varieties, one formed during walking along the substrate and the other formed when landing on the substrate after a bout of swimming (Fisher 1975a). Had these trace fossils been correctly interpreted, they would have highlighted some of the differences in behavior between *M. walchi* and its extant relatives. However,

Walther (1904), who first described the "death trails," did not recognize them as representative of two different types of behavior. He therefore illustrated them with a composite drawing that implicitly attributed both types to a single individual, stranded at the end of a walking trail. Subsequent workers (Abel 1935; Caster 1938, 1940) further embellished this scenario, concluding that we are seeing direct evidence of behavior identical to that of the extant species. Certain aspects of these interpretations are almost certainly in error (e.g., that *M. walchi* entered the Solnhofen lagoons on an annual mating migration exactly like that of living horseshoe crabs). Nevertheless, this work was influential and helped to sustain the general impression that the evolutionary history of horseshoe crabs is dominantly conservative.

The conventional interpretation of horseshoe crab history probably could not have survived if a considerable number of the newly described taxa had not shown a strong, general resemblance to living horseshoe crabs. Yet, with several notable exceptions (e.g., Raymond 1944; Eldredge 1974), most descriptions of new fossil material were not sufficiently detailed either to demonstrate identity or discover diversity. In part, this is probably because the expressed goals of most studies were more biostratigraphic and taxonomic than evolutionary. However, another important factor was problematic preservation. Horseshoe crab exoskeleton is essentially unmineralized and is thus more susceptible to postmortem compression and decomposition than is the exoskeleton of many other arthropod groups. As is common among arthropods, the ventral body surface is less well sclerotized than the dorsal surface, with the result that preservation of the limbs and associated ventral structures is extremely rare. This is aggravated in the case of horseshoe crabs (as distinct even from most other chelicerates) because their limbs do not extend laterally much beyond the margin of their prosoma or opisthosoma, and thus tend not to be exposed even when they are present within the matrix. In certain instances a great deal of information is available as a result of unusual preservational conditions (Stürmer and Bergström 1981) or special preparation techniques (Fisher 1975a, 1977), but in many other instances we have little more than molds of the dorsal surface of the body. This should by no means discourage further analysis, but it does place certain limits on the nature of the comparisons that have been made.

Despite its historical importance, *M. walchi* is not the best example to pick for demonstrating xiphosuran bradytely. It has often been considered more or less directly ancestral to Cretaceous and Cenozoic horseshoe crabs (Størmer 1952, 1955), but I have argued that it is phylogenetically, as well as morphologically, divergent (Fisher 1975a). As such, it is representative of not just one, but several sets of taxa that constitute radiations of horseshoe crabs which, although modest in diversity, depart significantly from the morphology typical of more plesiomorphic members of the order. These relatively divergent branches among horseshoe crabs include: advanced belinuraceans and all euproopaceans (Fig. 2o–y); most paleolimulids (Fig. 2A–D); a group of Triassic "mesolimulids" (Fig. 2E–G); and Triassic–Cretaceous mesolimulids (Fig. 2H–L). *Pringlia birtwelli* (Fig. 1E) and *Austrolimulus fletcheri* (Fig. 1F) should suffice as striking examples of some of the extremes in morphology to be found within these groups.

A better case for morphological and ecological conservatism can be made by comparing the living species and their closest relatives (Fig. 2P–U) to certain of the Triassic and Jurassic limulids (Fig. 1D and Fig. 2M–O). Although differences could certainly be enumerated, they are much less substantial than between the living species and the groups listed above as divergent. This interval of relative conservatism covers a period of about 200 million years. Stretching the case somewhat further, it could even be argued that a basically similar morphology and life style characterized a lineage extending back to the Devonian *Paleolimulus randalli* (Fig. 2k) and *Neobelinuropsis rossicus* (Fig. 2j). However, any attempt to include taxa older than about the Middle Devonian (ca. 360 million years) would involve comparisons with pseudoniscines and synziphosurans. These groups include such differences in body shape, intersegmental mobility, and limb design that it seems inappropriate to claim them as part of the manifestation of xiphosuran bradytely. I therefore consider it preferable to limit such claims to either the monophyletic infraorder Limuli-

cina or the lineage including taxa j, k, and M–U in Fig. 2 (referred to below as the "trunk lineage" of the Limulicina), depending on whether the clade or the lineage version of bradytely is preferred.

Phylogenetic Relationships

Having recognized the infraorder Limulicina as the most conspicuously bradytelic taxon within the Xiphosurida, it is now a straightforward matter to specify their closest relatives. Eldredge (1974; Eldredge and Plotnick 1974) has provided a compelling analysis of synziphosuran and pseudoniscine relationships, the results of which are incorporated, with only slight modification, into Fig. 2. The infraorder Pseudoniscina is portrayed as the sister of the Limulicina, and the suborder Synziphosurina as the sister of the suborder Limulina (Pseudoniscina + Limulicina). Looking on a somewhat broader scale, the Aglaspida have frequently been considered the sister-group of the Xiphosurida, but the evidence for this has been questioned recently (Briggs et al. 1979). Instead of a clear sister to the Xiphosurida, we have a series of groups, such as the chasmataspids (Caster and Brooks 1956), diploaspids (Størmer 1972), and others, whose relationships have not been adequately resolved. Other merostomes (e.g., the Eurypterida) and other chelicerates are presumably related at a still higher level, but phylogenetic inferences remain controversial (Bergström 1979).

The phylogenetic structure of the infraorder Limulicina itself will also be important in the following discussion. Most previous workers have tended to reconstruct an ancestor–descendant sequence based primarily on the overall order of stratigraphic occurrence of taxa. Woodward (1872) offers an early example of this, and both Størmer (1955) and Stürmer and Bergström (1981) repeat its essential features, though they include additional taxa. The proposed sequence is: Belinuracea → Euproopacea → Limulacea (Paleolimulidae → Mesolimulidae → Limulidae). In contrast, the interpretation of limulicine relationships adopted here (Fig. 2) preserves a much larger role for detailed comparative morphologic analysis. For instance, decisions regarding homology are based not just on gross topographic comparisons, but also on analysis of the relationship between topographic features and the metameric structure of horseshoe crabs. Elucidation of metameric structure has in turn involved studies of the soft anatomy and embryology of extant horseshoe crabs. Stratigraphic data can and should enter into phylogenetic analysis (Fisher 1980, 1982), but in this particular case, their inclusion does not significantly alter the pattern of relationships indicated by cladistic analysis of morphologic data (Fisher 1975a, 1981).

Species Diversity

As indicated above, the extant horseshoe crabs are generally described as comprising four species. A fifth, *Tachypleus hoeveni* was described by Pocock (1902) on the basis of preserved material but has not been recognized by subsequent workers (Waterman 1958).

An estimate of the species diversity of the Limulicina, through geologic time, is given at the right side of Fig. 2. This curve represents the minimum number of species-level lineages occurring in or passing through each third of the various geologic periods since the Upper Silurian. In addition to the species whose relationships are shown on the phylogenetic tree, the diversity profile includes information on a number of species that are considered valid, but whose morphology is not known well enough for phylogenetic analysis. This diversity profile differs from a count of the number of species known to occur within particular intervals in that it includes lineages inferred to have been present (through phylogenetic analysis), but not represented by known fossil material within those intervals. This procedure is particularly useful in the case of a group such as horseshoe crabs, where the number of occurrences per lineage per interval tends to be much less than one. A simple count of known, valid species would inevitably underestimate, to an unknown extent, actual species diversity. Although this is undoubtedly true also of the present diversity profile, the effect should be less pronounced because a lineage is not required to be preserved within an interval in order to be counted there.

As noted by Eldredge (1979), one indication that this diversity profile may be a reasonable estimate of the standing species number during much of the Mesozoic and Cenozoic is that it so closely approximates the present number of species.

The most remarkable feature of the pattern of horseshoe crab species diversity is that it is so low, and yet relatively constant. The only conspicuous departure from this pattern occurs during the Late Devonian and Carboniferous and is produced by the radiation of advanced belinuraceans and euproopaceans. It may be significant that this is the subclade of the Limulicina that, from a morphologic point of view as well, least deserves to be called bradytelic. It is also important to note that this is not a "pre-bradytelic" phase of more rapid evolution, but rather a departure from a pattern of bradytely developing simultaneously, and continuing subsequently, in a sister subclade.

Ecology

Among extant horseshoe crabs, the greatest amount of ecological information is available for *Limulus polyphemus*. Useful reviews of the recent literature have been provided by Shuster (1982) and Rudloe (1979). Terms such as generalist or specialist are usually difficult to apply in a way that simultaneously accounts for all aspects of an organism's ecology, but horseshoe crabs happen to offer a relatively uncomplicated situation. In most respects they have strong generalist tendencies. *Limulus polyphemus* is restricted to shallow, nearshore marine environments, but it occupies, consecutively throughout its ontogeny, a series of habitats within this zone. Its eggs hatch from "nests" buried high in the intertidal range of protected, sandy beaches. Juveniles remain year-round on intertidal mudflats, burrowing into the substrate at low tide and foraging during high tide. As they grow older they move progressively offshore (often to sandier substrates), but may return to the mudflat to forage during high tide. Adults tend to remain offshore, though generally at depths less than 50 m. In the late spring or early summer, they return to shallower water to reproduce. Development to sexual maturity requires 5–7 years for males and 7–9 years for females. Both sexes appear to live approximately 3–5 years after the attainment of sexual maturity.

The juvenile diet consists primarily of polychaetes. These are encountered by probing through the substrate with the chemosensitive chelae of the prosomal appendages and are captured by rapid excavation and pinching movements. As *L. polyphemus* grows larger, bivalves begin to constitute an important component of its diet. My observations of feeding in the wild have involved only live (though sometimes moribund) prey, but many feeding experiments have indicated that horseshoe crabs will eat a wide range of items (e.g., brine shrimp, chopped squid, or dead fish; French 1979; Kropach 1979; Brown and Clapper 1981).

Limulus polyphemus normally encounters relatively wide-ranging values for most environmental variables, and most of the capacity for accommodating these extremes appears to exist at the individual, physiological level. Except for populations in large estuarine systems such as the Chesapeake Bay, most horseshoe crabs live at normal marine salinities. However, juveniles on the mudflats and adults entering estuarine systems to breed encounter lowered salinities (Robertson 1970). Behavioral studies of the thermal preference of active individuals indicate an unusually broad temperature range (Reynolds and Casterlin 1979), and normal seasonally related stresses, such as winter on the mudflat, for juveniles, or temporary stranding of breeding adults during low tide, call for even greater thermal accommodation. Likewise, juveniles, in particular, and adults to a lesser extent occasionally experience sharply reduced availability of oxygen, moisture, and food (e.g., Johansen and Petersen 1975; Fields 1982). *Limulus polyphemus* also shows remarkable tolerance of extreme artificial perturbations of all these environmental variables (e.g., Mayer 1914; Pearse 1928; Fraenkel 1960). Although less directly relevant to the ecology of natural populations, this response is certainly consistent with conventional characterizations of an ecological generalist.

There have been relatively few studies of the ecology of the Indo-Pacific horseshoe crabs. However, most available reports emphasize their extreme similarity to *L. polyphemus* (e.g., Waterman 1953; Sekiguchi and Nakamura 1979; Shuster 1982). The most conspicuous difference is that *C. rotundicauda* is even more tolerant of reduced salinities (Annandale 1901; Sekiguchi and Nakamura 1979). Besides this, all living horseshoe crabs seem to be similar on a physiological level.

The ecology of fossil horseshoe crabs has been a matter of greater dispute, though only the issue of habitat has been routinely considered. One approach (e.g., Størmer 1955) involves an extremely literal reading of the record. The environments in which fossil horseshoe crabs are preserved are assumed to be those in which they lived. Many species are thus interpreted as brackish to fresh water. A contrasting approach (e.g., Caster 1957; Holland et al. 1975) is to interpret all brackish water or freshwater occurrences as individuals that died and were buried in the course of their annual mating migration, away from their normal marine habitat. These workers assume that all fossil horseshoe crabs were identical to the living species in behavior and habitat preferences. Their explanation for the predominance of brackish water to freshwater occurrences during much of horseshoe crab history is that the preservation potential of horseshoe crab body fossils and trails is higher in these environments than in a normal marine setting (Goldring and Seilacher 1971).

A shortcoming of the literal reading of the fossil record is that many fossil horseshoe crabs are known only on the basis of single specimens. With this sample size, any conclusions regarding habitat are probably risky. A more cautious approach would be to demand evaluation of evidence for postmortem transport and preservation bias and to distinguish clearly between those cases where strong inferences regarding habitat can be made and those where only tentative assignments are available. The main shortcoming of the interpretation suggested by Caster (1957) is that assumptions of behavioral stasis cannot provide a foundation for an empirical analysis of the history of behavior. The concept of behavioral stasis might be more useful if recast as a null hypothesis against which hypotheses of behavioral change could be tested; but even as such, it would be inappropriate in the context of a discussion of bradytely.

For our present purposes, the ideal approach for an ecological characterization of fossil horseshoe crabs is an analysis of evidence bearing on each species independently. Although still tentative and in need of further documentation, such a characterization can be offered for most limulicines. The "trunk lineage" of morphologically plesiomorphic taxa from *Neobelinuropsis rossicus* to the extant species inhabited nearshore marine environments and had a diverse behavioral repertoire (inferred from functional analysis), suggesting a relatively generalist ecology comparable to that of the living species. Advanced belinuraceans and euproopaceans were behaviorally generalized (at least one genus, *Euproops*, was probably even amphibious to some degree; Fisher 1979), but tended to be associated with a much more restricted range of aquatic environments. For instance, advanced euproopaceans, of which there are large samples from many localities, are restricted to the freshwater end of the spectrum of habitats available in Carboniferous coal swamps and are represented by relatively complete ontogenetic series (Fisher 1979). This pattern is unlikely to result from preservational bias, because contemporaneous brackish and marine facies are known to contain distinct taxa of horseshoe crabs (belinuraceans or paleolimulids). On the relative scale appropriate to such discussions, advanced belinuraceans and euproopaceans were ecological specialists. A similar characterization could be applied to Permian paleolimulids (Fig. 2B–D) and to certain Triassic forms (Fig. 2E–G), though smaller sample sizes make this inference less compelling. Finally, mesolimulids (Fig. 2H–L) appear to have been generalized in terms of habitat, but were rather specialized behaviorally, as noted above.

Distribution in Space and Time

As might be expected, the most detailed data on spatial distribution are available for *L. polyphemus*. It occurs along the east coast of North America from Maine to the Yucatan Peninsula, with a large hiatus along much of the Gulf Coast (Shuster 1979). In addition to this one large discontinuity, the species is broken up into a series of relatively discrete populations, organized around estuarine systems or coastal lagoon–shoal complexes. These provide the relatively sheltered beaches and intertidal flats that appear to be necessary for reproduction and juvenile development. Tagging experiments indicate considerable individual mobility (Baptiste et al. 1957), but immigration rates between consecutive populations are not high enough to overwhelm interpopulational morphologic differences.

Like *L. polyphemus,* the Indo-Pacific species are distributed discontinuously within their ranges (Shuster 1982). *Carcinoscorpius rotundicauda* and *Tachypleus gigas* occur from the Bay of Bengal, along the coast of Indonesia, to Borneo and the Philippines. During much of their life cycle the two species may be sympatric on a local as well as a regional scale, though reproduction takes place in separate habitats. *Tachypleus tridentatus* ranges from a small population on the coast of Borneo, where it occurs in relatively close proximity to *T. gigas* and *C. rotundicauda,* along the east coast of southeast Asia, to the east coast of Japan. Thus, horseshoe crab biogeographic ranges tend to be relatively extensive, and, with one major exception, the extant species are not sympatric (Sekiguchi and Nakamura 1979).

In most cases, geographic information is minimal for fossil horseshoe crabs. Of the 40 fossil species shown in Fig. 2, 12 are known from single specimens, and another 17 are known only from single localities. The only species for which occurrences are sufficiently numerous even to approach mapping a distribution are a few of the Carboniferous belinuraceans and euproopaceans. *Euproops danae,* for instance, occurs throughout much of the North American Carboniferous, from Illinois to maritime Canada, and has been reported from Europe. Likewise, *E. rotundatus* and *Belinurus reginae* are each known from a number of localities in Great Britain and Europe. While ranges of this magnitude are not implausible, given what is known about extant horseshoe crabs, there is insufficient information to determine whether they are typical of other fossil horseshoe crabs. The incidence of sympatry among the fossil species is also subject to sampling problems, but it can at least be estimated in the most favorable cases. Among limulicines it appears to be highest during the Carboniferous. On a regional level, horseshoe crab faunas associated with a coal swamp/deltaic complex were composed of one predominantly marine species (a paleolimulid or belinurid), one brackish "lacustrine" species (a belinurid; this element of the fauna has not been found in North America), one freshwater lacustrine species (a liomesaspid), and one or two species (euproopids) occupying small freshwater streams (and in at least some cases, the adjacent, moist litter of the forest floor). This pattern is repeated, with different species, several times along the formerly continuous tract of coal swamp development. The closest approach to local sympatry involved the pairs of euproopid species (e.g., *E. rotundatus* and *E. anthrax* in Great Britain). At the regional level, the relatively high degree of sympatry depended on the proliferation of taxa in nonmarine portions of the environmental spectrum. A very similar faunal pattern appears to have developed independently, during both the Permian and the Triassic, though in each case its reconstruction is based on much sparser data than are available for the Carboniferous. At no place in the fossil record of the Limulicina, with the possible exception of *Heterolimulus gadeai* and *Tarracolimulus rieki* (Romero and Via Boada 1977), is there demonstrable sympatry of predominantly marine species.

Analysis of species longevity suffers from the same sampling problems that plague the study of geographic distribution. Since most taxa occur at single localities and stratigraphic levels, few estimates of stratigraphic range are available. Even the living species are of no assistance since none of them are known as fossils. As with biogeography and faunal structure, the best data are from the Carboniferous, where several species (e.g., *B. reginae*, *B. koenigianus*, *E. rotundatus*) are recognized as having ranges of about 15–20 million years (Pruvost 1930; Van der Heide 1951). Although this is certainly longer than estimates of the average longevity of invertebrate species, it does not greatly exceed longevities typical of some groups of marine bivalves, reef corals, or forams (Stanley 1979). It is not clear, however,

how typical these horseshoe crab species longevities might be.

Intraspecific Variation

The most thorough study of genetic variation in horseshoe crabs is an electrophoretic analysis of allozyme loci in *Limulus polyphemus*. Looking at 19 loci in 6 population samples from the Gulf Coast of Florida to Massachusetts, Selander et al. (1970) found levels of polymorphism and heterozygosity that were comparable to or perhaps even greater than levels documented in a variety of other invertebrate and vertebrate groups. Actual measures of evolutionary rate are not available for most of these comparison groups, but a subjective consensus estimate would place most of them within the horotelic class. Based on this preliminary evidence, it seems unlikely that the low apparent rate of evolution of horseshoe crabs is caused by conspicuously low levels of genetic variation.

Morphological variation has also been studied most thoroughly in *L. polyphemus*. Shuster (1955, 1979, 1982) has documented patterns of size variation within and among populations, and Riska (1981) has provided an excellent multivariate analysis of variation in both size and shape. He finds within-locality coefficients of variation ranging from 7.31 to 11.52, well within the ranges known for a variety of other organisms (Riska 1981). In addition, an unusually large proportion (about one half to three fourths) of the total variance in individual characters is attributable to among-locality, as opposed to within-locality, variation. In this sense, horseshoe crabs seem more variable than most other groups for which comparable data are available (Riska 1981). Finally, Riska emphasizes that the differentiation among localities is due to variation in shape that is relatively independent of size differences. Available studies on the Indo-Pacific species suggest that they are characterized by comparable patterns of variation (Sekiguchi et al. 1976; Sekiguchi et al. 1978).

Ideally, similar studies of morphological variation might be undertaken on fossil horseshoe crabs. However, even for those species where adequate sample sizes are available, lack of control over the intensity of compressional deformation represents a serious problem. Furthermore, the effects of deformation are undoubtedly greatest on the continuous metric variables that are most likely to demonstrate intraspecific variation. Although approximate reconstructions can be handled routinely, fine scale analysis would be more difficult to justify.

Comparisons with Closely Related Groups

Characterization of the close relatives of the Limulicina, in terms of the issues discussed above, is at best difficult and incomplete. The only groups that are known well enough and related closely enough to provide relevant comparisons are pseudoniscines and synziphosurans. Comparison of either pseudoniscines to limulicines or synziphosurans to pseudoniscines + limulicines reveals that the former group has a smaller number of species, known, on average, from a smaller number of localities and individuals, and a much shorter whole-clade duration. This makes it even more difficult than usual to evaluate possible biases in the record. Nevertheless, general impressions may be useful for guiding future inquiries.

The species-level diversity of both synziphosurans and pseudoniscines is similar to what is known for the limulicines during most comparable intervals of their history. Both groups were probably similar to limulicines in overall physiology but do not appear to have invaded freshwater habitats. They occur in marine or marginal-marine (possibly even hypersaline) settings, often as rare elements associated with eurypterid faunas. In terms of diet, synziphosurans were probably similar to limulicines, but pseudoniscines were probably restricted more to small, soft-bodied prey. Their appendages do not appear to have been as well sclerotized as those of synziphosurans or limulicines, and the interophthalmic topography associated with the origin of extrinsic appendage musculature was not as well developed. In terms of both locomotor behavior and righting behavior, synziphosurans and pseudoniscines appear to have had less diverse repertoires than limulicines (Fisher 1982). In this sense they seem to have been ecologically less generalized than limulicines.

The fossil record of both synziphosurans and pseudoniscines is so sparse that we have little information on biogeographic ranges or species longevities. With regard to sympatry, a synziphosuran may occur together with a pseudoniscine (e.g., *Bunodes lunula* with *Pseudoniscus aculeatus*, in the Upper Silurian Oesel fauna, from the Baltic). However, as discussed by Eldredge (1974), most reported examples of sympatry between more closely related forms are probably questionable. In general, synziphosurans and pseudoniscines tend to be no better preserved than limulicines. Coupled with small sample sizes, this has prevented any meaningful study of phenotypic variation.

Discussion

Explanations for bradytely have been sought at a number of levels within the hierarchy of evolutionary processes and have attributed different significance to each of the general issues discussed above. It is therefore appropriate to examine the degree of congruence between horseshoe crab history, as characterized here, and the expectations associated with the competing, though not necessarily mutually incompatible, interpretations of bradytely.

Simpson (1944, 1953) was inclined to consider bradytely the long-term manifestation of an unusually low rate of morphologic change within species-level lineages (i.e., evolutionary species or series of chronospecies; see Fig. 3A). Ideally, this might be evaluated directly, on the basis of dense geographic and stratigraphic sampling of the ranges of representative species. However, the extremely sparse fossil record of xiphosurans virtually prohibits this approach. Those few species that do occur over a significant stratigraphic range do indeed show minimal amounts of morphologic change, but for two reasons this fails to provide compelling confirmation of Simpson's interpretation. In the first place, the species whose ranges can be traced are too few in number, too superficially studied, and too far removed phylogenetically from the "trunk lineage" of persistently plesiomorphic xiphosurans to adequately charac-

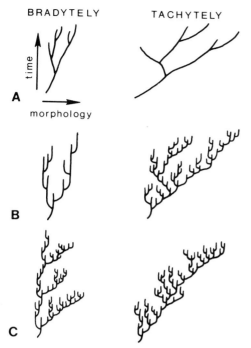

Fig. 3. Three hierarchic levels at which differences in measured evolutionary rate may arise (reproduced from Fisher, 1982, Text-fig. 4). See text for explanation.

terize the species-level pattern of change associated with bradytely. Secondly, confirmation of Simpson's interpretation requires a *comparison;* we must be able to show that the rates of change within species-level lineages belonging to bradytelic groups are *lower* than those that typify nonbradytelic groups. While not attempting to prejudge anything (since explicit comparisons of this sort have not yet been made), the likelihood that a certain degree of stasis is an attribute of many species in groups that are not normally considered bradytelic (Eldredge and Gould 1972; Gould and Eldredge 1977; Stanley 1979) should at least make us cautious about accepting Simpson's interpretation before both sides of the comparison are in hand.

A more indirect approach to evaluating Simpson's interpretation involves descending a level and considering factors related to the presumed dynamics, rather than just the gross pattern (or kinematics), of intraspecific (including interchronospecific) morphologic change. Do horseshoe crabs show features that we expect to be causally related to low rates of morphologic change? Certainly their generalist ecology and broad physiological tolerance are consistent with this interpretation, for reasons outlined by Simpson (1953). In addition, their population size, geographic distribution, and degree of variability (both morphologic and genetic) are compatible with low rates of transformation, though lack of variability is rejected as a cause (Simpson 1953). Details of phylogenetic history or the history of diversity are not seen as relevant to this approach. In summary, horseshoe crab history does not reject the idea that low rates of intraspecific change lie at the heart of bradytely, but neither does it present compelling support for this interpretation.

Simpson's (1944, 1953) approach to explaining bradytely is representative of a style of evolutionary thinking that Eldredge (1979) has referred to as "transformational," in contrast to an alternative, "taxic" style. The taxic interpretation of bradytely (Eldredge 1975, 1979; Eldredge and Cracraft 1980; Stanley 1975, 1979) tends to rely on the idea that significant morphologic change is associated primarily with speciation events (cladogenesis). It then interprets bradytely as the result of unusually low rates of speciation within a clade, and hence few opportunities for morphologic change (Fig. 3B). Once again, direct evaluation of this hypothesis, using horseshoe crab history, is difficult. Preservation of horseshoe crabs is so rare (i.e., there are so few fossil occurrences per species) that any measured "speciation rate" may actually reflect no more than a preservation rate. Even if the record is adequate for recognizing the existence of most species-level lineages and for giving an approximation of worldwide horseshoe crab diversity during an arbitrarily short interval of time, we have little idea of how many temporally successive species we are missing within species-level lineages.

Indirect evaluation of the taxic interpretation takes a variety of paths. Persistence of a group for long intervals at relatively constant, low diversity has been cited as an important correlate of low speciation rates (Eldredge 1975, 1979; Stanley 1975, 1979), not out of algebraic necessity, but rather on empirical and, to some extent, theoretical grounds. For instance, Stanley (1979) argues that most lineages evolve very slowly and that if a clade contains only a small number of lineages, the whole clade is apt to evolve very slowly. Such clades only have to survive to be considered bradytelic. The long history of low diversity shown by horseshoe crabs, particularly by the more conspicuously bradytelic Limulacea, is in evident accord with this view. In addition, Eldredge (1979) and Eldredge and Cracraft (1980) have discussed in some detail how eurytopy and well-developed dispersal abilities might be expected to be associated with broad geographic ranges, low incidence of sympatry, low extinction rates (high species longevity), low speciation rates, low incidence of morphologically divergent speciation, and consequent bradytely (again assuming an association between speciation and morphologic change). Most features of horseshoe crab history, including the *differences* between the histories of the various subgroups (e.g., Limulacea versus Belinuracea-Euproopacea, Limulicina versus Pseudoniscina), fit well with this model and thus corroborate the taxic interpretation of bradytely. However, horseshoe crabs fall short of directly demonstrating the primary role supposedly played by low rates of speciation.

A third approach to explaining bradytely, suggested in Fig. 3C (Fisher 1982), is that it is in part a function of the patterns of relationships among taxa comprising the bradytelic group. This approach operates at a higher level of organization within the hierarchy of evolutionary patterns than have the previous explanations. It differs also from most previous uses of phylogenetic information in that it does not concentrate on specific details of morphologic transformations, nor on diversity information abstracted from a genealogical context. Instead, it focuses on the comparative analysis of branching patterns, the distribution of longevities, and the relative degree of divergence from primitive conditions. Quantitative methods for comparing these aspects of phylogenetic trees have not really been developed, but some useful concepts and conventions may be gleaned from previous studies of other types of branching systems (e.g., Leopold and Langbein 1962; Leopold 1971; Honda 1971; Parker et al. 1971; Barker et al. 1973; McMahon 1975; McMahon and Kronauer 1976).

One of the most useful concepts concerns tree symmetry, which can be defined as the tendency for the two taxa that result from a bifur-cation to be similar in their characteristics of tree geometry. Symmetry/asymmetry relations might be taken as pertaining to individual species related by a single cladogenetic event, to average properties for the species belonging to the resulting clades, or to the properties of the entire clades. I envision three kinds of symmetry/asymmetry as being significant in the present context: morphologic, temporal, and cladistic. Morphologic symmetry at the level of individual species would be highest for daughter species that diverge equally from their common ancestral (parent) species and lowest for the situation of a single daughter species diverging substantially from a persistent and unchanging parent species. At other levels, this definition could be modified accordingly. Temporal symmetry refers to equality of longevity and again could be defined at any of the levels specified above. Cladistic symmetry refers to the tendency for the species resulting from a cladogenetic event to have equivalent subsequent histories of cladogenesis. It is highest when each species resulting from a bifurcation undergoes subsequent bifurcation itself.

Certain problems in developing a quantitative description of the geometry of phylogenetic trees can be illustrated by a more detailed discussion of cladistic symmetry. It can be measured by assigning Strahler order numbers (McMahon and Kronauer 1976) to the hierarchical system of "segments" that make up the tree and by determining branching ratios—i.e., the ratio between the number of segments of consecutive orders. Highly symmetrical trees have branching ratios near 2.0, whereas asymmetrical trees have branching ratios that may be much higher. Branching ratios may be averaged across the whole tree, or computed separately from consecutive low-order numbers (to reflect the small-scale symmetry of the tree) and higher order numbers (to reflect larger scale structure). Trees that are self-similar (i.e., whose small-scale neighborhoods would serve as an adequate model of their larger scale structure) have equal branching ratios for all order intervals. Although these conventions provide an index of cladistic symmetry, the range of values that the branching ratio can assume depends upon the size (i.e., number of lowest order segments) of the tree. This is a problem because it may frequently be necessary to compare trees of different sizes, either when comparing trees of unrelated groups or when examining the internal structure of a clade by comparing the structures of its constituent subclades. This problem might be dealt with by restricting descriptive statements to inequalities that are not affected by this size bias, by developing a size-independent index of symmetry, or by developing procedures for truncating trees so that they might be compared at similar effective sizes. Another problem is that previous studies of tree geometry have dealt with situations where it is possible to analyze each segment in the branching system, whereas studies of phylogeny will typically have only an incomplete sample of the original segments. It will therefore be necessary to consider the effects of incomplete sampling and to evaluate the likelihood that observed differences in cladistic symmetry have arisen through sampling error. These and other approaches to the study of tree geometry (e.g., length-probability histograms, McMahon and Kronauer 1976; fractal dimension, Mandelbrot 1977, 1982) remain intriguing, but largely unexplored within the context of phylogenetic studies.

Regardless of the technical problems associated with a fully quantitative analysis of the geometry of phylogenetic trees, it is possible to use some of these ideas in a more qualitative way to speculate on the relationship between bradytely and tree shape. The first of two characterizations of this relationship operates on a relatively descriptive level. It is based on the observation that during horseshoe crab history, when cladogenetic events appear to have been most morphologically asymmetrical, the longevity of the clade descended from the more plesiomorphic sister tends to be greater than the longevity of the clade descended from the more apomorphic sister. Examination of Fig. 2 shows that this is a fairly consistent pattern, and Fig. 3 should make it clear that it is not a trivial consequence of bradytely. Since Simpson (1953) remarks that other bradytelic groups have shown similar patterns, short-lived excursions into apomorphy among otherwise bradytelic clades do not appear to be an isolated phenomenon. Moreover, this pattern is important in that it implies that bradytely is not dependent on any clade-wide inability to *produce* evolutionary novelty. It explains bradytely in the sense that

it generates a clade history that shows conspicuously less apomorphy than might have developed had the more apomorphic subclades survived. As such, it appears to have been a factor in horseshoe crab bradytely.

However, differential longevity of plesiomorphic and apomorphic subclades has a serious shortcoming as an explanation of bradytely, in that it presently lacks a sufficient mechanism on an ecological level. As noted above, the plesiomorphic "trunk lineage" of horseshoe crabs is strongly eurytopic, and individuals probably showed excellent dispersal abilities. Eldredge and Cracraft (1980) have argued that this combination of features leads to high species longevity but low speciation rate. While the former attribute may contribute to high clade longevity, there is no necessary or direct causal relationship, and the latter attribute clearly works against high clade longevity. A possible link between ecological attributes and clade longevity could be postulated through a modification of Stanley's (1979) argument that adverse conditions will more readily suppress speciation than lead directly to extinction. In the special case of a eurytopic species in which the effectiveness of dispersal drops with the progressive demise of intervening local populations (this would obtain for horseshoe crabs, but not, for instance, for a species experiencing passive dispersal of long-lived planktonic larvae), the normally low probability of speciation may increase significantly with the onset of extinction. While the competitive interactions suggested by lack of sympatry might restrict the potential of incipient daughter species formed during the heyday of the parent species, a daughter species that originated during or shortly before the extinction of the parent would not be restricted to the same degree. The greater the number of incipient species formed in this way, the greater would be the likelihood that one of them would develop adaptations sufficient to deal with the ecological stress bringing about the extinction of the parent. A clade whose member species tend to respond to stress by speciating might tend to have greater clade longevity (regardless of species longevities and in spite of a normally low probability of speciation) than another clade in which stress suppresses speciation (even though this latter clade might have a normally higher probability of speciation). Nevertheless, this postulate is inadequate as a complete mechanism unless there is some basis for estimating stressed and unstressed speciation probabilities for the two types of clades.

An alternative approach to explaining the differential longevity of plesiomorphic and apomorphic clades of horseshoe crabs is that the most apomorphic clades (e.g., advanced belinuraceans and euroopaceans in the Carboniferous coal swamps) constituted radiations into environments that were relatively ephemeral in geologic time. This may have been an important factor in horseshoe crab bradytely, but it is unlikely that it represents a general explanation.

A second characterization of the relationship between bradytely and tree shape overlaps somewhat with the first, but focuses more on cladistic symmetry/asymmetry. It is based on the observation that, while the phylogenetic tree for limulicines as a whole shows moderate cladistic symmetry, the portion of the tree referring to the plesiomorphic "trunk lineage" (representing apomorphic daughter clades as single segments) is highly asymmetric. Few trees of other bradytelic groups are available for comparison, but Miles' (1977) analysis of lungfish phylogeny again shows extreme cladistic asymmetry. What would be the effect of extreme cladistic asymmetry in cases, such as these, where species longevities tend to be short relative to the entire history of the clade? At any one time, there would be relatively few species extant, and they would be relatively closely related. Their small number means that there would be relatively few genetic/morphological/ecological variants available to expand into any newly opened ecologic space. Their close relationship suggests that they are apt to be relatively similar, again resulting in a lack of variability at the interspecific level. Assuming that the ability to colonize a given "region" of newly opened ecologic space varies considerably according to the properties of a species, and that opportunities appropriate for any particular genetic/morphological/ecological variant arise relatively rarely, a narrow range of interspecific variability minimizes the probability that a successful colonization will occur. This involves little more than an extension, to the clade level, of the much older argument that lack of variability at the population level is a

cause of bradytely. In other words, cladistically asymmetric clades may evolve slowly because at any one point in time they have most of their evolutionary eggs in one cladistic basket and thus do not have the interspecific variability that would promote moving in new directions as unpredictable opportunities arise.

It should be emphasized that the low interspecific variability that is attributed here to cladistically asymmetric groups is more profound than that entailed by the "low species diversity" (i.e., low number of species) cited by Eldredge (1979) and Stanley (1979) as an important factor in explaining bradytely. For geometric reasons, if species extinction probabilities are more or less constant, cladistically asymmetric groups do tend to have low standing crop diversities, but the pattern of relationship among the species that make up the standing crop introduces an additional element.

It is also important to consider what ecological factors might lead to cladistic asymmetry. The pattern itself seems extremely improbable at first, since it implies that only one (in the extreme) of the most recently originated lineages is susceptible to further bifurcation. We of course have no idea, until further empirical research and modeling have been completed (e.g., Savage, 1983), whether this pattern occurs in nature more or less frequently than would be predicted by a random model of evolution, but in any case, a deterministic explanation of the pattern would be worthy of consideration. One possibility would be "directed speciation" as discussed by Grant (1963) and Stanley (1979), where ecologic gradients define the opportunities for speciation. Another possibility is that adaptive improvement or the biomechanical constraints relevant to a given morphological transition could generate an "adaptive archipelago" that would be followed by the speciational history of a clade. In each of these cases we may expect asymmetry at some level, either local or more global, depending on the nature and strength of the linearizing constraint. However, this will not necessarily result in slow evolution, since in these cases the properties of species, in a sense, *create* the opportunities for further speciation, rather than species being exposed to such opportunities by essentially random processes.

Another ecological situation that would lead

to cladistic asymmetry is the one hypothesized above, involving eurytopy, competitive exclusion, and increased probability of speciation associated with incipient extinction. Standing crop diversity would tend to remain low, and daughter species would tend to exclude or be excluded by their parent species. One of the few opportunities for the development of a larger standing crop of species, as well as some degree of cladistic symmetry, would be a major vicariance event such as may have separated North American and Eurasian limulids. As argued above, I would expect this mode of development of cladistic asymmetry to be associated with unusually low evolutionary rates. Given the ecological circumstances postulated, it would not be surprising to find low rates of intraspecific morphologic change, low rates of speciation, and low incidence of morphologically divergent speciation as well. However, the potential contribution to bradytely made by cladistic asymmetry is independent of any particular model of intraspecific morphologic change (as long as it is not both rapid and directed) and is independent of any assumption about how morphologic change is distributed relative to speciation events.

In summary, none of the explanations for bradytely considered here appear to be mutually incompatible or definitively rejected by the data. Support seems least strong for attributing a dominant role to low rates of intraspecific change, not because there is any reason to think such rates might be high, but rather because they are probably low enough even in many horotelic groups that it would be difficult to consider them responsible for the difference between horotely and bradytely. Choosing among the competing explanations of bradytely may prove difficult, not only because of inadequacies in the data, but also because they all tend to be associated with similar ecological contexts. It may be, in fact, that several processes operate simultaneously, at different levels, reinforcing one another. Each of the explanations of bradytely deserve much additional study, but the analysis of patterns of relationship and morphologic change seems particularly intriguing if only because it has not been generally considered. While comparisons of the geometry of phylogenetic trees may initially seem grossly epiphenomenal, they could also be the source of significant generalizations and a valid level of process study.

Literature

Abel, O. 1935. Vorzeitliche Lebensspuren. Jena: Gustav Fisher.

Annandale, N. 1901. The habits of Indian king crabs. Rec. Ind. Mus. 3:294–295.

Baptist, J. P., Smith, O. R., Ropes, J. W. 1957. Migrations of the horseshoe crab, *Limulus polyphemus* in Plum Island Sound, Massachusetts. U.S. Dept. Int. Fish. Wildl. Ser. Spec. Sci. Rept. Fisheries No. 220:1–15.

Barker, S. B. Cumming, G., Horsefield, K. 1973. Quantitative morphometry of the branching structure of trees. J. Theor. Biol. 40:33–43.

Bergström, J. 1979. Morphology of fossil arthropods as a guide to phylogenetic relationships, pp. 3–56. In: Gupta, A. P. (ed.), Arthropod phylogeny. New York: Van Nostrand Rheinhold Co.

Bleicher, M. 1897. Sur la découverte d'une nouvelle espèce de limule dans les marnes irisées de Lorraine. Bull. Soc. Sci. Nancy 14:116–126.

Briggs, D. E. G., Bruton, D. L., Whittington, H. B. 1979. Appendages of the arthropod *Aglaspis spinifer* (Upper Cambrian, Wisconsin) and their significance. Palaeontology 22:167–180.

Brown, G. G., Clapper, D. L. 1981. Procedures for maintaining adults, collecting gametes, and culturing embryos and juveniles of the horseshoe crab, *Limulus polyphemus* L., pp. 268–290. In: Committee on Marine Invertebrates (ed.), Laboratory animal management: marine invertebrates. Washington, DC: National Acad. Press.

Caster, K. E. 1938. A restudy of the tracks of *Paramphibius*. J. Paleont. 12:3–60.

Caster, K. E. 1940. Die sogenannten "Wirbeltierspuren" und die *Limulus*-Fährten der Solnhofener Plattenkalke. Paläont. Zeitschr. 22:12–29.

Caster, K. E. 1957. Problematica, pp. 1025–1032. In: Ladd, H. S. (ed.), Treatise on marine ecology and paleoecology, Vol. 2, Paleoecology. Geol. Soc. Amer., Mem. 67.

Caster, K. E., Brooks, H. K. 1956. New fossils from the Canadian–Chazyan (Ordovician) hiatus in Tennessee. Bull. Amer. Paleont. 36:157–199.

Desmarest, A. G. 1822. Histoire naturelle de crustacés fossiles: les crustacés proprement dits. Paris.

Eldredge, N. 1974. Revision of the Suborder Synziphosurina (Chelicerata, Merostomata), with remarks on merostome phylogeny. Am. Mus. Novit. No. 2543:1–41.

Eldredge, N. 1975. Survivors from the good old, old, old days. Nat. Hist. 84:60–69.

Eldredge, N. 1979. Alternative approaches to evolutionary theory. Bull. Carn. Mus. Nat. Hist. 13:7–19.

Eldredge, N., Cracraft, J. 1980. Phylogenetic patterns and the evolutionary process. New York: Columbia U. Press.

Eldredge, N., Gould, S. J. 1972. Punctuated equilibrium: an alternative to phyletic gradualism, pp. 82–115. In: Schopf, T. J. M. (ed.), Models in paleobiology. San Francisco: Freeman, Cooper.

Eldredge, N., Plotnick, R. 1974. Revision of the pseudoniscine merostome genus Cyamocephalus Currie. Am. Mus. Novit. No. 2557:1–10.

Fields, J. H. A. 1982. Anaerobiosis in Limulus, pp. 125–131. In: Bonaventura, J., Bonaventura, C., Tesh, S. (eds.), Physiology and biology of horseshoe crabs: studies on normal and environmentally stressed animals. New York: Liss.

Fisher, D. C. 1975a. Evolution and functional morphology of the Xiphosurida. Unpubl. Ph.D. Diss., Harvard U.

Fisher, D. C. 1975b. Swimming and burrowing in Limulus and Mesolimulus. Fossils and Strata 4:281–290.

Fisher, D. C. 1977. Functional significance of spines in the Pennsylvanian horseshoe crab Euproops danae. Paleobiology 3:175–195.

Fisher, D. C. 1979. Evidence for subaerial activity of Euproops danae (Merostomata, Xiphosurida), pp. 379–447. In: Nitecki, M. H. (ed.), Mazon Creek fossils. New York: Academic Press.

Fisher, D. C. 1980. The role of stratigraphic data in phylogenetic inference. Geol. Soc. Amer., Abst. Prog. 12:426.

Fisher, D. C. 1981. The role of functional analysis in phylogenetic inference: examples from the history of the Xiphosura. Amer. Zool. 21:47–62.

Fisher, D. C. 1982. Phylogenetic and macroevolutionary patterns within the Xiphosurida. Proc. Third N. Amer. Paleont. Conv. 1:175–180.

Fraenkel, G. 1960. Lethal high temperatures for three marine invertebrates: Limulus polyphemus, Littorina littorea and Pagurus longicarpus. Oikos 11:171–182.

French, K. A. 1979. Laboratory culture of embryonic and juvenile Limulus, pp. 61–71. In: Cohen, E. (ed.), Biomedical applications of the horseshoe crab (Limulidae). New York: Liss.

Goldring, R., Seilacher, A. 1971. Limulid undertracks and their sedimentological implications. N. Jb. Geol. Paläont., Abh. 137:422–442.

Gould, S. J., Eldredge, N. 1977. Punctuated equilibria: the tempo and mode of evolution reconsidered. Paleobiology 3:115–151.

Grant, V. 1963. The origin of adaptations. New York: Columbia U. Press.

Holland, Jr., F. D., Erickson, J. M., O'Brien, D. E. 1975. Casterolimulus: a new late Cretaceous generic link in limulid lineage. Bull. Amer. Paleont. 67:235–249.

Honda, H. 1971. Description of the form of trees by the parameters of the tree-like body: effects of the branching angle and the branch length on the shape of the tree-like body. J. Theor. Biol. 31:331–338.

Johansen, K., Petersen, J. A. 1975. Respiratory adaptations in Limulus polyphemus (L.), pp. 129–145. In: Vernberg, F. J. (ed.), Physiological ecology of estuarine organisms. Columbia: U. South Carolina Press.

Kristoffersen, Y. 1978. Sea-floor spreading and the early opening of the North Atlantic. Earth Planet. Sci. Letters 38:273–290.

Kropach, C. 1979. Observations on the potential for Limulus aquaculture in Israel, pp. 103–106. In: Cohen, E. (ed.), Biomedical applications of the horseshoe crab (Limulidae). New York: Liss.

Leopold, L. B. 1971. Trees and streams: the efficiency of branching patterns. J. Theor. Biol. 31:339–354.

Leopold, L. B., Langbein, W. B. 1962. The concept of entropy in landscape evolution. U.S. Geol. Surv. Prof. Pap. 500-A:1–20.

Mandelbrot, B. 1977. Fractals: form, chance, and dimension. San Francisco: Freeman.

Mandelbrot, B. 1982. The fractal geometry of nature. Salt Lake City: Freeman.

Mayer, A. G. 1914. The effects of temperature upon tropical marine animals. Pap. Tortugas Lab. 6:1–24.

McMahon, T. A. 1975. The mechanical design of trees. Sci. Amer. 233:92–102.

McMahon, T. A., Kronauer, R. E. 1976. Tree structures: deducing the principle of mechanical design. J. Theor. Biol. 59:443–466.

Miles, R. S. 1977. Dipnoan (lungfish) skulls and the relationships of the group: a study based on new species from the Devonian of Australia. Zool. J. Linn. Soc. London 61:1–328.

Parker, H., Horsfield, K., Cumming, G. 1971. Morphology of distal airways in the human lung. J. Appl. Physiol. 31:386–391.

Pearse, A. S. 1928. On the ability of certain marine invertebrates to live in diluted sea water. Biol. Bull. 54:405–409.

Pocock, R. I. 1902. The taxonomy of recent species of Limulus. Ann. Mag. Nat. Hist. 9:256–266.

Pruvost, P. 1930. La faune continentale du terrain houiller de la Belgique. Mém. Mus. Roy. Hist. Nat. Belg. 44:105–282.

Raymond, P. E. 1944. Late Paleozoic xiphosurans. Bull. Mus. Comp. Zool. 94:473–508.

Reeside, J. B., Harris, D. V. 1952. A Cretaceous horseshoe crab from Colorado. J. Wash. Acad. Sci. 42:174–178.

Reynolds, W. W., Casterlin, M. E. 1979. Thermoregulatory behavior and diel activity of *Limulus polyphemus*, pp. 47–59. In: Cohen, E. (ed.), Biomedical applications of the horseshoe crab (Limulidae). New York: Liss.

Riek, E. F. 1968. Re-examination of two arthropod species from the Triassic of Broodvale, New South Wales. Rec. Aust. Mus. 27:313–321.

Riska, B. 1981. Morphological variation in the horseshoe crab *Limulus polyphemus*. Evolution 35:647–658.

Robertson, J. D. 1970. Osmotic and ionic regulation in the horseshoe crab, *Limulus polyphemus* (Linnaeus). Biol. Bull. 138:157–183.

Romero, A., Via Boada, L. 1977. *Tarracolimulus rieki*, nov. gen., nov. sp., nuevo limulido del Triasico de Montral-Alcover (Tarragona). Cuad. Geol. Iber. 4:239–246.

Rudloe, A. 1979. *Limulus polyphemus:* a review of the ecologically significant literature, pp. 27–35. In: Cohen, E. (ed.), Biomedical applications of the horseshoe crab (Limulidae). New York: Liss.

Savage, H. M. 1983. The shape of evolution: tree topology. Biol. J. Linn. Soc. 20:225–244.

Sekiguchi, K., Nakamura, K. 1979. Ecology of the extant horseshoe crabs, pp. 37–45. In: Cohen, E. (ed.), Biomedical applications of the horseshoe crab (Limulidae). New York: Liss.

Sekiguchi, K., Nakamura, K., Sen, T. K., Sugita, H. 1976. Morphological variation and distribution of a horseshoe crab, *Tachypleus gigas,* from the Bay of Bengal and the Gulf of Siam. Proc. Jap. Soc. Syst. Zool. 12:13–20.

Sekiguchi, K., Nakamura, K., Seshimo, H. 1978. Morphological variation of a horseshoe crab, *Carcinoscorpius rotundicauda,* from the Bay of Bengal and the Gulf of Siam. Proc. Jap. Soc. Syst. Zool. 15:24–30.

Sekiguchi, K., Sugita, H. 1980. Systematics and hybridization in the four living species of horseshoe crabs. Evolution 34:712–718.

Selander, R. K., Yang, S. Y., Lewontin, R. C., Johnson, W. E. 1970. Genetic variation in the horseshoe crab (*Limulus polyphemus*), a phylogenetic "relic." Evolution 24:402–414.

Shuster, Jr., C. N. 1955. On morphometric and serological relationships within the Limulidae, with particular reference to *Limulus polyphemus* (L.). Unpubl. Ph. D. Diss., New York U.

Shuster, Jr., C. N. 1962. Serological correspondence among horseshoe "crabs" (Limulidae). Zoologica 47:1–8.

Shuster, Jr., C. N. 1979. Distribution of the American horseshoe "crab," *Limulus polyphemus* (L.), pp. 3–26. In: Cohen, E. (ed.), Biomedical applications of the horseshoe crab (Limulidae). New York: Liss.

Shuster, Jr., C. N. 1982. A pictorial review of the natural history and ecology of the horseshoe crab *Limulus polyphemus,* with reference to other Limulidae, pp. 1–52. In: Bonaventura, J., Bonaventura, C., Tesh, S. (eds.), Physiology and biology of horseshoe crabs: studies on normal and environmentally stressed animals. New York: Liss.

Simpson, G. G. 1944. Tempo and mode in evolution. New York: Columbia U. Press.

Simpson, G. G. 1953. The major features of evolution. New York: Columbia U. Press.

Stanley, S. M. 1975. A theory of evolution above the species level. Proc. Nat. Acad. Sci. 72:646–650.

Stanley, S. M. 1979. Macroevolution: patterns and process. San Francisco: Freeman.

Størmer, L. 1952. Phylogeny and taxonomy of fossil horseshoe crabs. J. Paleont. 26:630–639.

Størmer, L. 1955. Arthropoda 2: Merostomata, pp. 4–41. In: Moore, R. C. (ed.), Treatise on invertebrate paleontology, Part P. Geol. Soc. Amer. and U. Kansas Press, Lawrence, Ks.

Størmer, L. 1972. Arthropods from the Lower Devonian (Lower Emsian) of Alken an der Mosel, Germany. Part 2: Xiphosura. Senk. Leth. 53:1–29.

Stürmer, W., Bergström, J. 1981. *Weinbergina,* a xiphosuran arthropod from the Devonian Hunsrück Slate. Paläont. Zeitschr. 55:237–255.

Van der Heide, S. 1951. Les arthropodes du terrain houiller du Limbourg meridional. Med. Geol. Sticht. Maastrict. Ser. C, iv 3:1–84.

van der Hoeven, J. 1838. Récherches sur l'histoire naturelle et l'anatomie des Limules. Leyde.

Walther, J. 1904. Die Fauna der Solnhofener Plattenkalke bionomisch betrachtet. Jenaische Denkschr. Med. Nat. Gesell. 11:135–214.

Waterman, T. H. 1953. Xiphosura from Xuong-Ha. Amer. Sci. 41:292–302.

Waterman, T. H. 1958. On the doubtful validity of *Tachypleus hoeveni* Pocock, an Indonesian horseshoe crab (Xiphosura). Postilla 36:1–17.

Woodward, H. 1867. On some points in the structure of the Xiphosura, having reference to their relationship with the Eurypteridae. Geol. Soc. London Quart. J. 23:28–37.

Woodward, H. 1872. Sixth report on fossil Crustacea. Geol. Mag. 9:563–569.

23
Peripatus as a Living Fossil

Michael T. Ghiselin

Department of Invertebrate Zoology, California Academy of Sciences, San Francisco, CA 94118

Peripatus is sometimes considered a living fossil because it and other onychophorans are transitional between Arthropoda and Annelida and retain a large number of archaic features (Bouvier 1905, 1907; Zacher 1933; Cuénot 1968). Although *Peripatus* was originally described as a leg-bearing slug (Guilding 1826), it looks more like a caterpillar or a soft-bodied centipede with claws at the end of its stubby appendages. Hoyle and del Castillo (1979) object to calling onychophorans living fossils on the grounds that they have certain "unique features," but this only means that divergent evolution has occurred. The annelidan character of the eyes (Eakin and Westfall 1965) is but one example of a conservative feature. Except for the mode of reproduction, the group is quite homogeneous. There are about 70 species in two families. The Peripatidae, which include *Peripatus* in the strict sense, are viviparous, with a true placenta, or else have a yolky egg that develops internally. The Peripatopsidae are sometimes oviparous but may have internally developing young with various amounts of provisioning of the embryo via secretions, but no true placenta.

Although biogeography shows that the group is very old, there is little if anything of a fossil record. Modern onychophorans are all terrestrial, soft-bodied organisms whose habitat is not conducive to fossilization. A marine worm, *Xenusion,* from the Lower Cambrian is perhaps an annelid. It nonetheless possessed stubby projections suggestive of onychophoran "lobopod" appendages (see Heymons 1928; Häntschel 1962). Whittington (1978; see also Hutchinson 1969) has reexamined the lobopod *Aysheaia pedunculata* from the Middle Cambrian Burgess Shale. This marine animal bears some general resemblances to the modern onychophorans, but not enough to suggest more than its belonging to a clade along with tardigrades and perhaps pentastomids. A more likely possibility is the Middle Pennsylvanian Mazon Creek *Helenodora inopinata,* which could have been marine, freshwater, or terrestrial (Thompson and Jones 1980).

Relationships

The onychophorans are generally placed in a phylum by themselves, mainly in order to avoid the issue of their relationships. However, the available evidence is consistent with treating Onychophora as part of a strictly monophyletic phylum Arthropoda, albeit one that evinces much parallelism. If Manton (1977) is correct with respect to her thesis that the onychophorans are the sister-group of the myriapods and insects, then their closest relatives are the most diverse and abundant group of terrestrial animals. Various authors (e.g., Hoyle and Williams 1980) have challenged this view.

Perhaps the onychophorans branched off earlier. It is conceivable that they are closest to Tardigrada + Pentastomida.

Ecology

Onychophorans occupy a niche resembling that of centipedes, except that they are slow-moving animals restricted to moist situations (Lavallard et al. 1975). They feed upon small arthropods, which they can pursue into cracks and crevices by virtue of their flexible bodies. Poorly resistant to desiccation, they are very hygrophilic, though they avoid air that is completely saturated with water vapor (Bursell and Ewer 1950). Basically these animals are generalists with respect to feeding habits and specialists with respect to habitat. Some ecological divergence is represented by two species, ranked as monotypic genera and not closely related, of blind, troglobiotic forms (Kemp 1914; Peck 1975). The group is, therefore, "relictual" in one ecological sense.

Fig. 1. A living onychophoran, photographed by M. Gosliner in Tsitsikoma National Park, South Africa.

distribution implies that the group has not evolved significant differences since the breakup of Gondwanaland. This is true even of the modes of reproduction. Of course local speciation and adaptations of a minor nature have occurred, and some minor changes (perhaps the evolution of troglobiotic forms is an example) cannot be ruled out.

Distribution in Space

The biogeography of onychophorans indicates that the group is very old and that it has long existed in a condition of stasis. Poorly suited to passive dispersal, they provide excellent examples of the effects of continental drift (see Clark 1915; Brues 1923; Gravier and Fage 1926; Zacher 1933; Vachon 1954; Brinck 1956). We do not possess a good, modern phylogenetic classification of the group, but the taxa are sufficiently close to natural to allow some cautious historical inferences. The Peripatidae are a circumtropical group. Those with a true placenta are widespread in the Caribbean, Central America, and northern South America, and have one genus in a small area in western Africa. Their sister-group occurs in the eastern Himalayas, the Malay Peninsula, Sumatra, and Borneo. The Peripatopsidae are austral. They show connections between South America and South Africa on the one hand, and South Africa, Australia, Tasmania, and New Zealand on the other, with a distinct group in New Guinea, New Britain, and Ceram. This

Variability

Because onychophorans are inconvenient to collect, hard to culture, and characterized by a long generation time, they have not yet been the object of genetic research. However, their phenotypic variability has been well documented, and it evidently has a genetic basis. Color patterns can be highly variable (Wheeler 1898). Of greater interest is the variation in the number of limbs as a function of size, sex, and species. Comparison of different species suggests that there is a positive correlation between the number of limbs and the size of the body, one which would be closer, perhaps, if data were not expressed in terms of length. In all onychophorans, the males are smaller than the females, especially in the viviparous forms. Evidently, the sexual dimorphism is most pronounced in the species that attain the greatest bodily size (as is true of many dimorphic animals), but this point has not been adequately investigated. Males are also rarer in collections. This is due mostly, if not exclusively, to differential mortality, since clutches *in utero* have a sex ratio of approximately 1 : 1. The adaptive significance for this difference between the sexes has been explained as resulting from sexual selection (Ghiselin 1974). Fertilization can occur long after copulation, and the anatomical arrangement is such that intense sperm competition is likely (see Manton 1938). The males are in a position to maximize fitness by reproductive activity, frequent impregnation, and production of a large volume of sperm rather than by growth and survival. Indeed, males may even be born sexually mature (Rucker 1900). This pattern somewhat resembles that described for viviparous surfperch by Warner and Harlan (1982).

In most species, the males have fewer legs than the females do. The number of legs is congenital, at least as a rule. For some species it has been asserted that at birth the males are smaller than the females, but this requires confirmation. Males and females both vary in the number of legs, with males, as one would expect, tending to have fewer legs (Campiglia and Lavallard 1973; Lavallard and Campiglia 1975). There are some interesting sex-correlated patterns of variability of the sort that should have a genetic basis. Lavallard and Campiglia (1973) studied 683 specimens of *Peripatus acacioi* and found that 3.15% of the males and 1.89% of the females had more legs on one side than the other.

We lack a good, modern study of variability in space within what are known to represent reproductive populations. Nonetheless, the pattern as currently understood is such as one would expect from the normal process of speciation and modest divergence produced and maintained by selection.

Comparative Aspects

In general, the onychophorans provide a spectacle of persistent stasis. They have not changed much in a long period of time. Neither do they show any evidence of the sort of adaptive radiation so obvious in other lineages of arthropods. They have undergone a substantial amount of speciation, but the main outcome has been their maintaining themselves in a peculiar niche. The troglobiotic forms are specialists— minor variants upon a common theme.

We should not infer, however, that the onychophorans are incapable of undergoing extensive structural and functional change. This is contradicted by remarkable diversity in their reproduction, which might be compared to that of monotremes, marsupials, and placentals taken together; that is to say, the whole class Mammalia. The analogy with fishes is instructive (Wourms 1981). Here viviparity has frequently evolved, and it occurs in the "living fossil" *Latimeria*. On the other hand, the reproductive changes seem to have occurred very long ago, to judge from the biogeographical evidence, so a long period of stasis for the group still holds. The reason for stasis may have something to do with the ecology of the group, or with the limitations of functional anatomy. However, it does not seem attributable to some general lack of lability, especially an unspecified peculiarity of the "genetic program."

Literature

Bouvier, E.-L. 1905. Monographie des Onychophores. I. Ann. Sci. Nat. Zool. (9)2:1–383.

Bouvier, E.-L. 1907. Monographie des Onychophores. II. Ann. Sci. Nat. Zool. (9)5:61–318.

Brinck, P. 1956. Ch. 1. Onychophora. A review of the South African species, with a discussion on the significance of the geographical distribution of the group, pp. 7–32. In: Hanstrom, B. et al. (eds.), South African animal life, Vol. 4. Almquist & Wiksell Stockholm.

Brues, C. T. 1923. The geographical distribution of the Onychophora. Amer. Nat. 57:210–217.

Bursell, E., Ewer, D. W. 1950. On the reactions to humidity of *Peripatopsis mosleyi* (Wood-Mason). J. Exp. Biol. 26:235–253.

Campiglia, S., Lavallard, R. 1973. Contribution à la biologie de *Peripatus acacioi* Marcus et Marcus. II. Variations du poids des animaux en fonctions du sexe et du nombre des lobopodes. Bol. Zool. Biol. Mar. N.S. 30:499–512.

Clark, A. H. 1915. The present distribution of the Onychophora, a group of terrestrial invertebrates. Smithsonian Misc. Coll. 65(1):1–25.

Cuénot, L. 1968. Les onychophores. Traité Zool. 6:3–37.

Eakin, R. M., Westfall, J. A. 1965. Fine structure of the eye of *Peripatus* (Onychophora). Z. Zellf. Mikr. Anat. 68:278–300.

Ghiselin, M. T. 1974. The economy of nature and the evolution of sex. Berkeley: U. California Press.

Gravier, C., Fage, L. 1926. Remarques sur la distribution géographique des Péripates, pp. 725–727. In: Compte Rendu de la 49e Session, Association Francaise pour l'Avancement des Sciences, Grenoble, 1925.

Guilding, L. 1826. Mollusca carribbaeana. Zool. J. 2:437–444.

Häntschel, W. 1962. Trace fossils and problematica, pp. 177–245. In: Moore, R. C. (ed.), Treatise on invertebrate paleontology, Part W. Geol. Soc. Amer. and U. Kansas Press, Lawrence, K.S.

Heymons, R. 1928. Über Morphologie und verwandtschaftliche Beziehungen des *Xenusion auerswaldae* Pomp. aus dem Algonkium. Z. Morphol. Ökol. Tiere 10:307–329.

Hoyle, G., del Castillo, J. 1979. Neuromuscular transmission in *Peripatus*. J. Exp. Biol. 83:13–29.

Hoyle, G., Williams, M. 1980. The musculature of *Peripatus dominicae* and its innervation. Phil. Trans. Roy. Soc. London (B)288:481–510.

Hutchinson, G. E. 1969. *Aysheaia* and the general morphology of the Onychophora. Amer. J. Sci. 267:1062–1066.

Kemp, S. 1914. Onychophora. Zoological results of the Abor expedition, 1911–1912. Rec. Ind. Mus. 8:471–492.

Lavallard, R., Campiglia, S. 1973. Contribution à la biologie de *Peripatus acacioi* Marcus et Marcus. I. Pourcentage des sexes et variations du nombre des lobopodes dans un échantilonage de plusieurs centaines d'individus. Bol. Zool. Biol. Mar. N.S. 30:483–498.

Lavallard, R., Campiglia, S. 1975. Contribution à la biologie de *Peripatus acacioi* Marcus et Marcus (onychophore). V. Étude des naissances dans un élevage de laboratoire. Zool. Anz. 195:338–350.

Lavallard, R., Campiglia, S., Parisi Alvares, E., Valle, C. M. C. 1975. Contribution à la biologie de *Peripatus acacioi* Marcus et Marcus. III. Étude

descriptive de l'habitat. Vie et Milieu (C)25:87–118.

Manton, S. M. 1938. Studies on the Onychophora. IV. The passage of spermatozoa into the ovary in *Peripatopsis* and the early development of the ova. Phil. Trans. Roy. Soc. London (B)228:421–441.

Manton, S. M. 1977. The Arthropoda: habits, functional morphology, and evolution. Oxford: Clarendon Press.

Peck, S. B. 1975. A review of the New World Onychophora with a description of a new cavernicolous genus and species from Jamaica. Psyche 82 (3–4):341–358.

Rucker, A. 1900. A description of the male of *Peripatus Eisenii* Wheeler. Biol. Bull. 1:251–259.

Thompson, I., Jones, D. S. 1980. A possible onychophoran (*Helenodora inopinata*, new genus, new species) from the Middle Pennsylvanian Mazon Creek beds of Northern Illinois, U. S. A. J. Paleont. 54:588–596.

Vachon, M. 1954. Répartition actuelle et ancienne des Onychophores. Rév. Gén. Sci. Pures Appl. 61:300–308.

Warner, R. R., Harlan, R. K. 1982. Sperm competition and sperm storage as determinants of sexual dimorphism in the dwarf surfperch, *Micrometrus minimus*. Evolution 36:44–45.

Wheeler, W. M. 1898. A new *Peripatus* from Mexico. J. Morphol. 15:1–8.

Whitman, E. O. 1891. Spermatophores as a means of hypodermic impregnation. J. Morphol. 4:361–406.

Whittington, H. B. 1978. The lobopod animal *Aysheaia pedunculata* Walcott, Middle Cambrian, Burgess Shale, British Columbia. Phil. Trans. Roy. Soc. London, (B)284:165–197.

Wourms, J. P. 1981. Viviparity: the maternal-fetal relationship in fishes. Amer. Zool. 21:473–515.

Zacher, F. 1933. Onychophora. Hb. Zool. 3(2):79–138.

24
Neopilina, Neomphalus, and *Neritopsis,* Living Fossil Molluscs

Roger L. Batten

Department of Invertebrates, The American Museum of Natural History, New York, NY 10024

The living fossil mollusc examples used here represent three different phylogenetic patterns. The neopilinid monoplacophorans have a disjunct time pattern that separates the living species from the Cambro-Devonian tryblidiid (limpet) monoplacophorans—a gap just short of 430 million years—with the extinction of the last tryblidiid *Archaeophiala* in the Lower Devonian. The Silurian *Pilina unguis* (Lindström 1880) from the middle Silurian of Gotland is nearly identical (except for shell thickness) to *Vema (Vema) ewingi* Clarke and Menzies 1959 (Fig. 1).

The second example, that of *Neritopsis radula* (Linné 1758), illustrates a very long-lived species (it first appeared in the Paris Basin Eocene some 50 million years ago) that belongs to a family that flourished in the Late Paleozoic and Mesozoic. The rift limpet *Neomphalus fretterae* (McLean 1981) is an example of an arachaic member, without a fossil record, of the order Mesogastropoda. In this case, the internal anatomy is transitional between the primitive order Archeogastropoda and the more advanced Mesogastropoda. The shell of the living species is very highly convergent on modern advanced mesogastropod limpets (the calyptraeids).

The primary problem with studying molluscs, particularly the gastropods and other univalves, is the presence of relatively few character complexes, so that multivariate analyses are not feasible; with few characters to study, it is difficult to recognize synapomorphies, homologs, or convergences. Even so, most taxonomic schemes of living molluscs seem to be based on shell features, particularly on the generic and specific levels. It is also to be observed that most of the organ systems are plesiomorphic.

Neritopsis radula is Long Lived

Neritopsis radula, widely distributed in the Indo-Pacific bioprovince, is uncommon, and the sole survivor of the genus that first appeared in the Middle Triassic (225 million years ago) (see Fig. 2). There are a total of about 100 species of *Neritopsis* recognized, 7 from the Triassic, 80 from the Jurassic and Cretaceous of the Tethyan province, 11 from the Early Cenozoic, and one from the Miocene of central Europe, in addition to *N. radula. Neritopsis radula* first appeared in the Eocene, and fossil populations are scarce and essentially identical to those of today in morphologic variation. *Neritopsis* was almost certainly derived from a morphotype similar to *Trachydomia imbricata* Batten 1979 (a Permian Tethyan species from southeast Asia). The family Neritopsidae (Neritacea, Archeogastropoda) is composed of 14 genera worldwide in distribution from the Middle Devonian to Recent, but with 11 genera confined to the Late Paleozoic to Early and Middle Trias-

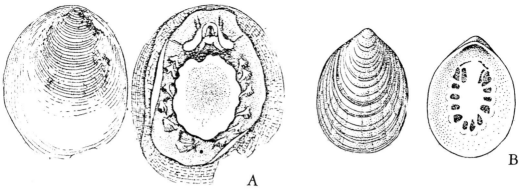

Fig. 1. (A) *Vema (Vema) ewingi* (Clarke and Menzies), 1959, ×5.3. (B) *Pilina unguis* (Lindström 1880), ×0.7. These figures are from Knight et al. (1960).

sic. Only one genus—*Neritopsis*—is known from post-Cretaceous time. By Permian time there was a very rapid increase in diversity of the neritopsids, particularly in the Tethys Sea. Most species are less than a geological period in duration, and most were probably specialists because they are short-lived and restricted geographically. It is difficult to document all of the speciation events, but in North America alone 40 species, mostly of *Naticopsis,* appeared from Mississippian to Middle Permian. There are at least 40 *Naticopsis* species in the Tethyan

Carboniferous through Jurassic from North Africa eastward to China. In North Africa and Sicily there are Permian species not found elsewhere. By Cretaceous time only *Damsia* (besides *Neritopsis*) remained, and is found at only a few Tethyan localities in France and Germany (see Batten 1979; Cossman 1925).

Today *N. radula* is found throughout the Indo-Pacific region where it is confined to small, isolated populations probably as a specialist grazer. It ranges from the sub-littoral at Mauritius to 200 m off Hawaii in sands and on hard surfaces. The well-fitted operculum and rounded shape indicates that the organisms lived on hard substrate in a current and/or tidal zone (Lindsey, personal communication 12/81).

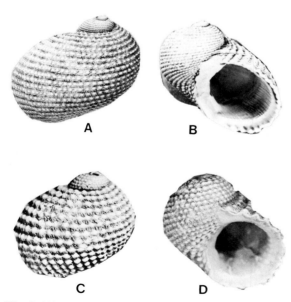

Fig. 2. (A) *Neritopsis radula* (Linné 1758), ×1.6, side view. (B) Same specimen, apertural view. (C) *N. radula* from the Miocene of Lapugy, Hungary, ×1.8, side view. (D) Same specimen, ×1.8, apertural view.

Neomphalus fretterae, a Primitive Mesogastropod Without a Firm Taxonomic Home

This Galapagos Rift limpet is, anatomically, a very primitive mesogastropod or advanced archeogastropod but not assignable to any living taxon of either order. The dibranchiate gills are generally associated with the archeogastropods but are also found in the more primitive mesogastropods, the Valvatacea. The position of the gonads, the rhipidoglossate radula, the anteriorly looped intestines, the overlapping esophageal pouches, and the radular diverticulum are basic archeogastropod features.

The shell, however, is highly convergent on the living *Cheilea cepacea* (Broderlip 1834, a calyptraed limpet) (Fig. 3) in many details, such as the arcuate (but much smaller) muscle platform under the apex, the shell shape, detailed ornament pattern, and in the shell structure (very high-angled and narrow multiple-layered crossed-lamellae). *Neomphalus fretterae* is convergent on the calyptraeids in having a long filamentous (but dibranchiate) gill over the head (the mesogastropods have a monofilamentous gill), incurrent channels are enclosed in the mantle cavity to the left of the neck, the monocardian heart is in the same central position, and they are filter feeders. The calyptraeids have, however, a taenioglossate radula, a single left kidney, a glassy style in the stomach, and other digestive features found only in the advanced mesogastropods. *Neomphalus fretterae* shares anterior epipodial neck lobes with the primitive mesogastropod taxa, the Cyclophora-

cea and Viviporacea, as well as a dystenoid nervous system, lack of a pretentacular snout, and a tentacle serving as a copulatory organ. Most important, *N. fretterae* is a monotocardian with a single right kidney and a glandular genital duct. It probably should be considered a separate, isolated archaic mesogastropod taxon, surviving by adaptation to a specialized habitat, much like the neritacean archeogastropods (Fretter et al. 1981:361). McLean (1981) suggests that *N. fretterae* should be assigned to the extinct Early Paleozoic order Macluritina based on the speculation that members of that order were filter feeders.

The paradox is that there is no similar limpet shell in any archeogastropod taxon (including the Macluritina) or lower Mesogastropoda groups living today or in the fossil record, and the primitive internal anatomy is not related to

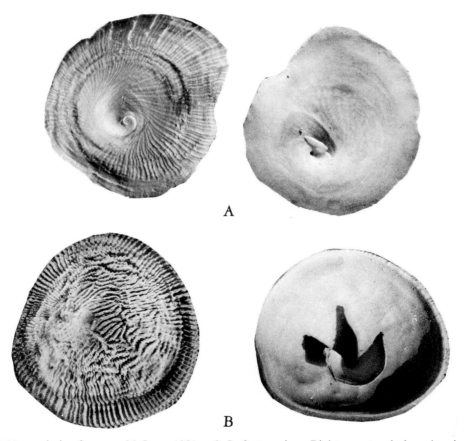

A

B

Fig. 3. (A) *Neomphalus fretterae* McLean 1981, ×2. Left, top view. Right, apertural view showing the small crescent-shaped muscle platform. (B) *Cheilea cepacea* (Broderlip 1834), ×2. Left, top view; the irregular elongate mounds are probably caused by differential growth on irregular substrate. Right, apertural view showing the very large muscle platform.

any single living gastropod group. What is certain is that without the internal anatomy, the shells would be identified as belonging to the shallow- and deep-water calyptraeids.

Neomphalus fretterae occurs in what has to be the most specialized ecologic niche of the marine ecosystem: in the thermal springs chemosynthetic system in the Galapagos Rift (Corliss et al. 1979). They feed exclusively on sulfur-producing bacteria and are found in large populations. They also occur in the East Pacific Rift just south of Baja California (MacDonald and Luyendyk 1981).

Neopilina, the Classic Living Fossil

The tryblidiid monoplacophoran *Neopilina galatheae* Lemche 1957 was first recognized in 1956 in a bottom sample from the Galathea expedition oceanic station #716 (9° 23′N, 89° 32′W, in 3590 m) collected May 16, 1952, as a metameric, nontorted, limpet-like mollusc related to the gastropods and the chitons (Amphineura). The notion that the group of Cambro-Devonian cap-shaped, multiple muscle-scarred limpets were not torted gastropod limpets but rather an untorted ancestor halfway between the annelid worms and the molluscs was speculated by Wenz (1940) and expanded on by Knight (1952).

For the most part, the 25 or so known occurrences of the neopilinids are in a variety of depths and habitats ranging from the gravelly bottom of the Cortez Bank off San Diego in 174 m to the green-gray muddy bottom of the Peru-Chile Trench in 6489 m (McLean 1979). There are two genera, two subgenera, and seven species. Lemche and Wingstrand (1959) originally reported that *Neopilina galatheae* lived in a soft, dark clay. This has always been a puzzle since the shape of the shell and the large foot is associated with a rocky or hard substrate habitat. All subsequent nontrench finds have been from rocky or gravelly bottoms. All living species are generalist deposit feeders, and the gut always contains diatoms, radiolaria, pelagic foraminiferans, or sponge spicules. The geographic range cannot be accurately defined, at present, because only *N. galatheae* and *Vema*

(*Vema*) *ewingi* are known from more than one locality. They are found mostly in the eastern Pacific, but isolated occurrences are known worldwide. Much more collecting must be done before a picture of their zoogeographic distribution can be constructed.

The two genera are recognized based on *Neopilina* having five pairs of gills and *Vema* having six pairs; the four species of *Neopilina* and three species of *Vema* demonstrate the consistency of this important character complex. Shell structure also appears to be generically important, with *Neopilina* having a very thin inner nacreous layer and a very thick outer sphaerulitic (complex prismatic) layer that is not visible through the thick periostracum.

Vema also has a thick nacreous inner layer, but the dominant layer is composed of large, blocky, simple prisms that form, in some species, a quincuncial pattern visible through the thin periostracum. The radula does not conform to the generic separation based on gill number and shell structure, but rather to absolute shell size (McLean 1979); for example, *N. velerons* and *V.* (*V.*) *hyalina* are the smallest sized species in each genus and they have the most similar first laterals and first marginal teeth.

Generically, there are also six pairs of nephridia with corresponding nerve ganglia and two pairs of gonads as opposed to two pairs of nephridia (but with eight pairs of ramifying masses) in the chitons which also have 2–80 pairs of gills and four pairs of nerve cords and a single gonad. Apomorphic features used to recognize species are size, shell ornament patterns, height of shell, apex features, and details of radular morphology.

Monoplacophora Classification

The Monoplacophora were one of the more successful of a group of early molluscs that originated in the Cambrian radiation. Along with the cap-shaped tryblidiids are the order Cyrtonellida (= Cyclomya), which are coiled or highly conical, curved shells represented by 7 families with 39 genera in the Cambrian and 18 post-Cambrian genera that became extinct in the Middle Devonian. The order Tryblidiida consists of 3 families with 12 Cambrian genera and 14 post-Cambrian genera, 12 of which were extinct by the Lower Devonian (see Runnegar and Jell 1976 for a full discussion of the fossil Monoplacophora; and McLean 1979 for a discussion of living Monoplacophora).

Class Monoplacophora Knight 1952
Order Tryblidiida Lemche 1957 (= Tergomya Horny 1965)
Family Neopilinidae Knight and Yochelson 1958
Neopilina Lemche 1957
Type species: *N. galatheae* Lemche 1957, Nature V.159, p. 413
Important features: relatively thick shells and periostraca and five pairs of gills.
Neopilina galatheae Lemche 1957
Type locality: off Central America, 9° 23'N, 89° 32'W, 2500–3591 m.
Important features: radial and concentric ribbing mostly in the early stages; moderately developed periostracum; shell moderately thick, shell low, apex just barely projects over the anterior margin; shape is almost circular, and the size averages 30 mm in length and width and 10 mm high.
Record: known from a single locality off Cedros Island, Mexico, 27° 52'N, 115°, 2739 m. Fourteen specimens were collected.
Bottom condition: not published in the original report.
Neopilina adenensis Tebble 1967
Important features: moderately large shell (10.7 mm long by 10.1 mm wide and 3.0 mm high), is moderately thick and has prominent, closely packed, concentric, irregular ribbing with faint radial threads; the apex is hooked but not overlapping the margin; the postoral tentacles are branched.
Record: a single specimen from off Oman, 13° 50'N, 50° 47'E in 3950–3000 m.
Bottom condition: not reported.
Neopilina bruuni Menzies 1968
Important features: relatively large shell (15.0 mm long by 13.0 mm wide), the irregular concentric ribbing is weakly formed; obscure radial threads, present only in early stages; outer shell layer irregular, rectangular prisms visible through the thin periostracum; the pointed small apex slightly overhangs, but is not near anterior margin.
Record: known from a single specimen in the Milne-Edwards Deep of the Peru Trench, 8° 54'S, 80° 41'W in 4825 m.
Bottom condition: not reported.
Neopilina oligotropa Rokop 1972
Important features: the small oval shell is 3.0 mm long, 2.5 mm wide, and 0.9 mm high, the apex is short, broad, and the apex does not overhang the margin; the radial threads are dominant, forming reenforcement nodes with the discontinuous concentric threads; the shell is very thin and discoid in shape.
Record: one specimen found in mid-Pacific, 680 miles north of Hawaii, 30° 05N, 156° 11'W in 6065–6079 m in oligotrophic water.
Bottom condition: red clay, manganese nodules associated with gastropods and bivalves, not specified.
Vema (Clarke and Menzies) 1959
Type species: *Neopilina* (*Vema*) *ewingi* Clarke and Menzies 1959
Important features: the shells are thin and the periostracum is also very thin; there are six pairs of gills; the ornament is not well developed and the shell structure of the thick outer layer consists of elongate simple prisms.
Vema (*Vema*) Clarke and Menzies 1959
Important features: large-sized shells and faint ornament.
Vema (*Vema*) *ewingi* Clarke and Menzies 1959
Type locality: the Peru-Chile Trench, 7° 35'S, 81° 24'W in 5821 m.
Important features: the thick shells have fine radial threads with concentric threads only on the early stages that seem to be defined by outer layer prisms; periostracum thick; gill lamellae composed of five to seven elements.
Record: eight other localities all within the Peru-Chile Trench and within 5 degrees of latitude and 2 degrees of

longitude of the type locality in 5817–6489 m.

Bottom condition: soft green muds with diatoms and a large fauna.

Vema (Vema) bacescui Menzies 1968

Important features: reticulate ornament, but rather faint; the shell is ovoid and large, 28 mm long, 16 mm wide with a pointed apex slightly overhanging and not protruding over anterior margin; post-oral tentacles with multiple branches.

Record: one specimen from the Milne-Edwards Deep, 8° 44′S, 80° 45′W, in 6000 m.

Bottom condition: not reported.

Vema (Laevipilina) McLean 1979

Type species: V. (Laevipilina) hyalina McLean 1979, Contr. Sci. Nat. Hist. Mus.-L.A. #307, p. 11.

Important features: small shells (adult 3 mm in length); no shell ornament; hexagonal shell structure prisms, apex blunt not protruding over margin; reduced number of gill lamellae.

Vema (Laevipilina) hyalina McLean 1979

Important features: same as subgenus

Record: Santa Rosa–Cortes Ridge eight specimens, two specimens from 32° 59′N, 119° 33′W in 374–384 m; two specimens from 32° 25′N, 119° 13′W in 229 m; and four specimens from 32° 58′N, 33° 05′W in 388 m of water.

Bottom condition: rocky terrain with a diverse fauna.

The Question of Metamerism

With the discovery of *N. galathea* came the striking, and unexpected, observation that it was segmented (according to Lemche in the original description) with repeated organs, thus linking the annelid worms to the molluscs as potential ancestors. In fact, in contrast to the annelids, there is only partial or pseudo-metamerism, and we must assume that the chitons are a sister-group of the Monoplacophora.

The Class Monoplacophora is composed, in part, of the cap-shaped limpets that are bilaterally symmetrical with repeated organs and repeated lateral-pedal muscles that form multiple, paired scars on the shell. Wenz (1940) conceived that the tryblidiid limpets were unlike the torted gastropod limpets such as the patellids and more like the chitons because the tryblidiids had eight pairs of discrete muscle scars that correspond to the eight plates of the chitons; in essence, he thought of a monoplacophoran shell as a fused set of eight chiton plates.

Neopilinids: Out Group Comparisons

The neopilinids have a docoglossate radula shared with the patellid limpets and the chitons. It is believed that since the Paleozoic tryblidiids were shallow-water rock dwellers with a habitat similar to that of the chitons and patellids of today; the docoglossate condition implies convergence. The crossed-lamellar wall structure of the patellids suggests no relationship with the chitons or the neopilinids. The eight pairs of latero-pedal muscles in both the chitons and the tryblidiid (limpet) monoplacophorans are serial repeats but nonetheless permit a sister-group comparison. The patellids are torted and therefore gastropods, thus far removed from any relationship with the neopilinids.

With so few data, speculation can be (and is) seemingly endless regarding the origin of the family Neopilinidae. The most speculative and least potentially documentable is that the Devonian tryblidiids managed to enter the deep ocean environment. Since deep ocean deposits are virtually unknown in the fossil record, we cannot substantiate this. Analyses of deep ocean faunas (Valentine 1973:466) tend to nullify the concept that the deep ocean houses primitive remnants—"the home of living fossils." Taylor and Forester (1979:411) have shown that cold-water biofacies tend to be similar to each other regardless of latitude or depth. However, since shallower water biofacies have failed to yield any monoplacophorans since the Devonian, we must keep the deep habitats in mind.

There is an important difference between the neopilinids and the Paleozoic tryblidiids: By and large the Paleozoic species are an order of magnitude thicker. This reflects the higher energy Paleozoic environments and also accounts for the unlikelihood that the monoplacophorans were in shallow-water deposits in post-Devonian time, for the thicker shells would certainly have been preserved somewhere and collected.

Literature

Batten, R. L. 1979. Permian Gastropods from Perak, Malaysia. Part 2. The trochids, patellids and neritids. Amer. Mus. Novit. #2685.

Cesari, P., Guidastri, R. 1976. Contributo alla conoscenza dei Monoplacophori recenti. Conchiglie, Anno 12, 1112:223–250.

Corliss, J. B., Dymond, J., Gordon, L. I., Edmond, J. M., Herzen, R. P. Von, Ballard, R. D., Green, K., Williams, D., Bainbridge, A., Crane, K., Andel, T. H. 1979. Submarine thermal springs on the Galapagos Rift. Science 203:1073–1083.

Cossman, M. 1925. Essais de paleoconchologie comparée, V.13, pp. 82–96. Paris.

Filatova, Z. A., Vinogradova, N. G., Moskalev, L. I. 1968. Molluscs of the Cambro-Devonian class Monoplacophora found in the North Pacific. Nature 220:1114–1115.

Fretter, V., Graham, A., McLean, J. 1981. The anatomy of the Galapagos rift limpet, *Neomphalus fretterae*. Malacologia 21 (122):337–361.

Knight, J. B. 1952. Primitive fossil gastropods and their bearing on gastropod classification. Smithsonian Misc. Coll. Vol. 117, #13.

Knight, J. B., Batten, R. L., and Yochelson, E. L. 1960. Part I. Mollusca. *In* Moore, R. C. Ed. Treatise on Invertebrate Paleontology., Geol. Soc. Amer., I169–I351.

Lemche, H. 1957. A new deep-sea mollusc of the Cambro-Devonian class Monoplacophora. Nature 179:413–416.

Lemche, H., Wingstrand, K. G. 1959. The anatomy of *Neopilina Galatheae* Lemche, 1957., Galatheae Rept. 3:9–71.

Linsley, R. 1978. Shell form and the evolution of the gastropods. Amer. Scien. 66:432–411.

MacDonald, K. C., Luyendyk, B. P. 1981. The Cretaceous of the East Pacific rise. Sci. Amer. 244 (5):100–118.

McLean, J. H. 1979. A new monoplacophoran limpet from the continental shelf off southern California. Contr. Sci., Nat. Hist. Mus. Los Angeles #307.

McLean, J. H. 1981. The Galapagos rift limpet *Neomphalus:* relevance to understand the evolution of a major Paleozoic-Mesozoic radiation. Malacologia 21 (1–2):291–336.

Menzies, R. J. 1968. New species of *Neopilina* of the Cambro-Devonian class Monoplacophora from the Milne-Edwards Deep of the Peru-Chile Trench. Mar. Biol. Assoc. Ind., Proc. Sym. Mollusca, Sym. Ser. 3:1–9.

Runnegar, B. Jell, P. A. 1976. Australian Middle Cambrian Mollusca and their bearing on early molluscan evolution. Alcheringa 1(2):109–138.

Taylor, M. E., Forester, R. M. 1979. Distribution model for marine isopod crustaceans and its bearing on early Paleozoic paleogeography and continental drift. Geol. Soc. Amer. Bull. 90 (4), 401–413.

Valentine, J. W. 1973. Evolutionary Paleoecology of the Marine Biosphere. Prentice-Hall, Inc. Englewood Cliffs, N.J. 511 pp.

Wenz, W. 1940. Ursprung und frühe Stammesgeschichte der Gastropoden. Arch. Molluskenk. 72:1–110.

25
Pleurotomaria: Pedigreed Perseverance?

Carole S. Hickman

Department of Paleontology, University of California, Berkeley, CA 94720

Introduction

Living pleurotomariids or slit shells provide a rare glimpse at the biology of remnants of a primitive and predominantly Paleozoic group of gastropods. Discovery of the first living pleurotomariid (Fischer and Bernardi 1856) generated considerable scientific excitement, in precisely the same pattern that later attended the discovery of *Neopilina galatheae* (Lemche 1957).

Because of their rarity, large relative size as gastropods, striking form and color patterns, and restriction to small habitat islands in the mysterious deep sea, pleurotomariids also captured the interest of amateur malacologists and conchologists, reinforcing the mystique and generating an interesting tendency to describe and illustrate (sometimes in color) each new individual specimen that emerged from the deep.

Although there are now 15 described living species, and their population sizes are probably greater than once estimated (Hickman 1976), *Pleurotomaria* remains poorly understood biologically. The question of how and why it has survived total replacement by alternative innovations in gastropod evolution is of considerable interest. The data presented below provide the basis for evaluation of some aspects of this question.

Morphology and Anatomy of Living *Pleurotomaria*

Shell

The shell of *Pleurotomaria* is typically dextral, conispiral, and trochoid in form, with a flattened base that may be either umbilicate or nonumbilicate (Fig. 1A). The most conspicuous feature of the shell is the prominent deep labral emargination or slit in the outer lip at or near mid-whorl and the selenizone (slit-band, fasciole, or cicatrix) or spiral trace of the slit on the shell. The slit is of particular anatomical significance because it reflects the primitive gastropod condition (and original molluscan organization) of paired bipectinate ctenidia in an elongate, deep mantle cavity. The slit is functionally the site of elimination of the paired convergent exhalant streams of water flowing through the mantle cavity.

Shell sculpture standardly consists of a combination of closely spaced spiral and axial elements that intersect to produce a cancellate pattern, frequently forming small raised beads. Pleurotomariids tend to be brightly colored, usually with irregular blotches of reddish pigments on light-colored grounds. Similar patterns are reported from Cretaceous *Pleurotomaria* (Kanie et al. 1980). Adult sizes range from moderate to large. Although one of the

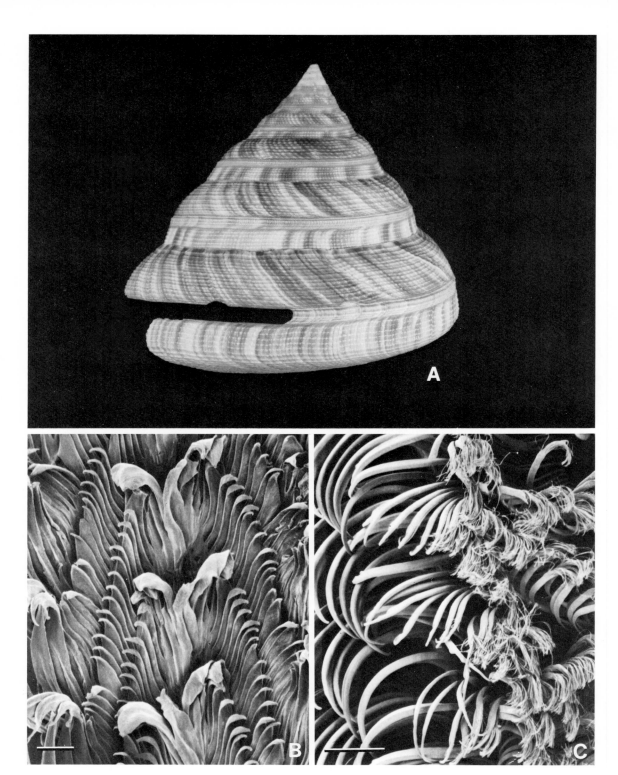

Fig. 1. Morphology of living *Pleurotomaria*. (A) Shell of *Pleurotomaria hirasei* Pilsbry 1903 from Japan. The specimen, 87.5 mm high, is in the collection of the Department of Invertebrate Zoology, California Academy of Sciences, Cat. No. 029229 (*ex* Stanford University). (B) Central complex of the assymmetric radula of *Pleurotomaria midas* Bayer 1965. The specimen is from the University of Miami, Rosenstiel Institute of Marine Science, Gerda Sta. G-10-16. Bar = 200 μm. (C) Marginal tooth complex of same radula figured in 1B, with hooked and "bristle" teeth. Bar = 200 μm.

living species, *Pleurotomaria rumphii* Schepman 1879, is known to attain adult sizes as great as 190 mm in diameter (Kuroda 1955), several species are generally less than 50 mm in diameter as adults.

Bayer (1965) illustrates and discusses 14 of the 15 living species and is the best single source of data on comparative morphology.

The shell ultrastructure of *Pleurotomaria* has not been studied in detail. There is a thick inner nacreous layer coupled with outer complex prismatic layers. The nacre is composed of stacked crystals, as is true of all gastropod nacre (Wise 1970; Batten 1972).

Anatomy of the Animal

The initial discovery of "living" *Pleurotomaria* was, in fact, only a shell that found its way into a fish trap via a hermit crab occupant (Dance 1969). Subsequent specimens containing animals provided the basis for early anatomical observations (Dall 1889; Bouvier and Fischer 1899; Woodward 1901). Most of the early collections of live animals were not fixed and preserved adequately for detailed anatomical examination, and descriptions emphasized gross external features and the radula. The best anatomical descriptions to date are those of Fretter (1964, 1966), who examined and compared three of the western Atlantic species on the basis of single preserved specimens. She found the major important anatomical features to be identical.

The most distinctive features of the anatomy of *Pleurotomaria*—those that set it most clearly apart from more advanced marine prosobranch gastropods—are related to the deep mantle cavity and primitive paired condition of the ctenidia. The ctenidia are bipectinate, i.e., the ctenidial leaflets arise from both sides of a central axis. There are, accordingly, four rows of filaments in pleurotomariaceans (as opposed to two in trochaceans, which have a single bipectinate ctenidium; and one in mesogastropods and neogastropods, in which the single ctenidium is monopectinate). Paired inhalant currents circulate over the ctenidial surfaces where respiratory exchange occurs across the hemocoelic networks, converge, and exit at the slit.

The ctenidia in *Pleurotomaria* lack an afferent membrane, and ctenidial circulation is inferred to be less efficient than in more advanced gastropods. Although the mantle cavity of *Pleurotomaria* is deep, the ctenidia occupy only the anterior portion, leaving a poorly ventilated postbranchial region containing the right and left kidney openings and accessory pallial glands. The hypobranchial glands, the organs of mucous consolidation responsible for cleansing the mantle cavity by trapping and binding suspended sediment, are, although paired, less extensively developed in *Pleurotomaria* owing to the greater space occupied by the ctenidia.

Other primitive characters in pleurotomariacean anatomy include pairing of the osphradia, auricles (receiving blood from the paired efferent branchial vessels), and kidneys, and the negligible degree of separation of ganglia from nerve trunks in the nervous system. There is but a single columellar muscle, but it is large, suggesting a double origin.

Operculum

The operculum of *Pleurotomaria* is chitinous and polygyrous and is remarkable in its small size relative to the aperture. The size is related to the fact that the animal retracts deeply into the shell so that closure occurs well behind the slit.

Radula

The pleurotomariid radula (Figs. 1B, 1C) is of the rhipidoglossan grade that characterizes most of the primitive gastropod families that traditionally have been grouped within the Archaeogastropoda. The rhipidoglossan grade is characterized by division of the radula into three major fields of food-preparing and food-gathering teeth: a central field of robust rachidian and lateral teeth, with two fields of finer marginal teeth on either side (Hickman 1980; Morris and Hickman 1981). The number of marginal teeth is highly variable. It ranges from many thousands of teeth to fewer than a hundred, with inferred total loss in a few families. Living Pleurotomariidae fall somewhere in the middle of the range with respect to degree of development of the marginal tooth fields.

A unique feature of the pleurotomariid radula is the unusual form of one group of marginal teeth, which terminates in bundles of chitinous filaments that have been called "bristles." On the basis of these, the radula has been termed hystricoglossate rather than rhipidoglossate by some authors (Hyman 1967). The bristle teeth are, however, only one of a number of unique features that are superimposed on the rhipidoglossan plan. Other unique features of the radula include the unusually great number of teeth in the central complex (26 on either side of the 7 central teeth), their peculiar lamellate form, and the unusual sinusoidal cusp-row configuration. The largest and most formidable teeth in the radula are the two sets of 13 hooked inner marginal teeth (Fig. 1C).

An additional character of interest in the pleurotomariid radula is its asymmetry, which may be either right- or left-skewed (Hickman 1981). The function of the asymmetry is to accommodate the hooked inner marginal tooth rows alternately, as in a zipper, during withdrawal of the radula and while it is stored in the radula sac (Hickman 1981; Morris and Hickman 1981). Asymmetry cannot be interpreted as a primitive feature of the rhipidoglossan radula because it has evolved to solve the same functional problems in at least five major archaeogastropod clades (Hickman 1981).

What is important about the radula of *Pleurotomaria* is that, aside from rhipidoglossan plan, it has very little in common with any other living gastropods, including the two other extant pleurotomariacean families of more recent geologic appearance. Although highly specialized, it is inferred to represent a variation on rhipidoglossan design that was established early in gastropod evolution.

Phenotypic Variation

Little is known about phenotypic variation in living pleurotomariid populations. With the exception of large museums, most institutions lack specimens. And even in major collections there are seldom more than two or three specimens of any species. Some species are known from single specimens. *Pleurotomaria (Perotrochus) hirasei* Pilsbry 1903 has been recovered sufficiently frequently by Japanese commercial fishermen that hundreds of specimens exist, although pleurotomariids are primarily in private collections and not accessible for analysis of variation. Species with which I am familiar are not highly variable in any aspect of shell morphology.

Both ornamentation and shell form may, however, change dramatically during ontogeny, and it is therefore important to exercise caution in examining shells of widely differing size or age. Bayer (1965: 746–756) provides a detailed discussion of ontogenetic variation in *Pleurotomaria (Perotrochus) midas*.

Fossil *Pleurotomaria* and the Time Scale

Pleurotomaria DeFrance 1826 appeared in the Jurassic and is based on the conispiral trochiform *Trochus anglicus* Sowerby 1818. Because the shell of the type species is nodose, there has been a proliferation of names for post-Jurassic non-nodose taxa. It must be noted that some authorities have assigned the living species, on the basis of differences in shell characters, to *Perotrochus* Fischer 1885; *Entemnotrochus* Fischer 1885; and *Mikadotrochus* Lindholm 1927 (Knight et al. 1960; Bayer 1965). Because living pleurotomariid species are so conservative anatomically and because the shell differences, such as the presence or absence of an umbilicus and depth of the slit, are superimposed on a conservative shell form, I prefer to recognize these names as subgenera of *Pleurotomaria* (Hickman 1976).

The important point is that, regardless of nomenclature, living *Pleurotomaria* is the last vestige of a group of conispiral gastropods with the deep labral emargination that clearly reflects a common set of primitive organizational features. Furthermore, these features (the deep mantle cavity, paired bipectinate ctenidia, and paired convergent exhalant streams of water exiting at the slit) predate the Mesozoic appearance of the genus. The Pleurotomariacea is an ancient superfamily that appeared during the Late Cambrian, and the Paleozoic radiation is primarily a radiation of conispiral deeply slit shells that is conservative with respect to basic shell geometry. Most of the variation occurs in ornamentation patterns (i.e., variation in the complex temporal and spatial programming of calcium carbonate deposition by the mantle).

Pleurotomariacean Phylogeny

Pleurotomariacean ancestry is not clear from the fossil record. These organisms may be derived from bellerophontaceans (with which they share the labral emargination, paired ctenidia, and other vestiges of internal bilateral symmetry) by abandoning isostrophic coiling (and thus bilateral symmetry of the shell) in favor of conispiral coiling.

Because some bellerophontaceans have multiple symmetrically placed muscle scars and others have single paired scars, there has been heated debate as to whether the animals were monoplacophorans or gastropods (i.e., whether or not they were torted) (Harper and Rollins 1982). The only relevant point for our discussion is that the pleurotomariaceans were most likely derived from *gastropods,* with which they shared torsion and an anterior mantle cavity.

Two other living families are usually classified as pleurotomariaceans: the Scissurellidae and the Haliotidae. Although both families share a number of primitive features with the Pleurotomariidae, both are such highly modified experiments in primitive gastropod evolution that their derivation is not clear. Scissurellids are essentially microgastropods, while haliotids have adopted the limpet mode of life with accompanying major changes in coiling geometry and anatomy.

If both families are derived from the Pleurotomariidae, as their point of appearance in the geologic record suggests (see Fig. 2), it is interesting that the apparently canalized group has indeed been capable of producing evolutionary novelty. It has been postulated, however, that scissurellids are neotenously derived from within the Fissurellacea (Batten 1975) or from a Paleozoic pleurotomariacean ancestor (McLean, in press).

Diversity

Sampling the literature for an indication of species-level diversity suggests that the number of valid pleurotomariacean taxa is something on the order of 1500 species, most of which are Paleozoic. Approximately 85 species have been described from the Cenozoic, with maximum

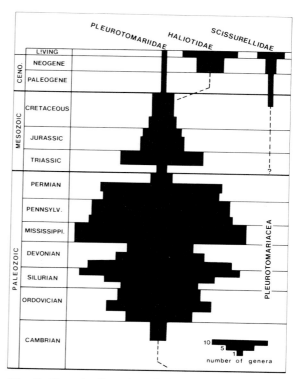

Fig. 2. Generic diversity of pleurotomariacean gastropods over Phanerozoic time. Compiled from Knight et al. (1960) and other sources.

diversity in the Eocene (Hickman 1976). None of the Cenozoic species are known from large populations of specimens, including the 15 living species.

The generic level diversity is summarized in Fig. 2, which clearly indicates that the Paleozoic was the heyday of the group.

Ecology

Living pleurotomariids are restricted to relatively deep water (>200 m), low latitudes, and hard substrates—the tops of submarine banks, steep talus slopes, or vertical rock walls as in the Tongue of the Ocean (see Hickman 1976 for a review of the Cenozoic ecology of the genus).

Pleurotomaria lives below the level of primary production, and both gut contents and the surfaces on which it occurs suggest that it obtains its nutrition primarily from animal sources (Woodward 1901; Thiele 1935; Yonge 1973; Hickman 1976 and unpublished observations).

It is not, however, a specialist carnivore in an active predatory sense, but rather a generalist "grazer" on sessile animal turfs consisting of sponges, hydroids, tunicates, alcyonarians, gorgonians, etc. It is the only large, mobile, epifaunal invertebrate in some of these bathyal habitats.

Although population densities are not high, the alleged rarity of living *Pleurotomaria* is partially a function of its preference for vertical and overhanging rock faces and the attendant difficulties of sampling such habitats with conventional gear.

Pleurotomariaceans have occupied a broader range of depths and environments over geologic time, although they have never been prominent in areas of high inferred concentrations of suspended fine particles with which the primitive mantle cavity is ill prepared to deal. Paleozoic and Mesozoic species are common in shallow-water communities. Species abundances during the Paleozoic are greatest in carbonate, and particularly reef, environments: regions of low inferred turbidity (Lindström 1884; Batten 1958, 1966; Ahern 1972). They appear to reach maximum abundance on fore-reef talus slopes (Ingles 1963; Wolfenden 1958). In some depositional environments with higher inferred concentrations of fine suspended sediment, dwarfing has been noted in Paleozoic pleurotomariaceans (Thomas 1940; Batten 1958).

Pleurotomaria is not without enemies. Multiple breakage and repair is a universal feature on modern pleurotomariid shells. The breaks are primarily minor chipping of the aperture and slit, however, and they appear to represent attempts to nibble at the mantle edge or epipodium rather than attempts to crush or capture the entire animal. Disruptions of color patterns are frequently associated with breakages.

Geographic Distribution

Biogeography of living *Pleurotomaria* is related to the somewhat unusual restriction of species to "islands" of rocky substrate in deep water. On a global scale, species exist in disjunct, highly localized populations with one major center off eastern Japan (Okutani 1963) and a second in the western Atlantic (Bayer 1965).

On a local scale, populations may be further broken up and isolated topographically, as in the case of species that are restricted to the tops of submarine banks and insular shelves between 31° and 35°N latitude off eastern Japan (Okutani 1963).

The pattern of restriction began to emerge in the Eocene, with distinct clusters of species in the Tethyan region, northern Europe, the Atlantic Coastal Plain, and the northeastern Pacific (Hickman 1976). The northeastern Pacific cluster represents a particularly interesting rare occurrence of at least three species that are associated with thin sedimentary interbeds and limestone lenses in oceanic basalts that have been interpreted as a chain of submarine banks that was subsequently rafted against and sutured to the North American Plate (Hickman 1976).

Summary

In basic shell form and anatomy, living *Pleurotomaria* remains faithful to a primitive grade of organization that has been successively replaced over Phanerozoic time by a series of innovations in gastropod evolution. Although sculptural variation and minor variations in coiling geometry and apertural form have served to define many pleurotomariacean species, there is an underlying anatomical conservatism to the group that remains intact. The 15 living species are narrowly restricted to islands of firm substrata in the deep sea, where they are the predominant epifaunal generalist carnivores on encrusting invertebrate communities. Although the superfamily has occupied a broader bathymetric and habitat range over geologic time, it has never been invasive of turbulent environments where suspended fine sediment concentrations are high. There is no basis for believing that *Pleurotomaria* has retreated to the deep sea or that it is hanging on in the modern world in a poorly adapted state. Following editorial exhortations to remain objective, I leave the broader evolutionary evaluations to the reader along with the caveat that there are many things we still need to document about the basic biology of these gastropods in order to understand their pattern of evolution.

Literature

Ahern, K. 1972. Patterns of Paleozoic gastropod diversity. U. California, unpub. Masters Thesis.

Batten, R. L. 1958. Permian Gastropoda of the southwestern United States. 2. Pleurotomariacea: Portlockiellidae, Phymatopleuridae and Eotomariidae. Amer. Mus. Nat. Hist. Bull. 114(2):153–246.

Batten, R. L. 1966. The Lower Carboniferous gastropod fauna from the Hotwells Limestone of Compton Martin, Somerset. Palaeont. Soc. Mon. Vol. 119 (Publ. 509), Vol. 120 (Publ. 513).

Batten, R. L. 1972. The ultrastructure of five common Pennsylvanian pleurotomarian gastropod species of eastern United States. Amer. Mus. Novit. 2501.

Batten, R. L. 1975. The Scissurellidae: are they neotenously derived fissurellids? (Archaeogastropoda). Amer. Mus. Novit. 2567.

Bayer, F. M. 1965. New pleurotomariid gastropods from the western Atlantic, with a summary of the Recent species. Bull. Mar. Sci. 15:737–796.

Bouvier, E. L., Fischer, H. 1899. Reports on the results of dredging . . . in the Gulf of Mexico and the Caribbean Sea, and on the east coast of the United States, 1877 and 1880 . . . XXVIII. Etude monographique des pleurotomaries actuels. Bull. Mus. Comp. Zool. Harv. 32(10):193–249.

Dall, W. H. 1889. Reports on the results of dredging . . . in the Gulf of Mexico (1877–78) and in the Caribbean Sea (1879–80), by the U.S. Coast Survey steamer "Blake" . . . XXIX. Report on the Mollusca. Part II. Gastropoda and Scaphopoda. Bull. Mus. Comp. Zool. Harv. 18:1–492.

Dance, S. P. 1969. Rare shells. London: Faber and Faber.

Fischer, P., Bernardi, A. C. 1856. Description d'un pleurotomarie vivant. J. Conchyliol. 5:160–166.

Fretter, V. 1964. Observations on the anatomy of *Mikadotrochus amabilis* Bayer. Bull. Mar. Sci. 14(1):172–184.

Fretter, V. 1966. Biological investigations of the Deep Sea. 16. Observations on the anatomy of *Perotrochus*. Bull. Mar. Sci. 16(3):603–614.

Harper, J. A., Rollins, H. B. 1982. Recognition of Monoplacophora and Gastropoda in the fossil record: a functional morphological look at the bellerophont controversy. Abstr., N. Amer. Paleont. Conv. III. J. Paleont. 56 Suppl. 2:12.

Hickman, C. S. 1976. *Pleurotomaria* (Archaeogastropoda) in the Eocene of the northeastern Pacific: a review of the Cenozoic biogeography and ecology of the genus. J. Paleont. 50(6):1090–1102.

Hickman, C. S. 1980. Gastropod radulae and the assessment of form in evolutionary paleontology. Paleobiology 6(3):276–294.

Hickman, C. S. 1981. Evolution and function of asymmetry in the archaeogastropod radula. Veliger 23(3):189–194.

Hyman, L. H. 1967. The invertebrates: Mollusca I (Vol. VI). New York: McGraw-Hill.

Ingles, J. C. 1963. Geometry, paleontology and petrography of the Thornton reef complex, Silurian of northeastern Illinois. Amer. Assoc. Petrol. Geol. Bull. 47(3):405–440.

Kanie, Y., Takahashi, T., Mizuno, Y. 1980. Color patterns of Cretaceous pleurotomariid gastropods from Hokkaido. Sci. Rept. Yokosuka City Mus. 27:37–42.

Knight, J. B., L. R. Cox, A. M. Keen, R. L. Batten, E. L. Yochelson, and R. Robertson. 1960. Systematic descriptions, pp. 169–331. In: Moore, R. C. (ed.), Treatise on invertebrate paleontology, Part I. Mollusca 1. Geol. Soc. Amer. and U. Kansas Press, Lawrence, Ks.

Kuroda, T. 1955. A new *Pleurotomaria* from Japan with a note on a specimen of *P. rumphii* Schepman collected from Taiwan. Venus 18:211–221.

Lemche, H. 1957. A new living deep-sea mollusc of the Cambro-Devonian Class Monoplacophora. Nature 179:413–416.

Lindström, G. 1884. On the Silurian Gastropoda and Pteropoda of Gotland. Kongl. Svenska Vetenskap-Akad., Handl. 19(6).

McLean, J. H. In press. On the possible derivation of the Fissurellidae from the Bellerophontacea. Malacologia.

Morris, T. E., Hickman, C. S. 1981. A method for artificially protracting gastropod radulae and a new model of radula function. Veliger 24(2):85–90.

Okutani, T. 1963. Preliminary notes on molluscan assemblages of the submarine banks around the Izu Islands. Pacif. Sci. 17:73–89.

Thiele, J. 1935. Handbuch der systematischen Weichtierkunde. Teil 4:1023–1154. Jena: Fischer.

Thomas, E. G. 1940. Revision of the Scottish Carboniferous Pleurotomariidae. Geol. Soc. Glasgow Trans. 20:30–72.

Wise, S. 1970. Microarchitecture and mode of formation of nacre (mother-of-pearl) in pelecypods, gastropods, and cephalopods. Ecolg. Geol. Helv. 63(3):776–697.

Wolfenden, E. B. 1958. Paleoecology of the Carboniferous reef complex and shelf limestones in northwestern Derbyshire, England. Geol. Soc. Amer. Bull. 69(7):871–898.

Woodward, M. F. 1901. The anatomy of *Pleurotomaria beyrichii* Hilg. Quart. J. Micr. Sci. 44:215–268.

Yonge, C. M. 1973. Observation of the pleurotomariid *Entemnotrochus adansoniana* in its natural habitat. Nature 241(5384):66–68.

26
The Giant Creeper, *Campanile symbolicum* Iredale, an Australian Relict Marine Snail

Richard S. Houbrick

Department of Invertebrate Zoology (Mollusks), National Museum of Natural History, Smithsonian Institution, Washington, DC 20560

Introduction

During the Early Tertiary, prosobranch gastropods of the family Campanilidae Douville 1904 comprised an extensive group of many large-shelled species that were common in the Tethys Sea. Some species such as *Campanile giganteum* (Lamarck 1804), the type-species of the genus *Campanile*, attained a length of 1 m and are among the largest gastropods on record. The family is represented today by a single living species from southwestern Australia, *Campanile symbolicum* Iredale 1917.

Description

The shell is large, heavy, turreted and elongate, reaching 244 mm in length, comprising of 25–30 flat-sided whorls that become weakly inflated on penultimate and body whorls (Fig. 1). The spire is concave, each whorl sculptured with a presutural spiral cord producing a weak keel at the base of the whorl. Early whorls have a nodulose subsutural spiral cord. Nodules are elongated axially and are sometimes weak or absent. Penultimate and body whorls are smooth, lacking sculpture. The suture is distinct. The protoconch is smooth, with one-and-a-half whorls. The body whorl is round in cross section with the anterior siphonal canal in the center. The aperture is triangular-fusiform, at a 45° angle to the shell axis, and one-fourth to one-fifth the shell length. The anterior canal is distinct, moderately short, and slightly twisted to the left of the shell axis. The columella is short, concave, and twisted slightly to the left at the anterior canal. A slight fold at the columellar base does not continue into the aperture. The outer lip is thin, sinuous, smooth, and forms a deep sinus where it joins to the body whorl. The periostracum is calcified, pitted, and weakly cancellate. The operculum is corneous and paucispiral, with a subcentral nucleus. Jaws are thick. The radula is taenioglossate, short, moderately stout, and about one-tenth the shell length. The rachidian tooth has a broad, platelike cutting edge forming a central, triangular cusp flanked on each side by a tiny, blunt denticle.

Unique anatomical features include: (1) a deeply ciliated pedal gland around entire sole of foot; (2) papillae around entire edge of mantle; (3) short, thick snout; (4) very long columellar muscle; (5) short, oval bipectinate osphradium; (6) modification of hypobranchial gland into tiny leaflets adjacent to anus; (7) elaboration of transverse interior folds of distal end of pallial oviduct into rounded filaments forming large albumen gland; (8) saclike seminal receptacle located inside pericardium and occurring in males and females, but more fully developed in latter sex; (9) paired salivary glands and ducts anterior to nerve ring; (10) paired buccal pouches in

Fig. 1. Shell of *campanile symbolicum* Iredale, 120 mm long. Apertural (A), side (B), and dorsal (C) views showing conical shell and details of aperture and outer lip.

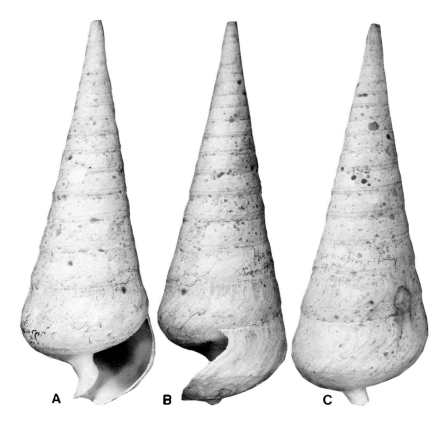

anterior esophagus; (11) a pit located in the sorting area of the stomach bearing spirally arranged leaflets; and (12) a mixture of loose and condensed neural elements including dialyneury and zygoneury. The possibility of protandry and the presence of chalazae joining individual egg capsules within the spawn mass are also unusual features among the Cerithiacea. Taken together, these characters distinguish *Campanile* from other cerithiaceans and indicate a taxonomic separation at the familial level.

Ecology

Campanile symbolicum lives subtidally in sandy patches on algae-covered rocky platforms. The snails are found in large populations in depths of 1–4 m and, although epifaunal, may be partially buried adjacent to rocks. They appear to be most active at night when they crawl about leaving long trails in the sand. *Campanile* is an algal feeder having a complex stomach and producing tiny fecal pellets comprised of fine detrital particles. Spawn is deposited in large, jellylike, crescent-shaped masses that contain an average of 4000 eggs and closely resemble the spawn of some opisthobranchs and polychaetes. Egg capsules are joined by chalazae and are deposited in a continuous, spirally coiled strand (Houbrick 1981). Embryonic stages from early cleavage embryos to advanced veligers may occur within a single spawn mass. Embryonic shells are smooth, comprise one-and-a-half whorls and lack a sinusigera notch. A free veliger stage is unknown, and although there may be a short, demersal larval stage, development appears to be direct. Spawn masses are deposited on algae, grass blades, or rocks and are neutrally buoyant. The developmental biology has not been studied.

Present Geographic Range

Campanile symbolicum is confined to southwestern Australia, from Geraldtown south to the Recherche Archipelago and Great Australian Bight. It occurs as a fossil in the early Pleistocene of the Eucla Basin, South Australia (Ludbrook 1971).

Relationships

The exact placement of this relict snail within the framework of the superfamily Cerithiacea and its relationship to the fossil species have been controversial. It is frequently assigned to the family Cerithiidae Fleming 1828. A recent study (Houbrick 1981), which described anatomical, embryological, opercular, and radular characters, established it unequivocally as a member of the superfamily Cerithiacea and supported the recognition of a separate family, Campanilidae, to accommodate it.

The relationship of *Campanile* to other cerithiacean groups is difficult to assess. Several anatomical features such as the short, distally placed, bipectinate osphradium, anterior position of salivary glands and ducts relative to the nerve ring, and the complex, spirally arranged leaflets in the sorting area of the stomach are reminiscent of some neogastropods. The presence of a calcified periostracum is known in some rissoids and epitoniids but is not common among mesogastropods. The general physiognomy and ecology of *Campanile* is closest to members of the Potamididae and Cerithiidae, but *Campanile* is probably related to them only distantly.

Evolutionary History

The family Campanilidae Douville 1904 was well represented in the Tertiary by numerous large-shelled species and has a fossil record extending back to the Maestrichtian of the Upper Cretaceous (Cossmann 1906:73). The most comprehensive study of this group was by Delpey (1941) who presented a thorough history of the nomenclature and traced the fossil lineage. The family underwent widespread adaptive radiation in the Tertiary and was especially well represented by numerous species in the Tethys Sea during the Eocene. Delpey (1941:20–21)

recognized three subgenera: *Diozoptyxis* Cossmann 1896, *Campanilopa* Iredale 1917, and *Campanile* Fischer 1884, but the group was probably more generically diverse and is represented in New World deposits by the endemic Miocene genus *Dirocerithium* Woodring and Stenzel 1959. It is apparent that the family comprises several supraspecific categories that differ from the living species herein assigned to the genus *Campanile*. The entire fossil assemblage is in need of a thorough revision before the exact number of genera, fossil species, and their lineages may be understood. I conservatively estimate that there are about 40 fossil species. Some fossil campanilids had elaborate parietal, palatal, and columellar folds that approach in complexity those observed in the Neriniidae. *Campanile symbolicum* differs from many of the fossils in lacking these folds; however, there is considerable interspecific variation in the presence, placement, and number of these folds as well as in shell sculpture of the fossils (Delpey 1941). Some fossil species lack columellar folds, and the shell of the living species does not differ substantially from those of these extinct species. The pitted surface of the shells of some Eocene species noted by Wrigley (1940:111) resembles the pattern seen on the calcified periostracum of the living species.

The exact causes for the virtual extinction of this group are unknown, but the Messinian crisis at the end of the Miocene was undoubtedly an important factor. The ecology of the living species indicates that the campanilids were probably herbivores and occupied the same trophic niche in Tethyan shallow-water ecosystems as do large snails of living strombid genera such as *Strombus, Lambis,* and *Pterocera.* The Strombidae appeared in the Late Eocene–Early Miocene and flourished during the Pliocene and Early Pleistocene (Abbott 1960:33) when the Campanilidae began to disappear. Competition between these two trophically similar groups may also have been a factor in the extinction of the Campanilidae.

Literature

Abbott, R. T. 1960. The genus *Strombus* in the Indo-Pacific. Indo-Pacific Mollusca 1(2):33–146.
Cossmann, M. 1896. Catalogue illustré des coquilles

fossiles de l'Éocene des environs de Paris faisant suite aux travaux paléontologiques de G. P. Deshayes. Ann. Soc. Roy. Malac. Belg. 31:1–94.

Delpey, G. 1941. Histoire du genre *Campanile*. Ann. Paléont. 24:3–25.

Fleming J. 1828. Mollusks. Encyclopedia Brittanica, Supp. 4–6, Vol. 3, Part 1. Edinburg.

Houbrick, R. S. 1981. Anatomy, biology and systematics of *Campanile symbolicum* Iredale with reference to adaptive radiation of the Cerithiacea (Gastropoda:Prosobranchia). Malacologia 21(1–2): 263–289.

Ludbrook, N. 1971. Large gastropods of the families Diastomatidae and Cerithiidae (Mollusca:Gastropoda) in southern Australia. Trans. Roy. Soc. S. Aust. 95:29–42.

Wrigley, A. 1940. The English Eocene *Campanile*. Proc. Malac. Soc. Lond. 24(3):97–112.

27
Diastoma melanioides (Reeve), a Relict Snail from South Australia

Richard S. Houbrick

Department of Invertebrate Zoology (Mollusks), National Museum of Natural History, Smithsonian Institution, Washington, DC 20560

Introduction

Diastoma melanioides (Reeve 1849) is the sole survivor of a long lineage of gastropods in the genus *Diastoma* Deshayes 1850, family Diastomatidae, and is not well known in the literature. This group was common in the Tethys Sea during the Tertiary and was most abundant during the Eocene. The living species is confined to shallow-water habitats in southern Australia. The taxonomic assignment of the group and its relationship with other cerithiaceans was based on shell characters until recently, when I examined the soft parts of the one extant species. The operculum, radula, and internal anatomy provide the additional characters necessary for a proper assignment and allow a more informed judgment about the relationships of the diastomids to other cerithiaceans. A more detailed account about the anatomy and systematics of *Diastoma* may be found in Houbrick (1981:589–621).

Description

The living survivor, *Diastoma melanioides* (Reeve) has an elongate shell averaging 39 mm in length, comprised of 10–13 convex whorls (Fig. 1). The smooth protoconch is one-and-a-half whorls. Adult whorls have an overall fine cancellate sculpture. The aperture is tear-shaped, wide at the base, and about one-third the length of the shell. The anterior siphonal canal is virtually undefined and exists only as a wide shallow notch at the anterior of the shell. No anal canal is present. The suture is deeply incised. There is a slight median oblique fold on the columella that extends into the aperture where it ends about halfway up the body whorl. The outer lip is thin, smooth, moderately curved, and slightly separated from the body whorl at the suture where a distinct, narrow, sutural ramp appears, especially in older specimens. The operculum is thin, corneous, ovate, paucispiral, and has an eccentric nucleus. The radula is taenioglossate and typically cerithioid in layout.

The anatomy of *Diastoma* differs from that of other cerithiaceans by the following diagnostic features: (1) a highly extensible snout and large cephalic hemocoel; (2) paired, wormlike salivary glands that run through the nerve ring; (3) loose connection of anterior and mid-esophagus to walls of cephalic cavity by numerous long muscle strands; (4) the simple, generalized arrangement of the spermatophore bursa and seminal receptacle in the outer lamina of the pallial gonoduct; (5) a large, complex ovipositor on the right side of the mesopodium and its associated large, mucus-producing, glandular inner incubatory chamber; (6) length of the labial and buccal nerves and innervation of the incubatory chamber by the mesopodial ganglion;

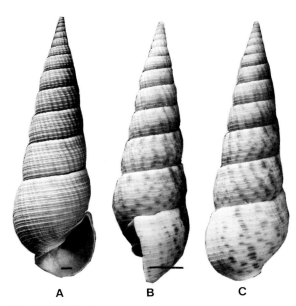

Fig. 1. Shell of *Diastoma melanoides* (Reeve), 49.7 mm long. (A) Apertural view with shell whitened to show sculptural details. (B) Side view showing outer lip and sutural ramp. (C) Dorsal view showing color pattern.

and (7) absence of the osphradium, ctenidium, and hypobranchial gland from the posterior portion of the mantle cavity.

Ecology

Few ecological observations have been made on the living species *Diastoma melanoides*. It lives in sandy, shallow-water habitats in depths of 1–5 m where it burrows in clean sand associated with grass beds and algae. Anatomical evidence shows that it is herbivorous and probably feeds on the microalgae, diatoms, and detrital particles in the sand.

Paired snails have been collected in June and large ova observed in the ovaries of individuals collected in July. This indicates that the reproductive season may begin in early winter. The spawn is unknown, but females have a large ovipositor on the right side of the foot leading into a spacious cavity that produces much mucus. This structure may be a brood pouch or a large mucoid-jelly gland that produces a jelly-string mass similar to those observed in other ceriths. The smooth protoconch and lack of a

sinusigera notch on its outer lip point to a direct development.

Present Geographic Distribution

The genus *Diastoma* was most abundant in species during the Eocene. Most of these underwent extinction in the Tethys Sea after the Miocene. The genus spread to the Pacific Ocean where it remained in the south Australian region after the closure of the Tethys in the Miocene.

The living species is confined within the 19–20°C warm, temperate isotherms of the Great Australian Bight. Museum records show a distribution from Cheyne Beach, near Albany, Western Australia, to Streaky Bay, South Australia.

Relationships

Study of the anatomy of the living species establishes it unequivocally in the superfamily Cerithiacea (Houbrick 1981). Anatomical and shell characters indicate a close relationship with members of the widespread marine families Cerithiidae and Potamididae. Several anatomical features and the shell morphology suggest a relationship with members of the large, tropical freshwater family Melaniidae (Houbrick 1981). Characters derived from a study of the living species and the fossil record indicate that *Diastoma* differs enough from other cerithiid groups to merit familial status. Its exact relationships with the families cited above remain problematic until more comparative anatomical information is available on the numerous genera comprising these families. The most likely candidates as sister-groups among marine prosobranchs are members of the widespread tropical genera *Cerithium* Bruguière and *Rhinoclavis* Swainson (Family Cerithiidae). These successful groups comprise numerous species found throughout the warm, shallow-water habitats of the tropics and were abundant and widespread throughout the Tertiary, especially during the Eocene (Houbrick 1974:35; 1978:16). Species in these genera have always been more abundant and widespread than *Dias-*

toma species and may have eventually outcompeted them for similar niches in the tropics, leaving a few *Diastoma* species in outlying temperate habitats where they were less successful.

The other possible sister-group, at least on morphologic grounds, is the freshwater genus *Melanoides* Oliver, family Melaniidae (= Thiaridae, Melanopsidae), found throughout the tropics. Members of this genus share a remarkable similarity in shell morphology to *Diastoma,* as the living species' name, *melanioides,* bears testimony. While these may indeed be the true sister-group, and they do share many anatomical features with the living *Diastoma* species, not enough is known about other genera in the family Melaniidae and closely related cerithiid genera such as *Diala* A. Adams and *Finella* A. Adams to reach any conclusions at this time.

Members of the family Potamididae, a brackish water group, also bear some anatomical and conchological resemblance to *Diastoma.*

Evolutionary History

The family Diastomatidae comprises a number of fossil species that have elongate, turreted shells of convex whorls, cancellate sculpture, and ovate apertures with sinuous outer lips. The anterior siphonal canal is present only as a wide, shallow notch and is virtually undefined from the base of the outer lip. Former varices are normally present. The family, as originally defined by Cossmann (1894:322–323), embraced a number of fossil genera some of which were relatively small-shelled species and included the genus *Diastoma* Deshayes. The concept and limits of the family were recently narrowed by Houbrick (1981).

The genus *Diastoma* comprises a compact group of moderately sized, distinctive-looking snails, most of which are extinct. It is represented by about 13 extinct taxa and one living species, all having a very similar shell morphology. Cossmann (1894) was unaware of the existence of the south Australian living species, as were many subsequent workers. Few American workers had seen the European fossils, and the genus was known largely through illustrations. Thus, workers like Dall (1889:258) were misled

by Cossmann's (1906:174) inclusion of rissoid groups in the family. They assigned other small-shelled groups to the genus such as *Alaba* Adams and Adams and *Bittium* Gray, overlooking the differences in scale between these groups and the type-species of *Diastoma.* As a result, the genus was expanded in scope to include various unrelated small-shelled groups that have convergent shell features. This distorted both the familial and generic concepts into unnatural polyphyletic groups (see Houbrick 1981:609–612 for a detailed account of the history of these taxa). Recent publications have erroneously referred several small-shelled snails such as *Bittium varium* Pfeiffer to *Diastoma.* The genus *Diastoma,* however, is a relatively circumspect group, as pointed out by Ludbrook (1971).

Diastoma melanoides is not common in museum collections and is little known by malacologists; consequently, the full extent of shell variability is not known. There is some minor variation in the number of axial riblets and the cancellate sculpture of specimens I examined (Houbrick 1981:602). The Recent species differs from the Tethyan fossils, and especially from the type-species of the genus *Diastoma costellata* Lamarck 1804, in that the latter has a more pronounced sutural ramp and lacks the columellar fold. The sculpture of most fossil species is more rugose than in the living species. The shell morphology of the Australian fossil species is nearly identical with that of the living species. Because of these differences, the Australian fossils and the Recent species were assigned to a new genus, *Neodiastoma* Cotton 1932, but I regard these differences as trivial and do not recognize generic separation.

According to Cossmann (1889:176), the genus *Diastoma* may have originated in the Senonian (Late Cretaceous) but the assignment of fossils from this period to the genus is debatable. The genus *Diastoma, sensu stricto,* first appeared in the Paleocene of Europe but is best known from the Eocene of the Paris Basin. It has also been found in the Upper Oligocene (Cossmann 1889). The genus occurred in the Tethys Sea throughout most of the Tertiary, but was less common, at least in species diversity, after the Eocene. Ludbrook (1971:31) recorded the genus from the Tertiary of Egypt, the East Indies, and North and South America. The taxonomy of this genus is not well known; consequently, some of these citations may be erroneous. Three unequivocal *Diastoma* species occur in the Australian Miocene, Early Pliocene, and Pleistocene (Ludbrook 1971:31; 1978:112).

Literature

Cossmann, M. 1894. Revision sommaire de la faune du Terrain oligocène marin aux environs d'Étampes. J. Conch. 41:297–363.

Cossmann, M. 1906. Essais de paléoconchologie comparée, 7:261. Paris. Chez l'auteur.

Dall, W. H. 1889. Reports on the results of dredging, under the supervision of Alexander Agassiz, in the Gulf of Mexico (1877–78) and in the Caribbean Sea (1879–80), by the U.S. Coast Survey Steamer "Blake," Lieut.-Commander C. D. Sigsbee, USN, and Commander J. R. Bartlett, USN, commanding. Report on the Mollusca. Part 2. Gastropods and Scaphopoda. Bull. Mus. Comp. Zool. 18(29):1–492.

Houbrick, R. S. 1974. The genus *Cerithium* in the western Atlantic. Johnsonia 5(50):33–84.

Houbrick, R. S. 1978. The family Cerithiidae in the Indo-Pacific. Part 1. The genera *Rhinoclavis, Pseudovertagus* and *Clavocerithium*. Mon. Mar. Moll. 1:1–130.

Houbrick, R. S. 1981. Anatomy of *Diastoma melanioides* (Reeve 1849) with remarks on the systematic position of the Family Diastomatidae (Prosobranchia:Gastropoda). Proc Biol. Soc. Wash. 94(2):598–621.

Ludbrook, N. H. 1971. Large gastropods of the families Diastomidae and Cerithiidae (Mollusca:Gastropoda) in southern Australia. Trans. Roy. Soc. S. Aust. 95(1):29–42.

Ludbrook, N. H. 1978. Quaternary Mollusks of the western part of the Eucla Basin. Geol. Surv. W. Aust. Bull. 125:1–286.

28
The Relict Cerithiid Prosobranch, *Gourmya gourmyi* (Crosse)

Richard S. Houbrick

Department of Invertebrate Zoology (Mollusks), National Museum of Natural History, Smithsonian Institution, Washington, DC 20560

Introduction

Gourmya gourmyi (Crosse 1861) is a rare, relatively large, stout, and distinctive-looking prosobranch that lives on subtidal coral reefs in a restricted part of the southwest Pacific. It has largely been ignored in the scientific literature and is not well represented in museum collections. This species is the sole survivor of a Tethyan lineage that can be traced back to the Eocene of the Paris Basin. The shell morphology of the living form is virtually identical with those of its ancestors. Nothing was known about the animal or its ecology until recently, when I was able to examine a few preserved animals from New Caledonia and the Chesterfield Islands, Coral Sea. Prior to this only the shell was known. My study (Houbrick 1981) provided information about anatomy, the radula, and operculum that established *Gourmya* as a valid genus in the family Cerithiidae.

Description

Adults have a stocky, heavy shell about 44 mm long and 19 mm wide, and are comprised of about 8.5 moderately inflated, nearly smooth whorls having an apical angle of 40–45° (Fig. 1).

The body whorl is obese and has a thick, smooth varix at the edge of the outer lip. The suture is incised and straight. The ovate aperture is about one-third the shell length and has a concave columella. The anal canal is deeply incised, bordered with a columellar fold that extends into the shell aperture. The siphonal canal is tubular, straight and located at the mid-anterior of the shell. The outer lip is smooth, beginning at the middle of the body whorl and extending, hooklike, over the anterior siphonal canal, but not fused to it. A thin brown periostracum covers the shell. The operculum is ovoid, corneous, paucispiral, and has an eccentric nucleus. The radula is taenioglossate, moderately short, and typically cerithiacean in morphology.

Unique anatomical features are: (1) six to seven siphonal ocelli located on the inner edge of the inhalant siphon; (2) a short sperm-collecting gutter, large kidney-shaped spermatophore bursa, and seminal receptacle at the proximal end of the outer lamina of the pallial oviduct; (3) paired salivary glands that lie mostly anterior to the nerve ring but pass partially through it; (4) a large esophageal gland formed of two longitudinal strips; and (5) a stomach with a very large chitinous gastric shield and a sorting area with numerous serially arranged leaflets of unknown function.

Fig. 1. Shell of *Gourmya gourmyi* (Crosse), 63 mm long. (A) Apertural view showing outer lip crossing anterior siphonal canal. (B,D) Dorsal and anterior views showing central position of tubular anterior canal. (C) Side view showing outer lip in line with plane of shell axis.

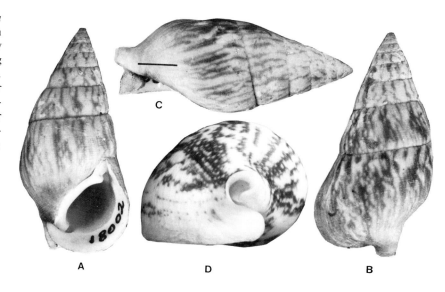

Ecology

Gourmya gourmyi is a relatively rare animal that lives subtidally from 5 to 30 m on rocky substrata associated with coral reef drop-offs. Living animals are collected clinging tightly to algae-covered rocky surfaces along high-energy reef slopes. The unusual, centrally placed anterior siphonal canal and the large ovate aperture with hooklike extension of the outer lip over the siphonal canal allow the snail to withdraw its head and clamp down on the rocks without any gaps between the aperture edge and the foot. Communication with the external environment is maintained via the siphonal canal by means of the siphonal ocelli and sensory receptors of the siphonal mantle edge. The thickened varix along the outer lip serves to strengthen the body whorl and protects the snail from crushing predators. Stomach anatomy and contents, fecal pellet analysis, and the radular ribbon suggest that *Gourmya* is a grazer of coarse algal substrates. Nothing is known of its reproductive biology, spawn or larvae.

Present Geographic Range

This species is restricted to the New Hebrides (Cernohorsky 1978:53), New Caledonia,

the Chesterfield Islands, and Marion Reef, in the Coral Sea. For a map and exact locality data see Houbrick (1981:8–10).

Relationships

Several members of the genera *Pyrazus* Montfort and *Terebralia* Swainson, in the large estuarine cerithiacean family Potamididae show an interesting shell convergence with *Gourmya,* especially in the physiognomy of their apertures. This similarity has led some authors (Sowerby 1865; Tryon 1887) to refer mistakenly the living species to potamidid genera. While some *Terebralia* species are frequently found clamped to mangrove roots in much the same manner as *Gourmya* clamps on its rocky substratum, their anatomy is quite different, and it is clear that this similarity is convergent and does not represent close relationship.

Open pallial gonoducts, aphallic males, and the production of spermatophores are shared characters with other members of the Cerithiacea. The distinctive conchological and anatomical features of *Gourmya* establish its standing as a separate genus and indicate that it is probably closely related to the genera *Cerithium* Bruguière, *Pseudovertagus* Vignal, and *Rhinoclavis* Swainson, all very diverse, widespread groups since the Tertiary.

Evolutionary History

The genus *Gourmya* Fischer 1884 was proposed for the living species, *Gourmya gourmyi*, the type-species of the genus. *Gourmya* first appeared during the Eocene and was present in the Tethys Sea until the end of the Miocene when it disappeared. The genus undoubtedly reached the Indo-Pacific via the Tethys before its closure in the Late Miocene and appears as a fossil in the Pliocene of Java (Cossmann 1906:68–69).

The shell morphology of the living species is virtually identical to that of its ancestors. The genus was not diverse in species numbers: there are five fossil species known. An Eocene fossil, *Gourmya romeo* (Bayan 1870), is strikingly like the extant species and was designated by Fischer (1884) as the "genoplesiotype" of *Gourmya*. *Gourmya romeo* is well figured by Bayan (1870; plate 9, Fig. 5) and Cossmann (1906; plate 1, Figs. 8–9), and both authors noted its resemblance to the living species. An Oligocene species, *Gourmya ocirrohoe* (d'Orbigny 1850) is figured by Vignal (1897:69–70) and Cossmann (1906:69). There are two species cited from the Miocene of France, *Gourmya klipsteini* (Michilotti 1847), and *Gourmya geminatum* (Grateloup 1832 [*non* Sowerby 1816]). The latter species lacked an aperture and is assigned to the genus with some doubt. The Pliocene species from Java, *Gourmya parungpontengense* (K. Martin 1899), confirms the presence of the genus in the Indo-Pacific in the Late Tertiary. It resembles the living species and appears to be closely related to it (Martin 1899:204, Cossmann 1906:69). I believe this fossil species may be the direct ancestor of *Gourmya gourmyi*.

The lineage of this group may clearly be traced back to the Eocene, and there is little doubt that *Gourmya gourmyi* is closely related to the extinct species in the genus. The ecology of the living species suggests that members of this genus exploited an adaptive zone different from those of other cerithiid genera. Most cerithiids live in subtidal or intertidal shallow water environments where they have undergone extensive adaptive radiations into sandy lagoons, grass beds, and the rocky intertidal. The environmental niche of *Gourmya*, high-energy outer reef drop-offs, is unusual and may explain why the group never became as geographically widespread as other cerithiids. In terms of species numbers, and compared to the other large cerithiid groups, the genus was not a successful group.

Literature

Bayan, F. 1870. Études faites dans la collection de l'École des Mines sur des fossiles nouveaux ou mal connus. Premier Fascicule, Mollusques Tertiaires. Paris. F. Savay.

Cernohorsky, W. O. 1978. Tropical Pacific marine shells. Sydney. Pacific Publications.

Cossmann, M. 1906. *Essais de Paleoconchologie comparée,* Vol. 7. Paris. Chez l'auteur.

Grateloup, J. D. S. 1832. Tableau (suite du) des coquilles fossiles qu'on rencontre dans les terrans calcaires tertiares (faluns) des environs de Dax, department des Landes; par M. Grateloup, membre honoraire. 5em Article. Actes Soc. Linne. Bordeaux 5(29):263–282.

Houbrick, R. S. 1981. Some aspects of the anatomy, reproduction and early development of *Cerithium nodulosum* (Bruguière) (Gastropoda, Prosobranchia). Pacif. Sci. 24(4):560–565.

Martin, K. 1899. Die Fossilien von Java. Samml. Geol. Reichs-Mus. Leiden. Neue Folge 1(6–8):133–221.

d'Orbigny, A. 1850. Prodrome de paléntologie stratigraphique universelle des animaux mollusques et rayonnés faisant suite au cours elementaire de paléontologie et de géologie stratigraphiques. Paris. Victor Masson.

Sowerby, G. B. 1865. Cerithium. In: Reeve, L. A. (ed.), Conchologia Iconica: or Illustrations of the shells of molluscan animals. Vol. 15, 20 plates + index (no pagination). Reeve and Co., London.

Tryon, G. W. 1887. Cerithium, pp. 127–149. Manual of conchology; structural and systematic; with illustrations of the species. Philadelphia. Acad. Nat. Sci.

Vignal, L. 1897. Note sur le *Cerithium* (*Gourmya*) *ocirrhoe* A. d'Orbigny. J. Conch. 45:69–70.

29
Neotrigonia, the Sole Surviving Genus of the Trigoniidae (Bivalvia, Mollusca)

Steven M. Stanley

Department of Earth and Planetary Sciences, The Johns Hopkins University, Baltimore, MD 21218

Introduction

The bivalve mollusk genus *Neotrigonia* holds a special place in the history of evolutionary biology. When it was discovered in 1802, naturalists had believed for many years that the family to which it belongs, the Trigoniidae, had disappeared at the end of the Mesozoic Era, about 65 million years ago (Gould 1968). Thus, *Neotrigonia* has been regarded as a "living fossil" in the sense that it is the sole surviving genus of a once flourishing and widely distributed family. In Upper Cretaceous deposits (ranging from about 100 million years old to 65 million years old) approximately 20 genera of trigoniids are recognized (Cox *et al.* 1969). In early Cenozoic deposits, however, only one trigoniid genus has been found; this is *Eotrigonia,* the apparent ancestor of the living genus.

The Living Species and Their Geographic Occurrences

Since the discovery of the first living trigoniid species, additional species have been encountered. All occupy water adjacent to Australia, and all are now assigned to the genus *Neotrigonia.*

Habe and Nomoto (1976) recognized six living species of *Neotrigonia,* including the spe-

cies *N. kaiyomaruae,* which they named from a single distinctive specimen collected off Western Australia. The other species and their geographic ranges are:
1. *Neotrigonia margaritacea* (Lamarck 1804) (New South Wales, Victoria, Tasmania)
2. *Neotrigonia uniophora* (Gray 1847) (New South Wales, Queensland, North Australia, Western Australia)
3. *Neotrigonia strangei* (A. Adams 1854) (New South Wales)
4. *Neotrigonia bednelli* (Verco 1912) (South Australia, Western Australia)
5. *Neotrigonia gemma* (Ireland 1924) (New South Wales)

Morphology

Tevesz (1975) found *N. margaritacea* and *N. gemma* to be quite similar to one another in general anatomy. They are non-siphonate species that possess filibranch gills on which the ciliary tracts resemble those of the freshwater Unioacea, a group that on other morphological grounds as well appears to share a common ancestry with the Trigoniidae (Newell and Boyd 1975). The most distinctive anatomical feature of *Neotrigonia* is its large, muscular foot, which is shaped like the letter *L,* but possesses a small heel in addition to the long, anteriorly directed toe (Figs. 1D, 1E).

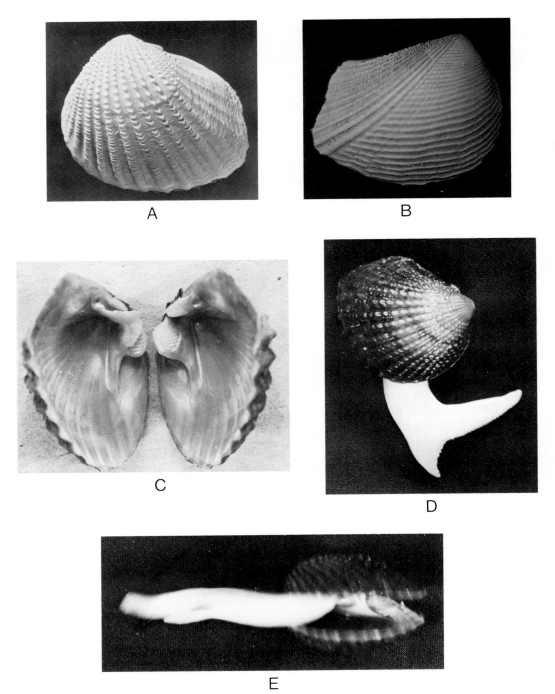

Fig. 1. Trigoniidae of the Cenozoic. (A) Left-lateral view of *Neotrigonia margaritacea* (Lamarck), ×1.14 (Westernport, Victoria, Australia). (B) Right-lateral view of *Eotrigonia subundulata* (Jenkins), ×1.4 (Brit. Mus. Nat. Hist. L 42330). (C) Oblique view of the interior of *Neotrigonia margaritacea*, showing the massive hinge teeth with secondary dentition. (D) and (E) Extension of the large foot of *Neotrigonia margaritacea* for jumping (×1); the hinge teeth maintain valve alignment at the wide angle of gape.

In shell form, the living species of *Neotrigonia* are all quadrate and slightly longer than high, with distinct radial ribs surmounted by tubercles or scales (Fig. 1A). The large hinge teeth, which bear secondary dentition, are characteristic of the Trigoniidae in general (Fig. 1C); these structures serve to maintain valve alignment at the unusually wide angles of gape required for protrusion of the large, muscular foot (Stanley 1977).

Neotrigonia does not closely resemble Mesozoic Trigoniidae. It does exhibit the nacro-prismatic shell structure characteristic of the group, and also the truncated posterior shell region and complex hinge teeth; on the other hand, its overall shell outline and ornamentation are not closely duplicated in any Mesozoic species of the family.

Preceding *Neotrigonia* in the Cenozoic fossil record of Australia is *Eotrigonia*, which is recognized in deposits ranging in age from Early Eocene to Miocene (Fig. 1B). The shell of *Eotrigonia* displays an oblique array of ridges quite unlike the radial ribbing of *Neotrigonia* (Fig. 1A). *Neotrigonia* overlapped slightly in time with *Eotrigonia*; it has been recognized as far back as the Oligocene series.

Ecology and Life Habits

The living species of *Neotrigonia* are all subtidal forms that occupy sea floors from slightly below low tide to at least 400 m in depth (Tevesz 1975; Habe and Nomoto 1976; Stanley 1977). They occupy sediments that range from poorly sorted sand to sandy mud. *N. margaritacea*, the form most commonly reported from depths less than 20 m, occupies the floors of tidal channels at Westernport, Victoria, where currents commonly range up to 5 knots (Stanley 1977).

Neotrigonia lives with the truncated posterior end of its shell more-or-less coincident with the sediment surface, although *N. margaritacea* often lives with this end of the shell slightly exposed and colonized by epibionts (Tevesz 1975; Stanley 1977).

As illustrated by behavioral studies of *Neotrigonia margaritacea* and *N. gemma* (Tevesz 1975; Stanley 1977), members of the genus *Neotrigonia* differ from nearly all other non-siphonate bivalves in being moderately rapid burrowers. This ability and the ability to jump when disturbed are conferred by the presence of the large, musclar foot.

Complex hinge teeth like those of the living *Neotrigonia* made their appearance within the Trigoniidae during the Triassic Period, which implies that the large, muscular foot was also present in the group by this time. In fact, it seems likely that the great success of the Trigoniidae during the Mesozoic Era was in large part fostered by the high degree of mobility conferred by the presence of this foot (Stanley 1977). A similar foot, functionally comparable adaptations of the hinge teeth, and the ability to jump and burrow rapidly are also all traits of modern cockles (Cardiidae). Coadaption among these features strengthens the inference that Mesozoic trigoniids were highly mobile.

Levels of Adaptation

There is no reason to believe that *Neotrigonia* or *Eotrigonia* has been in any way adaptively superior to the trigoniids of the Mesozoic Era. Perhaps most significant is the evidence that Mesozoic trigoniids were highly mobile animals. *Neotrigonia* is much more mobile than are extant members of bivalve families that are represented in the world today by larger numbers of non-siphonate burrowing species. These numerically successful non-siphonate families include the Astartidae, Crassatellidae, Carditidae, and Arcidae.

Predation is the dominant environmental agent limiting population sizes of most bivalve species in the marine realm today (Stanley 1973), and mobility is at a premium in escape from predation. Hence, it can be argued that Mesozoic trigoniids were in a very important way more highly adapted than were contemporary species of the Astartidae, Crassatellidae, Carditidae, and Arcidae (Stanley 1977).

Thus, it appears that the trigoniids did not become outmoded any more than did the highly mobile dinosaurs (Bakker 1980). Both of these groups were simply struck especially hard by the mass extinction that ended the Cretaceous Period. The trigoniids were quite diverse until the very end of the Cretaceous. Losses of marine taxa in the terminal Cretaceous mass extinction were concentrated in the tropical Tethyan biogeographic realm, and the Trigoniidae were presumably one of the groups that suffered severely simply because they were predominantly Tethyan in distribution. It appears that the survival into Cenozoic time of a small number of species—possibly only one—may have been a statistical accident.

One might wonder whether the failure of the trigoniids to diversify greatly during the Cenozoic Era argues against their being well-adapted animals in the modern world. In fact, the rate of adaptive radiation for bivalve taxa is characteristically low (Stanley 1979). *Neotrigonia,* having generated its five or so living species in perhaps 30 million years, has been expanding at close to the normal rate for adaptive radiation within the Bivalvia (Stanley 1977). The fact that *Eotrigonia,* the solitary genus of Early Cenozoic Age, suffered extinction during the Miocene Epoch may relate to climatic changes or other factors unrelated to adaptive success under stable conditions. The extinction of a single member genus is no measure of the level of adaptation of a family.

Geographic Patterns and the Survival of the Trigoniids

Fleming (1964) reviewed the biogeographic history of Austral trigoniids, noting that many Jurassic species of New Zealand and Australia resemble species of Europe, the Tethys, or the northern Pacific. On the other hand, endemic genera came to characterize the Austral region in Middle and Late Cretaceous times. In light of plate tectonic reconstructions unavailable when Fleming wrote on the history of the Austral trigoniids, we can now offer a reasonable explanation for this pattern. It was during the Cretaceous Period that the great southern continent Gondwanaland fragmented. Of its several fragments, South America, Africa, and peninsular India drifted northward, leaving Australia attached to Antarctica, which was positioned over the South Pole. As a result, endemism increased in and around Australia and Antarctica. Thus, the trigoniid species that survived the Cretaceous were positioned in an isolated biogeographic province far to the south of the family's Mesozoic center of distribution in the Tethys. Following the terminal Cretaceous mass extinction, there has been little opportunity for the Trigoniidae to reoccupy other regions.

Literature

Bakker, R. T. 1980. Dinosaur heresy—dinosaur renaissance. Amer. Assoc. Adv. Sci. Selected Symp. Ser. 28:351–505.

Cox, L. R., *et al.* 1969. Superfamily Trigoniacea, pp. 471–489. In: Moore, R. C. (ed.), Treatise on invertebrate paleontology, Part N, Mollusca 6, Bivalvia. Geol. Soc. Amer. and U. Kansas Press, Lawrence, KS.

Fleming, C. A. 1964. History of the bivalve family Trigoniidae in the southwest Pacific. Austral. J. Sci. 26:196–204.

Gould, S. J. 1968. *Trigonia* and the origin of species. J. Hist. Biol. 1:41–56.

Habe, T., Nomoto, K. 1976. A new species of the genus *Neotrigonia* from off Western Australia. Bull. Nat. Sci. Mus. Tokyo, Ser. A. 2:174–177.

Newell, N. D., Boyd, D. W. 1975. Parallel evolution in early trigoniacean bivalves. Amer. Mus. Nat. Hist. Bull. 154:53–162.

Stanley, S. M. 1973. Effects of competition on rates of evolution, with special reference to bivalve mollusks and mammals. Syst. Zool. 22:486–506.

Stanley, S. M. 1977. Coadaptation in the Trigoniidae, a remarkable family of burrowing bivalves. *Palaeontology* 20:869–99.

Stanley, S. M. 1979. Macroevolution: pattern and process. San Francisco: Freeman.

Tevesz, M. J. S. 1975. Structure and habits of the living fossil pelecypod *Neotrigonia.* Lethaia 8:321–327.

30
Is *Nautilus* a Living Fossil?

Peter Ward

Department of Geology, University of California, Davis, CA 95616

The genus *Nautilus* of the Nautiloidae family appeared in the Triassic Period, and its representatives have remained unchanged to this day. It is one of the most remarkable of nature's living fossils; it has been called, and is, "the Coelocanth of the invertebrates."

(Cousteau and Diole 1973)

At the present, for instance Miller (1951) restricts *Nautilus* to only the Recent species and all of the Tertiary forms previously assigned to *Nautilus* are placed in *Eutrephoceras* or *Cimomia*. In this interpretation I am in full agreement.

(Kummel 1956)

The two quotes above are representative of the controversy concerning the age of *Nautilus*, the last externally shelled cephalopod. If we accept the first age estimate, we would have to include *Nautilus* as a living fossil, the end member of a clade that has survived for a long time and undergone little morphologic change.

The second quote is more representative of specialists studying cephalopods. Although much disagreement exists about the actual age of the genus *Nautilus*, no modern worker extends the genus back to the Triassic. Cousteau and Diole did not simply invent their Triassic date for the origin of the genus, however. In the latest Triassic, a small number of involute, smooth-shelled nautiloids survived a major extinction of the diverse Triassic nautiloids and ammonoids. Like the ammonoids, the one or several species of coiled nautiloids that survived this crisis became the ancestral stock of a large radiation in the ensuing Mesozoic. Unlike the ammonites, however, which showed extraordinary morphologic plasticity in their Jurassic and Cretaceous radiations, the nautiloid species and subsequent genera evolving from the original *Cenoceras* complex of the Late Triassic and Early Jurassic were very homogeneous in form. Most post-Triassic nautiloids occupied a quite limited portion of the W,D,S shell coiling spectrum as defined by Raup (1967), being confined to globose to slightly depressed or compressed, involute shells. These particular shell shapes also appeared to have been little utilized by ammonites (Ward 1980). Other morphologic features of the shell, such as suture line and external ornament were, in general, simple or reduced. With a very limited variety of shell form and ornament, homeomorphy and convergence were common, as the same morphologic themes were used and reused (Miller 1947; Kummel 1956, 1964). The only discernible evolutionary trend within the Nautilaceae was a move toward higher proportion of compressed shell forms in the Tertiary, following the extinction of the ammonites (Ward 1980). *Nautilus*, the last descendant, is easily differentiable from the ancestral *Cenoceras* species of the Late Triassic on the basis of suture and cross section. A nonspecialist, however, holding the two shells, would probably find the differences quite slight.

If the two endpoints are not so dissimilar, what about the evolutionary road in between? The problem is to decide if the nautiloids have produced a few, long-ranging species, or a larger number of shorter-ranging species that have largely been unrecognized because of some combination of lack of distinctive morphology, poor fossil record, or tendency toward iterative evolution and homeomorphy. In this contribution I examine the geologic record of the post-Triassic nautiloids, and some distributional and ecological aspects of the extant species of *Nautilus* in order to decide if *Nautilus* does or does not qualify as a "living fossil."

Anatomical Similarity of *Nautilus* to Extinct Nautiloids

Nautilus Linnaeus is defined as follows in Kummel's (1964) *Treatise on Invertebrate Paleontology* (Part K): "Smooth, nautiliconic, involute to occluded; suture consisting of broad rounded ventral saddle, broad lateral lobe, small saddle in vicinity of umbilical wall, small saddle near umbilical seam, broad shallow dorsal lobe and annular lobe, siphuncle subcentral." Evidently, most of the differentiating features are related to the morphology of the suture line. Other Mesozoic and Cenozoic nautiloids had whorl section and ornament comparable to *Nautilus* (e.g., *Pseudonautilus, Hercoglossa, Cimomia, Aturoidea, Aturia*). All, however, can be differentiated on the basis of suture lines.

To the above description I can add a previously overlooked morphologic character that is found in the extant species of *Nautilus*, and may be useful in differentiating the genus from other Nautilaceae. The nepionic constriction in *Nautilus* and other externally shelled cephalopods is a shallow indentation found on the flanks of the juvenile shell, separating distinctly different shell microornament on its two sides. In nautiloids the nepionic constriction separates cross-hatched, unscarred ornament of the earliest shell regions from fine growth lines (but not cross-hatching) of the later shell. Even immediately after the nepionic constriction, this latter, growth-like ornament is commonly interrupted by tiny shell-breaks that have been rehealed,

leaving small scalloped-shaped scars in the growth-line pattern.

Some workers consider the nepionic constriction to mark the time of hatching from the egg (Naef 1928; Stenzel 1964), while others interpret the nepionic constriction to mark the point of migration of the juvenile nautilus from shallow to deeper waters (Eichler and Ristedt 1966). In any event, this feature appears to be present in all chambered cephalopods, and provides a point of comparison of various taxa. Shimanskii (1962) has examined the nauta (juvenile shell portion prior to deposition of nepionic constriction) of a variety of nautiloids. Among the coiled forms, the nauta of *Nautilus* is by far the largest, occurring at a shell diameter of approximately 25 mm in all of the extant species. I have recently examined nauta from species of *Aturia, Cenoceras, Eutrephoceras, Cymatoceras,* and *Hercoglossa* (Ward, in preparation). In all of these forms, the nauta are at most half the diameter of equivalent stages in *Nautilus*. *Nautilus* may be, at hatching, the largest of all cephalopods still extant, and could be one of the largest cephalopods hatching in the history of the class. In summary, *Nautilus* is quite similar in shell shape and lack of ornament to numerous past species. It can be readily differentiated, however, on sutural and developmental differences.

Phylogenetic Relationships

A major question still to be answered is the actual age of the genus *Nautilus*. Miller (1947, 1951) included only the currently extant species of nautiloids in *Nautilus*. Kummel (1956) followed Miller, but later (Kummel 1964) listed the range of the genus as Oligocene to Recent. Shimanskii (1962) has listed the range as Cretaceous to Recent.

The problem in interpreting the actual range of *Nautilus* is in the lack of nautiloid fossils reported from Pliocene and Pleistocene strata worldwide, and in the lack of specimens that can be assigned to *Nautilus* from the last Tertiary rocks, of Miocene age, that do have fossil nautiloids.

Numerous Mesozoic and Tertiary species have been placed in *Nautilus;* according to

Miller (1947, 1951) and Kummel (1956), all of these have differed from *Nautilus* on the basis of suture line. Soon after the publication of Kummel's 1956 monograph on the post-Triassic Nautiloids, however, a new nautiloid species from the Soviet Union, from Paleogene rocks, was described by Shimanskii (1957). This specimen, *Nautilus praepompilius* Shimanskii, appears quite similar in shell form and possibly suture to *N. pompilius,* the most common of the currently extant *Nautilus* species. Its last fossil representative, however, is from Lower Tertiary rocks. Kummel (1956) listed 20 species of nautiloids known from Miocene rocks, and none in either the Pliocene or Pleistocene. None of the Miocene specimens can be assigned to *Nautilus,* although several from Australia are close (Chapman 1914) (see Fig. 1).

If we accept an origin from *N. praepompilius,* we are left with a gap of 40 to 50 million years without record of the genus. If we assume that *Nautilus* evolved from one of the Miocene eutrephoceratids (Miller 1947), we still have a gap, but of about 5 million years. Until such time as well-dated Pliocene and Pleistocene rocks yield fossil nautiloids, or unquestioned specimens of *Nautilus* are found in Eocene through Miocene rocks, the controversy over the origin and age of *Nautilus* will remain.

It is not too surprising that no Pliocene or Pleistocene nautiloid fossils have yet been reported, since I doubt that anyone has taken the trouble to look for them. If we assume that Pliocene nautiloids were restricted to those approximate areas where *Nautilus* range today, we should search in Pliocene rocks of the Indo-Pacific. In contrast to North America and Europe, where Plio-Pleistocene rocks have been intensively studied, deposits in the southwest Pacific are poorly known and rarely collected.

Species-Level Diversity

The species-level diversity of the post-Triassic nautiloids has been tabulated most recently by Kummel (1956), with little addition of species since that time (Fig. 2).

The apparent restriction at the end of the Cretaceous is probably more apparent in the diversity diagram than real. Late Cretaceous diver-

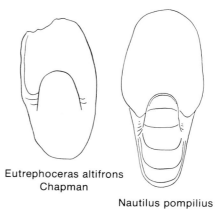

Fig. 1. Whorl profiles of *N. pompilius* and *Eutrepho-ceras altifrons* Chapman, a possible Miocene ancestor of the genus *Nautilus.*

sity is shown to be approximately 90 species, as compared with about 30 for the Paleocene. The Cretaceous species, however, are an aggregate number from a time interval of approximately 30 million years, as compared with about 5 million for the Paleocene. The highest diversity occurred in the Eocene, with more than 100 species described.

An interesting and as yet unresolved question concerns the fate of the nautiloids during the terminal Cretaceous extinctions. Some nautiloids did suffer. Except for one example from the Tertiary of Japan (which may be due to homeomorphy in *Eutrephoceras*), the highly diverse Cretaceous form *Cymatoceras* (59 Cretaceous species) and its offshoots *Heminautilus*

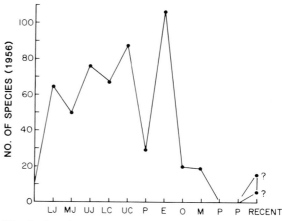

Fig. 2. Species level diversity of post-Triassic nautiloids. From Kummel 1956.

(6 sp.), *Deltocymatoceras* (2 sp.), *Epicymatoceras* (1 sp.), and *Syrionautilus* (1 sp.) do not extend beyond the Cretaceous. The available records, however, suggest that species of these genera were already rare or absent by Maestrichtian time. If anything, some nautiloid genera seem to expand across the Cretaceous–Tertiary border; *Eutrephoceras,* for instance shows a higher species level diversity in the Danian than it does in the Maestrichtian (Kummel 1956). Other genera, such as *Aturoidea* and *Aturia,* began their radiations in the latest Cretaceous, and continued into the Tertiary. We might well ponder why nautiloids seem so unaffected by the Cretaceous–Tertiary extinctions, as compared to the ammonites, which were completely wiped out. My feeling is that the differing developmental strategies between nautiloids and ammonites exemplify the major differences between the two groups. Ammonites produced numerous, small (1 mm) juveniles, which may have spent a long time in the plankton. Nautiloids, on the other hand, seem to have produced a few, large progeny. The demise of the ammonites may have been due to the ecologic collapse of the plankton at the end of the Cretaceous, affecting not so much the mature ammonites, but their young stages. Nautiloids may have escaped this through the production of deeply-hatching young, which started a nektobenthic, instead of a planktic or shallow-water developmental phase.

From the Eocene high nautiloid species-level diversity dropped markedly in the Oligocene and Miocene, both with about 20 species. By Miocene time only two genera, *Aturia* and *Eutrephoceras,* can be detected in the fossil record.

Present-Day Diversity of *Nautilus*

As in seemingly everything else relating to nautiloids, the number of species accepted in *Nautilus* is disputed. Three species are accepted by everyone: *N. pompilius* Linnaeus, characterized by an umbilical plug; *N. macromphalus* Sowerby, without an umbilical plug; and *N. scrobiculatus* Solander, which is more evolute than either *N. pompilius* or *N. macrompha-*

lus. Also accepted by some workers (myself included) is *N. stenomphalus* Sowerby, which is smaller than the others and has no umbilical plug. A fifth species, *N. repertus* Iredale, is also accepted in some quarters as valid. Other previously defined species, such as *N. moretoni* Willey and *N. alumnus* Iredale, are generally considered to be junior synonyms of *N. pompilius.*

Ecological Niches of Living Species

The species of *Nautilus* are generally found on deep fore-reef slope environments. In the two areas where I have studied *Nautilus* (*N. pompilius* in Fiji and *N. macromphalus* in New Caledonia and the Loyalty Islands), the greatest abundance of individuals appears to be in the 200–500 m depth range. The buoyancy system of *Nautilus* appears to be used simply for maintaining neutral buoyancy within these environments, rather than as a means of buoyancy-induced propulsion (Ward and Martin 1978; Ward 1979; Ward, Greenwald and Greenwald 1980). Hyponome-powered swimming, rather than buoyancy change, allows these species to forage. In the few observations of naturally occurring specimens that have been made in relatively shallow water (*N. macromphalus,* 5–50 m, New Caledonia and Loyalty Islands; *N. pompilius,* New Hebrides; Ward and Martin 1980), the *Nautilus* have been seen to move slowly over the bottom, dragging two tentacles over and just above the substrate as they swim backwards. Similar behavior is also widely reported from aquarium observations. I have never observed a naturally occurring *Nautilus* any distance up in the water column; rather, my impression is that members of the *Nautilus* genus are true nektobenthos, foraging on or just above the substrate.

Not much information is available about food preferences. Ward and Wicksten (1980) have published the only list of taxonomically identifiable gut contents (from hand-caught specimens of *N. macromphalus*). The specimens dissected showed a distinct specificity for one species of hermit crab (*Aniculus aniculus*), which is one of the only hermit crabs in the Indo-Pacific that does not bear commensal actinians on its gastropod-shell home. Other identified crop material was from a variety of crustacean groups, including xanthid crabs, raniid crabs, and galatheids. A fish bone was also identified. Also reported in this study was the behavior of *N. macromphalus* toward fresh crustacean molts, which are readily consumed. For these shallow-caught specimens of *Nautilus*, the picture is that of a generalist-scavenger. For deeper specimens, no information is known.

Perhaps the most striking aspect of the ecology of *Nautilus* is the very low reproductive potential that seemingly exists. *Nautilus macromphalus* and *N. pompilius* have now been successfully maintained in aquaria for long periods of time at several locations, and have produced numerous eggs in captivity (Martin et al. 1978; Mikami and Okutani 1977). From these observations it seems that about ten eggs are laid per female during each reproductive season. It is not yet known if the mature female *Nautilus* die after spawning, or produce eggs for several years. Even in the latter case, it appears that *Nautilus* produced far fewer eggs per individual than any other cephalopod. The eggs themselves are very large for cephalopods (up to 45 mm), which may account for the low fecundity. The smallest *Nautilus* yet observed (Davis and Mohorter 1973) were like miniature adults in external soft-part morphology. Because of these observations, the inference has been drawn that the young hatch after direct development, bypassing a free-living larval stage, and begin a nektobenthonic life similar to that of the adults immediately after hatching (Eichler and Ristedt 1966; Blind 1976). This type of developmental history would have to have profound influences on dispersal, and ultimately, on the evolutionary history of the species. If, as has been suggested, *Nautilus* bypasses any sort of planktonic phase, then dispersal of the species would be limited to benthic corridors of less than 750–800 m (the implosion depth of the shells) (Ward and Martin 1980; Ward, Greenwald and Rougerie 1980), since it appears that neither young nor adults swim any distance in the water column. In this situation, gene flow between populations separated by corridors of water deeper than 800 m could occur only during very rare, ''sweepstakes'' events, and for all practical purposes be nonexistent between populations separated by hun-

dreds of kilometers of deep water. Since most known habitats of *Nautilus* are widely scattered island groups separated by hundreds of thousands of kilometers of deep sea, I predict that gene-flow will be found to be low or nonexistent between most populations.

Species Distribution in Space and Time

Distribution of Modern Species

Nautilus are today found only in the western Pacific and perhaps parts of the Indian Oceans. Stenzel (1964) and Reyment (1973) have discussed the geographic ranges of the various species.

Three of the four recognized species appear to be restricted to one archipelago or island. These include *N. macromphalus* (New Caledonia and the neighboring Loyalty Islands), *N. scrobiculatus* (Soloman Islands and perhaps New Guinea), and *N. stenomphalus* (northern Australia). The most widely distributed species, *N. pompilius*, is found over a much wider range, from Fiji in the east to Australia and the Indian Ocean in the west. The highest diversity may be in the region of the Torres Straits, where drifted specimens of *N. pompilius* (*N. alumnus* Iredale), *N. pompilius* (*N. repertus* Iredale), and *N. stenomphalus* have been found.

Generic and Species Longevity

Range data for the post-Triassic nautiloid genera were derived from Miller (1947), Kummel (1956, 1964), and Shimanskii (1962). The mean range for the 26 genera listed by Kummel (1964, and excluding *Nautilus*) is 45 million years per genus. This type of longevity would place nautiloids in the same longevity range as bivalves, and is far less than generic longevity of ammonites and mammals (Stanley 1979).

In contrast to the relatively long average ranges for the nautiloid genera, the available data for species ranges suggests that they are far shorter. Kummel (1956) lists ranges in stages for a number of Mesozoic nautilid genera; for the genera *Cenoceras, Eutrephoceras, Pseudo-*

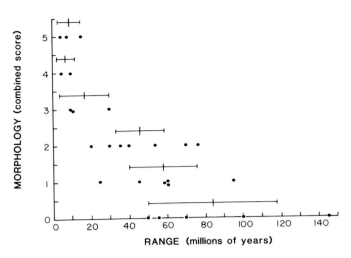

Fig. 3. Range of nautiloid genera in millions of years, plotted against morphologic complexity. One point was awarded for each morphologic departure from a smooth, involute, globular shell with straight suture. Points were separately awarded for: ribbing, nodes, keels, shell furrows, evolute shell, highly compressed shell, incised sutures. A high score indicates more complex morphology than a lower score. Above the data points for each score (0, 1, 2, 3, 4, 5) are plotted the mean value for that score, and 95% confidence intervals.

cenoceras, Pseudaganides, Paracenoceras, and *Cymatoceras,* 106 species are listed with stage ranges. Only 9% of these species ranged longer than one stage. In my own experience in the Lower and Upper Cretaceous of the North Pacific Province, I have found nautiloid species similarly to have ranges comparable to many of the coexisting ammonite species. Invariably, however, the nautiloids are a rare component of the fauna, and hence the observed range data may be due to inadequate sampling, rather than a very short geologic range. According to Kummel (1956), low number of specimens is characteristic of most nautiloid species.

Is there a connection between geologic range of the nautiloid genera, and their degree of identifiable morphology? In a very important paper, Schopf et al. (1975) found a correlation between extinction rate and degree of morphologic complexity for a variety of higher taxa. They concluded that evolutionary change in groups with complex morphology is more likely to be recognized than in morphologically simpler taxa, thus producing apparently shorter ranges per taxa. This may serve as the null hypothesis to the entire question of living fossils, and should be addressed for each possible group of supposed living fossils. I have examined this question by scoring post-Triassic nautiloid genera on the basis of the distinguishing morphology of each genus. The scoring has been kept as simple as possible by giving one point for each distinctive morphology that departs from a globular nautiloid shell with simple suture; the exercise simply allows a yes or no answer to the question of whether or not there is a correlation

between morphology and range in nautiloids. Range values are computed in millions of years, and are derived from first and last stage or epoch appearances and disappearances for species of each genus; if a nautiloid is recovered from any part of the stage, the age value for the entire stage is used, thereby possibly inflating the ranges to some degree.

The results of this exercise can be seen in Fig. 3. Genera with small scores include *Cenoceras, Eutrephoceras, Pseudocenoceras, Paracenoceras,* and *Cimomia.* All of these genera are typified by species with no ornament, involute inflated shells, and straight, or only slightly sinuous suture lines. At the other end of the morphologic scale are the genera *Carinonautilus, Obinautilus, Cymatonautilus, Epicymatoceras,* and *Tithonoceras.* All of these forms showed complex ornament, including combinations of ribs, keels, and depressed areas on the shell, as well as complex sutures, and in some cases highly evolute or strongly compressed shell outlines. As can be seen in Fig. 3, there does seem to be an inverse correlation between the aggregate scores for morphology and generic longevity.

Genotypic and Phenotypic Variation

Living populations of *Nautilus* have now been studied from the Philippines (Haven 1972; Hirano and Obata 1979), Fiji (Ward et al. 1977), New Caledonia (Ward and Martin 1978; Ward, unpublished), and Palau (Saunders and Spinosa 1978). In these populations there seems to be very little phenotypic variation between mature shells. The greatest degree of within-population variation is due to sexual dimorphism, with mature males having a slightly wider shell, as well as a greater shell diameter and total weight compared with females in the population. Not much work has been done on the degree of variation within extinct species. My own unpublished examinations of *Eutrephoceras dekayi* from the Cretaceous of the Western Interior, as well as the detailed examination of *Paracenoceras calloviense* by Tintant (1969) suggest that within-species variation may be similarly low. Much more work is needed, however, to resolve this question.

Of far more interest is the degree of between-population variability in separated populations of *N. pompilius*, especially in diameter, shell color pigments, degree of coloration, microornament of the shell, coiling geometry of the shell, and jaw morphology. Ward et al. (1977) first pointed out the great shell diameter differences between populations of *N. pompilius* captured live in Fiji and those from the Philippine Islands. Hirano and Obata (1979) looked at the shape of shells from two widely separated locales in the Philippine archipelago, finding subtle but significant differences in shell-coiling geometry, and concluded that the two populations are isolated by an ecological barrier, as earlier defined by Hamada (1977).

Museum collections of *N. pompilius* show subtle, yet distinctive differences between specimens from various island groups. Some, such as the specimens from Palau, and from various localities around Australia, are charac-terized by very large size. Others, such as the forms from Fiji, Tonga, and New Guinea, are characterized by very small size. Still others, intermediate in size, such as the *N. pompilius* from the New Hebrides, are differentiable by distinctive shell color and degree of coloration. The specimens from the New Hebrides have the same purplish-brown shell color that typifies newly captured *N. macromphalus*. In addition to this distinctive color, both *N. macromphalus* and *N. pompilius* from the New Hebrides have a higher proportion of their shells covered with stripes of pigment than can be found in other populations. Degree of coloration can be measured by photographing a shell from the side and measuring the area of color with a planimeter. In 16 specimens of *N. macromphalus*, mature shells were found to have 40% to 65% of the side view covered with pigment; 4 specimens of *N. pompilius* from the New Hebrides showed values of 32% to 45%. In contrast, 21 specimens of *N. pompilius* from Fiji showed coloration values of 6% to 30%. *N. pompilius* from the Philippines and Palau are intermediate between these two extremes.

I have measured mature shell diameters for *N. pompilius* maintained in the Australian National Museum, the United States National Museum, and my own collections of shells. Additionally, data of shell size of mature *N. pompilius* is available from Saunders and Spinosa (1978) for *Nautilus* from Palau, and from Hirano and Obata (1979), for specimens from the Philippines. These data, tabulated as mean diameters of mature specimens (Fig. 4), indicate that major differences in mature shell sizes exist between some of the populations, and that some of these between-population size differ-

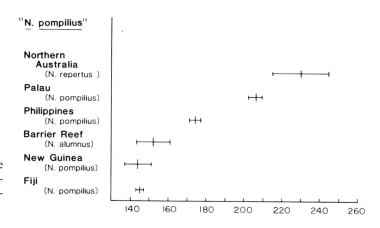

Fig. 4. Mean shell diameters of mature *"Nautilus pompilius"* from various, separated habitats in the Pacific. 95% Confidence intervals shown.

Table 1. Morphs of *"N. pompilius."*

Name previously used	Location	Distinguishing features
N. repertus, *N. pompilius*	Coastal Australia	Largest known *Nautilus,* low shell coloration
N. pompilius, N. aff. *N. pompilius*	Palau	Large size, distinct "chevron" pattern of shell ornament
N. pompilius	Philippines	Intermediate to large size, no chevron ornament
N. pompilius	New Hebrides	Intermediate size, virtually identical to *N. macromphalus,* except for presence of umbilical plug
N. pompilius, *N. alumnus,* *N. moretoni*	Great Barrier Reef, New Guinea	Small size, no chevron ornament
N. pompilius	Fiji, Tonga	Small size, low degree shell coloration
Other Species		
N. scrobiculatus	New Guinea, Solomon Islands	Intermediate to large size, wide umbilical opening, chevron pattern, square whorl profile, thin, yellowish color pattern
N. macromphalus	New Caledonia, Loyalty Islands	Intermediate size, open umbilicus, reddish-brown color, high degree shell coloration
N. stenomphalus	N. Australia	Smallest *Nautilus;* small, open umbilicus

ences are greater than the degree of within-population morphologic variation (Table 1).

Further research may show that the diameter and other morphologic differences between the isolated *N. pompilius* populations are phenotypic responses to slightly differing environmental challenges across the range of the species. However, in light of the low within-population variability, and possibility of low juvenile or adult dispersal potential for crossing the vast oceanic distance between the isolated populations of *N. pompilius,* the possibility that these differences between the various *"N. pompilius"* are genotypic, rather than phenotypic should be explored. If the various morphs of *N. pompilius* as currently constituted are actually separate, sibling species, then the number of species of *Nautilus* could be between 10 and 20 instead of the currently accepted three or four. Speciation events would be brought about by chance, "sweepstakes events," producing isolation. The absence of *Nautilus* from many of the smaller islands or island groups in the Pacific may indicate that such chance events only occur on the larger island groups or land masses. Just such an event may have been documented recently. Hamada et al. (1980) reported the capture in Japan

of a single *N. pompilius,* similar to those found in the Philippines. These workers concluded that this specimen represents a drift of the warm Kuroshiwo Current, which occasionally deposits tropical organisms in southern Japan. (Alternatively, this specimen may be part of an unrecognized population living in this area.)

Conclusion

Should *Nautilus* be considered a living fossil? The data at hand are insufficient to make a clear case either for or against. There *are* species in the latest Cretaceous and Early Tertiary that are morphologically similar to the currently extant species of *Nautilus,* and there *are* good range data that support the notion that post-Triassic nautiloid genera were characteristically long ranged. Finally, there was a marked diversity drop at the end of the Miocene. Evidence against includes the short geologic ranges of fossil nautiloid species, the great gaps in the record between the Cretaceous and Lower Tertiary *"Nautilus"* and those of the present day, the common homeomorphy in the group, and the evidence that nautiloid genera are seemingly susceptible to what I think of as the "Schopf Effect": that recognition of evolutionary longevity is influenced by degree of morphological complexity, tending to obscure the true rates or ranges involved in either highly ornamented, or unornamented forms.

Can evolutionary stasis actually be demonstrated, or is it an artifact of low morphologic complexity? My prejudice in the case examined here is that the nautiloids since the Triassic were capable of rapid radiations. However, the morphologies evolved have been constrained by two main factors: the need for effective swimming, and considerations of the buoyancy control system. In nautiloids, solutions to both of these problems have resulted in short body chambers (to allow straight-line retraction of muscles used in swimming, and to allow formation of new chambers with the siphuncle in a vertical orientation within the chambers, sitting atop the convex swell of the new septum). The necessity for effective swimming, and the need for a short body chamber have severely limited the use of ornament and the types of shell shapes available to the nautiloids, resulting in the repeated convergence of shape. Variability within the populations appears to be low. Speciation events, producing morphologic change, may have been typified by very slight changes in morphology. Perhaps these speciation events were common, but are obscured in the fossil record by small samples. Only large samples of nautiloids, carefully allocated from measured stratigraphic sections, would reveal that speciation occurred at all. Rather than being a prime example of a living fossil, the nautiloids may be examples of rapidly speciating organisms that change only slightly during each event, and return to the same form over and over. The result would be apparent stasis, but the actual history would be similar to that of any other rapidly speciating group—except that the net morphologic change over time would be small, rather than large.

Finally, what of the species of *Nautilus* currently extant? I believe that they arose from a eutrephoceratid ancestor in the latest Miocene of Australia, and may today be engaged in active radiation. In this view, the terminal Miocene extinctions of *Eutrephoceras* and *Aturia* are but one more of the many "crises" in the history of the chambered cephalopods. We may

be witnessing a new expansion, rather than final decline.

Literature

Blind, W. 1976. Die ontogenetische Entwicklung von *Nautilus pompilius* (Linne). Paleontographica, Abt A 153:117–160.

Chapman, F. 1914. New or little-known Victorian fossils in the National Mus. Proc. Roy. Soc. Victoria 27(N.S):350–361.

Cousteau, J., Diole, P. 1973. Octopus and squid. New York, Doubleday.

Davis, R. A., Mohorter, W. 1973. Juvenile *Nautilus* from the Fiji Islands. J. Paleont. 47:925–928.

Eichler, R., Ristedt, H. 1966. Isotopic evidence on the early life history of *Nautilus pompilius* (Linne). Science 153:734–736.

Hamada, T. 1977. Distributional and some ecological barriers of modern *Nautilus* species. Sci. Papers Coll. Gem. Education, U. Tokyo 27(2):89–102.

Hamada, T., Tanabe, K., Hayasaka, S. 1980. The first capture of a living chambered *Nautilus* in Japan. Sci. Papers Coll. Gen. Education, U. Tokyo 30:63–66.

Haven, N. 1972. The ecology and behavior of *Nautilus pompilius* in the Philippines. Veliger 15:75–80.

Hirano, H., Obata, I. 1979. Shell morphology of *Nautilus pompilius* and *N. macromphalus*. Bull. Nat. Sci. Mus., Ser. C 5(3):113–130.

Kummel, B. 1956. Post-Triassic nautiloid genera. Bull. Mus. Comp. Zool. 114:324–493.

Kummel, B. 1964. Nautiloidea-Nautilida, pp. K383–K466. In: Moore, R. C. (ed.) Treatise on invertebrate paleontology, (K) Mollusca 3. Geo. Soc. Am. and U. of Kansas Press, Lawrence, Ks.

Martin, A. W., Catala-Stucki, I., Ward, P. 1978. The growth rate and reproductive behavior of *Nautilus macromphalus*. N. Jb. Geol. Paleont. Abh. 156:207–225.

Mikami, S., Okutani, T. 1977. Preliminary observations on maneuvering, feeding, copulating and spawning behavior of *Nautilus macromphalus* in captivity. Venus 36(4):29–41.

Miller, A. K. 1947. Tertiary nautilids of the Americas. Geo. Soc. Am. Memoir 23.

Miller, A. 1951. Tertiary nautiloids of west-coastal Africa. Ann. Mus. Congo Belge, Ser. 8, Sci. Geol. 8:1–88.

Naef, A. 1928. Die Cephalopoden-Fauna und Flora des Golfes von Naepel. Zool. Stat. Neopel. 35(2).

Raup, D. M. 1967. Geometric analysis of shell coiling: coiling in ammonoids. J. Paleont. 41:43–65.

Reyment, R. A. 1973. Factors in the distribution of fossil cephalopods. Part 3. Experiments with exact models of certain shell types. Bull. Geol. Inst. U. Uppsala, N.S. 4:7–41.

Saunders, W. 1981a. A new species of *Nautilus* form Palau. Veliger. 24:1–7.

Saunders, W. 1981b. The species of living *Nautilus* and their distribution. Veliger. 24:8–17.

Saunders, B., Spinosa, C. 1978. Sexual dimorphism in *Nautilus* from Palau. Paleobiology 4(3):349–358.

Schopf, T., Raup, D., Gould, S., Simberloff, D. 1975. Genomic versus morphologic rates of evolution: influence of morphologic complexity. Paleobiology 1(1):63–70.

Shimanskii, V. 1957. New forms of the order Nautilida in the USSR. Materialy k "Osnovam Paleontologii" 1:35–41.

Shimanskii, V. 1962. Superorder Nautiloidea: general section. In: Ruzhentsev, V. (ed.), Fundamentals of paleontology 5:33–70.

Stanley, S. 1979. Macroevolution. San Francisco: Freeman.

Stenzel, H. B. 1964. Living *Nautilus*, pp. K59–K93. In: Moore, R. C. (ed.), Treatise on invertebrate paleontology, (K) Mollusca 3. Geol. Soc. Am. and U. Kansas Press, Lawrence, Ks.

Tintant, H. 1969. Un cas de dimorphisme chez les *Paracenoceras* du Callovien, pp. 167–184. In: Westermann, G. (ed.), Sexual dimorphism in fossil metazoa, and taxonomic implications. E. Schweizerbartsche Verlagsbuchhandlung.

Ward, P. 1979. Cameral liquid in *Nautilus* and ammonites. Paleobiology 5:40–49.

Ward, P. 1980. Comparative shell shape distributions in Jurassic-Cretaceous ammonites and Jurassic-Tertiary nautilids. Paleobiology 6(1):32–43.

Ward, P., Greenwald, L., Greenwald, O. 1980. The buoyancy of the chambered *Nautilus*. Sci. Amer. 243(4):190–203.

Ward, P., Greenwald, L., Rougerie, F. 1980. Shell implosion depth for living *Nautilus macromphalus* and shell strength of extinct cephalopods. Lethaia 13(2):182.

Ward, P., Martin, A. 1978. On the buoyancy of the pearly *Nautilus*. J. Exp. Zool. 205:5–12.

Ward, P., Martin, A. 1980. Depth distributions of *Nautilus pompilius* in Fiji and *N. macromphalus* in New Caledonia. Veliger 22(3):259–264.

Ward, P., Stone, R., Westermann, G., Martin, A. 1977. Notes on animal weight, cameral fluids, swimming speed, and color polymorphism of the cephalopod *Nautilus pompilius* in the Fiji Islands. Paleobiology 3:377–388.

Ward, P., Wicksten, M. 1980. Food sources and feeding behavior of *Nautilus macromphalus*. Veliger 23(2):119–124.

Note Added in Proof

Since the writing of this manuscript (1981), Dr. W. Saunders has conducted additional research on the *Nautilus* populations in Palau, Western Caroline Islands. Dr. Saunders considers the Palauan forms to be sufficiently distinct from *N. pompilius* to warrant separate species status (Saunders, 1981a). Saunders has also reviewed the other extant species of *Nautilus* (Saunders, 1981b.) Saunders and I are currently reviewing the *Nautilus* species problem through morphological and gel electrophoretic studies of the major populations.

31
The Bryozoan *Nellia tenella* as a Living Fossil

Judith E. Winston and Alan H. Cheetham

Department of Invertebrates, American Museum of Natural History, New York, NY 10024
Department of Paleobiology, National Museum of Natural History, Smithsonian Institution, Washington, DC 20560

Bryozoans are a group of invertebrates with a history dating from Early Ordovician time, soon after well-skeletonized invertebrates first appeared. Many living representatives of the tubular Bryozoa, or Stenolaemata, retain some characteristics of their early Paleozoic ancestors, generally as order-level characters (Boardman 1981). Stenolaemates of simple morphology, such as the stomatoporids, show relatively little change from Ordovician to Recent (Boardman and Cheetham 1973). The peculiar mode of reproduction of living stenolaemates, polyembryony, resulting in production of many genetically identical larvae, also suggests a conservatism that makes this group the logical place to look for a species that could be called a living fossil.

Unfortunately, homeomorphy, particularly in external characters, is common in the Stenolaemata. Paleozoic and post-Triassic stenolaemates traditionally have been studied by different techniques, making it difficult to recognize effects of convergence on apparently long-lasting characteristics (Boardman 1981). Wider application to the post-Triassic stenolaemates of techniques involving oriented sections (Boardman 1981) may well provide examples of living species with long stratigraphic ranges.

In the meantime, it is necessary to turn to the other group of calcified bryozoans, the Cheilostomata, to find a candidate that can be more readily evaluated. Much of the morphologic detail in this group is displayed on external surfaces or on the readily observed internal surfaces of their generally box-shaped skeletons.

The Cheilostomata apparently arose in Late Jurassic time (Pohowsky 1973). Early Cretaceous cheilostomes exhibit low diversity, having gone little beyond the morphologic simplicity of their soft-bodied ctenostome ancestors (Banta 1975; Cheetham 1975b). Some of this morphology has persisted to the present, but because of its simplicity, is difficult to evaluate for possible convergence. Rapid diversification ensued during Late Cretaceous time (Voigt 1959), culminating in the Maestrichtian (latest Cretaceous) in a number of stocks that had developed complex states of some morphologic characters while retaining simple states in others, apparently as a result of mosaic evolution (Boardman and Cheetham 1973). *Nellia tenella* is an example of one of these stocks, first recorded in the Maestrichtian of Jamaica, that may have persisted as a single species to the present, a duration of approximately 65–70 million years. Arguments for its persistence were given by Lagaaij (1969) and Cheetham (1968), and arguments against it by Schopf (1977).

The definition of a living fossil used here is a species that is "anatomically similar (bordering on identity) to a fossil species which occurs very early in the history of the lineage," (see

Introduction, this volume). According to evidence now available, the *Nellia tenella* stock appears to meet that definition, whether or not it retained sufficient genetic uniformity to have remained a single species throughout its 65–70 million-year range. Of course, we cannot know how much the soft-part morphology not reflected in the skeleton (e.g., number of tentacles) might have differed through this interval, but by analogy to living cheilostomes we have no reason to suspect large differences (Winston 1981).

Nellia tenella (Lamarck 1816) [Synonymy in Cheetham 1966:49] forms bushy whitish to glassy colonies on algae, sponges, and other substrata. Each colony or subcolony (Fig. 1A) arises from spreading stolons attached by straw-colored uncalcified, cuticular rootlets (see Fig. 1). Erect parts of colonies begin with three or four short, calcified kenozooidal segments (internodes) connected to each other by flexible joints composed of two parallel uncalcified, cuticular tubes. Above this initial portion are calcified autozooidal internodes shaped like four-sided prisms, with a row of autozooids on

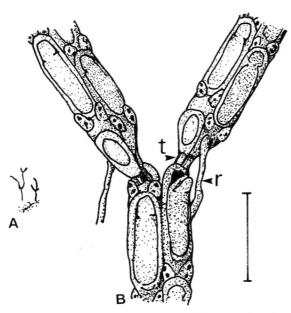

Fig. 1. (A) Lifesize colony of *Nellia tenella*. Redrawn after MacGillivray 1880. (B) Portion of a colony of *Nellia tenella* showing budding pattern (A–H in Fig. 2) and joints. Redrawn after Marcus 1939. r = Uncalcified cuticular rootlets; t = uncalcified cuticular tubes joining internodes. Scale = 0.5 mm.

each face. The autozooids of all faces are similar in morphology and orientation. Variation in their number (usually 1–12 per row) results in variable internode lengths (usually increasing distally). Each autozooidal internode is connected to one internode proximally by a cluster of three parallel organic tubes fitting into one large and two small sockets (Figs. 2F, 2G) and to two internodes distally by a pair of similar clusters (Figs. 2B, 2H). This method of articulation results in a pattern of regularly bifurcating branches. In more proximal internodes additional connections are formed by organic rootlets (Fig. 1B) that extend from frontal surfaces of autozooids along surfaces of internodes to the colony base.

Autozooids (Figs. 2A, 2E, 2H) are elongate rectangular in frontal outline, more variable in length than in width, both within and among internodes. Most of the frontal surface of a zooid is taken up by an uncalcified area that narrows slightly at the hinge of the operculum. This membranous area is surrounded by a smooth raised calcified border that widens proximally to form a gymnocyst of variable width. From the inner margin of the gymnocyst, on the proximal and lateral sides of the zooid, a narrow crescent-shaped calcified shelf (cryptocyst) extends under the membranous area. A pair of large opercular occlusor muscle scars, smaller parietal muscle scars, and communication pore plates are usually visible on the obliquely oriented vertical walls of the zooid (Figs. 2A, 2C, 2D, 2H). The lophophore has 11 or 12 tentacles.

A pair of small oval avicularia (rarely single or absent) is placed on each autozooid, near the proximolateral corners of the gymnocyst. The mandible is bluntly pointed and hinged on a complete skeletal bar; it covers about half of the avicularium surface and when closed is directed proximally. Mandibles may fail to develop on some avicularia, the surfaces of which may then remain membranous or calcify to leave only a minute pore (Fig. 2A). On the proximal end of the avicularium, and subequal in size to it, is a round rootlet pore.

Ovicells (Fig. 2E) are externally inconspicuous and hence have been described as vestigial (Harmer 1926). Internally, an ovicell forms a somewhat globular chamber deeply buried at the distal end of the maternal autozooid and

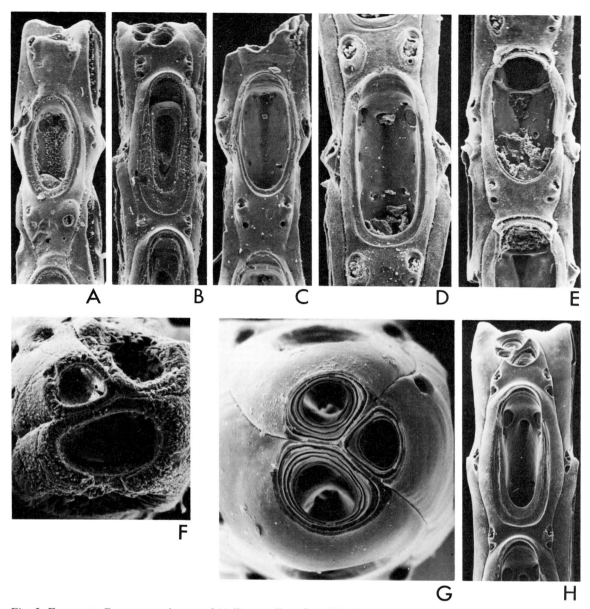

Fig. 2. Eocene to Recent specimens of *Nellia tenella* showing autozooids with preserved opercular occlusor muscle scars (A. D. H), parietal muscle scars (C, D, H), and ovicells (E); avicularia with mandibular areas occluded (A) or functional (B–E, H); rootlet pores on avicularian chambers (A–H); kenozooid apparently budded within autozooid (B); and articulation sockets at distal (B, H) and proximal (F, G) ends of internodes. (A, F) Eocene (Ypresian), Loeblich loc 4703, France; (A) USNM 337431, ×100; (B) Oligocene (Chickasawhay Fm.), St. Stephens, Alabama; USNM 337343, ×100. (C) Oligocene (Byram Fm.), Old Byram, Mississippi; USNM 337434, ×100. (D, E) Miocene Olsson loc 179, Dominican Republic; D, USNM 337435, ×150; (E) USNM 337436, ×100. (F) USNM 337423, ×300. (G, H) Recent, Johnson-Smithsonian Expedition, north of Puerto Rico; (G) USNM 337437, ×300; H, USNM 337438, ×100.

partly separated from it by a low ridge. At the surface, the ovicell roof is formed by a slight, narrow elevation of the gymnocyst of the distal zooid in which a crescent-shaped uncalcified area exposes an inner calcified layer.

Measurement of autozooids distal to the zone of change with which each internode begins (Boardman and Cheetham 1969:Fig. 7) are summarized from Cheetham (unpublished data). Those for Recent material are for four internodes of the same colony; those for fossils are for four disarticulated internodes from the same sample.

Nellia tenella, Recent, north of Puerto Rico
Zooid length ($n = 16$), $\bar{X} = 0.42$ mm, $s = 0.011$ mm
Zooid width ($n = 16$), $\bar{X} = 0.23$ mm, $s = 0.010$ mm

Nellia tenella, Miocene (Cercado Fm.), Dominican Republic
Zooid length ($n = 16$), $\bar{X} = 0.45$ mm, $s = 0.036$ mm
Zooid width ($n = 16$), $\bar{X} = 0.21$ mm, $s = 0.027$ mm

Nellia tenella, Oligocene (Chickasawhay Fm.), Alabama
Zooid length ($n = 16$), $\bar{X} = 0.45$ mm, $s = 0.019$ mm
Zooid width ($n = 16$), $\bar{X} = 0.18$ mm, $s = 0.015$ mm

Nellia tenella, Eocene (Crystal River Fm.), Florida
Zooid length ($n = 16$), $\bar{X} = 0.48$ mm, $s = 0.069$ mm
Zooid width ($n = 16$), $\bar{X} = 0.22$ mm, $s = 0.031$ mm

Nellia tenella, Cretaceous (Maestrichtian), Jamaica
Zooid length ($n = 16$), $\bar{X} = 0.46$ mm, $s = 0.043$ mm
Zooid width ($n = 16$), $\bar{X} = 0.17$ mm, $s = 0.025$ mm

Differences in zooid length among geologic ages are not significant ($F = 1.087$, $P > 0.25$), by two-level nested analysis of variance. Differences in zooid width are marginally significant ($F = 4.801$, $P < 0.025$), but show no trend in mean values with time. Differences among internodes within ages are highly significant for both length ($F = 18.074$, $P \ll 0.001$) and width ($F = 20.474$, $P \ll 0.001$). The pattern of varia-

tion in these measurements thus seems to reflect fluctuating stasis.

Other apparent consistencies among specimens of all geologic ages, from Maestrichtian to Recent include: (1) arrangement of autozooids in four similar series in internodes (generic character); (2) tripartite articulation of internodes (specific character); (3) morphology of cryptocyst and gymnocyst (generic character); (4) size, shape, orientation, and distribution of avicularia (specific characters); (5) rootlet pores (generic or higher level character); and (6) morphology of ovicell (specific character). An additional generic character, (7) similarly placed muscle scars and communication pore plates, can be seen in well-preserved specimens of Eocene to Recent age. The Cretaceous specimens are too recrystallized for these structures to be observed.

Harmer (1926) described within-species variation in living *N. tenella*. According to him, variety *quadrilatera* differed from the "typical" form in having (1) longer zooids with less raised margins; (2) ovicell less raised, with maternal zooids widened slightly at the proximal end; (3) larger colonies with longer internodes having up to 30 or more zooids on each face; (4) avicularia all functional, with more sharply pointed mandibles. This variety, which was first named by D'Orbigny (1851) for specimens from "mers de la Chine," was found at only four Siboga Expedition stations; at two of these it occurred with "typical or transitional" forms of *N. tenella*. It thus appears that *quadrilatera* represents an infrasubspecific variant or possibly a sibling species.

Two other living species of *Nellia* are known, *N. appendiculata* (Hincks) and *N. tenuis* Harmer. Both species share with *N. tenella* a colony form consisting of flexibly jointed quadriserial internodes arising from creeping stolons. Their zooids are also similarly arranged and oriented on the four faces of the internodes, have similar gymnocysts and cryptocysts, and bear similarly distributed, paired avicularia. They differ from *N. tenella*, however, in the following ways: (1) Their internodes are connected by sets of two, rather than three organic tubes; (2) Their zooids are wider (*Nellia appendiculata*, Recent, Australia; zooid length, $n = 16$, $\bar{X} = 0.49$ mm, $s = 0.024$ mm; zooid width,

$n = 16$, $\bar{X} = 0.39$ mm, $s = 0.044$; Cheetham, unpublished data); (3) The roofs of their ovicells are wider and somewhat more protuberant; and (4) Their avicularia are larger, diverge outward, and have an acute, hooked mandible. *N. appendiculata* and *N. tenuis* differ from each other mainly in relative lengths of their internodes and shapes of their avicularia.

At present *Nellia tenella* has a circumtropical/subtropical distribution, being reported from warm waters everywhere except Japan and Hawaii. Its flexibly jointed, rooted (cellariiform) colonies can cling to ephemeral substrata like sponges and algae, and so inhabit areas where primary substrata (sand or mud) are unsuitable for colonization by most bryozoans, and can remain intact on these or more stable substrata in areas of strong water movements. Its depth range is great (from low water to 1000 m), and it is apparently tolerant of fluctuating and somewhat reduced salinities (Lagaaij 1969).

Lagaaij (1969) investigated the distribution of *Nellia tenella* with regard to its paleoecological interpretation and concluded that the species was characteristic of inner neritic or open bay environments with slightly reduced salinity conditions. This was the conclusion reached also by Rucker (1967) who in a paleoecological analysis of bryozoans from Venezuelan shelf sediments found a "transitional" biofacies characterized by an almost completely unispecies assemblage of *Nellia tenella* occurring between the barren silty clays of the inner shelf and the bryozoan-rich calcareous sands of the outer shelf. While this interpretation may hold in most cases, it should be remembered that *Nellia tenella* has been reported to occur in some very cold and deep water (e.g., 404 m off Crozet Island, one of the Possession Islands, according to Busk 1884) and may also be common intertidally. In fact Ryland (1974) described "swards" of *Nellia tenella* occurring along with *Poricellaria ratoniensis* and *Synnotum aeqyptiacum* under intertidal rocks in Queensland, Australia, in what is apparently a very silty environment. *Nellia* is also known to occur, though not in such abundance, on coral reefs (Ryland 1974; Brood 1976). Accurate paleoecological interpretation must depend on analysis of remains of other organisms as well

as other bryozoans, and careful evaluation of the possibility of allochthonous deposition (the delicate colonies being present as fossils only as disarticulated internodes or internode fragments).

Knowledge of the biology and ecology of living *Nellia tenella* is limited. The bushy colonies of the species serve as attachment sites for other bryozoans with similar colony form, e.g., *Caulibugula, Crisia, Vittaticella, Savignyella* (Cook 1968). Ryland (1974) noted many pycnogonids inhabiting intertidal colonies in Queensland, though he did not investigate whether or not these arthropods were predators on the bryozoans as has been reported elsewhere. Of the basic aspects of its population biology (growth rates, reproductive season, longevity) nothing is known.

The phylogenetic position of *Nellia* was unclear for a long time. Harmer (1926) placed it in the Membraniporidae, but later authors (e.g., Osburn 1950; Bassler 1953; Cheetham 1966; Cook 1968) included it in the Farciminariidae, a cellularine family. A study of the evolution of zooidal morphology in poricellariid cheilostomes led Cheetham (1968) to the conclusion that a *Nellia* of the *tenella*- morphology might actually be ancestral to a whole group (including *Nellia, Mediosola, Curvacella, Rimosocella, Vincularia, Poricellaria*) to which it had previously been regarded as little related. *Nellia tenella* appears to most closely approach the primitive condition in that: (1) Its budding pattern and articulation of internodes are the same as in primitive poricellariids; (2) Its colonies contain only one type of non-ovicelled autozooid (are monomorphic) and the zooids are symmetrical in shape; (3) Ovicells are consistently developed and not limited to particular budding series; (4) Avicularia are of the same form and have the same general position as those in primitive poricellariids—adventitious, gymnocystal, paired, with pointed rostrum and pivotal bar and directed more or less proximally; (5) Colony form is cellariiform (erect, jointed, and attached to the substratum by rootlets rather than a calcareous base). The occurrence of *Nellia* of the *tenella* type along with primitive poricellariids in the Upper Maestrichtian of Jamaica (Cheetham 1968) makes the phylogenetic arguments advanced by Cheetham at least plausible.

On the basis of stratigraphic distributions and these morphologic criteria, Cheetham suggested that poricellariids could have evolved from *Nellia* in the Cretaceous through the development of the following new characteristic: (1) zooid asymmetry and differentiation of lateral and frontal series of zooids; (2) completion of the cryptocyst; and (3) restriction of ovicells to lateral zooids.

Vincularia possibly evolved from *Nellia* in the Paleocene, sharing with the poricellariids the development of zooidal asymmetry and dimorphism and restriction of ovicells to lateral zooids, but differing from them in: (1) retention of minimal calcification of the cryptocyst; (2) modification of the budding pattern; and (3) development of asymmetry of avicularia.

While close phylogenetic relationships within this "nelliid" group are suggested by evolutionary trends in a large number of morphologic characters at both the colony and zooid level, there are few such comparisons that can be made with possible ancestors of the group as a whole. Their highly regular system of internodes and joints is unknown in other Cretaceous stocks. Adventitious avicularia occur in a few Cretaceous stocks, but these are mostly ascophorans with more complex zooidal structure. Ovicells in Cretaceous stocks are differently constructed. These major differences suggest that the "nelliid" group may have evolved rather suddenly from one of the simpler anascan stocks that are so common in the Upper Cretaceous.

General trends in this "nelliid" group can be summarized based on the work of Cheetham (1968, 1973, unpublished data; Lagaaij 1968, 1969; Lagaaij and Cook 1973; Labracherie 1970, 1975). Trends in diversity (number of species) are shown in Fig. 3.

In the Late Cretaceous two lineages of *Nellia* were already present in the Caribbean (Cheetham 1968): (1) those of a "*ventricosa*" type, having shorter internodes, broader zooids and a greater degree of gymnocyst development (leaving less of the frontal surface uncalcified) and larger avicularia; and (2) those more clearly of the *N. tenella* type. During the Paleocene, representatives of the "*ventricosa*" group occurred with primitive poricellariids in the southeast United States (Canu and Bassler 1920, 1933). The greatest diversity in the genus ap-

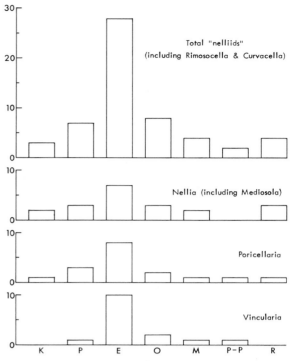

Fig. 3. Diversity of *Nellia* and its relatives over time. Vertical scale shows number of species known during a particular interval.

pears to have been in the Eocene, when at least six, possibly seven species were present: *N. tenella* and *N. ventricosa* (Cheetham 1966; Labracherie 1970), *N.* aff. *N. ventricosa* (Cheetham 1975a), *N. oscitans* (Cheetham 1963), *N. petila* and *N.* (subgenus *Mediosola*) *stricta* (Labracherie 1970), and *N.* sp. (Labracherie 1975). By the Eocene, *N. tenella* had achieved a virtually circumtropical distribution, i.e., both sides of the Atlantic (Lagaaij 1969) to the Indian Ocean (Labracherie 1975). The known distribution of the "*ventricosa*" lineage was even wider, extending from both sides of the Atlantic through the Indian Ocean (Labracherie 1975) to the Pacific (Cheetham 1975a).

The extensive Miocene records of *Nellia tenella* show that the species had clearly become circumtropical/subtropical. At that time it still occurred in Europe and the Mediterranean, from which as of Pliocene time, it was no longer known, giving it essentially its Recent distribution (Lagaaij 1969). The other two living species of *Nellia* are apparently both members of the "*ventricosa*" lineage (Cheetham 1966), although no records are yet known for that group in Pliocene or Pleistocene time. One species, *N. tenuis,* has a wide distribution, from the Indo-Pacific (Harmer 1926) to Pacific Panama (Osburn 1950) and the Caribbean (Osburn 1940). The other, *N. appendiculata,* is known only from southern Australia.

Diversity among related lineages also reached a maximum in Eocene times. The genus *Rimosocella* is limited to the Eocene (Cheetham 1960, 1968). The genus *Curvacella* occurred from Mid-Eocene to Late Oligocene (Labracherie 1970).

Diversity in *Vincularia* was also highest in the Eocene, with possibly as many as ten species represented (Cheetham 1966, 1973). Principal component analysis (Cheetham 1973) of morphologic variation in *Vincularia* populations from Early Eocene to Mid-Oligocene showed a pattern of decreasing polymorphism. Within one lineage (A) frontal zooids changed very little over time, but lateral zooids became first larger and more asymmetrical, then smaller and more elongate; therefore eventually lateral and frontal series became less morphologically distinct. *Vincularia,* which had a Caribbean-Atlantic distribution during the Paleocene and Eocene, had spread by the Miocene to India ("*Nellia*" *kutchensis* Tewari and Srivastava 1967) and the Pacific (Lagaaij 1969; "*Nellia* aff. *oculata*" of Brown 1964). The genus is apparently extinct; its last recorded occurrence being from New Hebrides Pliocene or Pleistocene sediments (Cheetham 1979).

Poricellaria first occurred with *Nellia* in the Upper Cretaceous of Jamaica (Cheetham 1968). The greatest diversity in this genus was also apparently in the Eocene, with at least eight species represented at that time. Principal component analysis of morphologic variation in *Poricellaria* populations from Late Cretaceous to Recent (Cheetham 1973) showed two main trends. In lineages C and D (*Poricellaria* s.s.) all zooids increased in size at a very rapid rate (becoming larger than those of lineages A and B ever became), but lateral and frontal zooids did not differentiate morphologically, remaining monomorphic. These lineages became extinct by the end of the Eocene. In lineages A and B (*Diplodidymia*) lateral zooids changed more rapidly (though following the same allometric trends) than frontal zooids so that in living pop-

ulations measurements of the two series of zooids do not overlap. From its inception in the Caribbean, *Poricellaria* reached Atlantic Europe during the Paleocene and the Indo-West-Pacific during the Eocene. After the Eocene all lineages except B underwent a rapid decline to extinction. Lineage B had a circumtropical/subtropical distribution (excluding Australia and New Zealand) in the Miocene. Its Recent distribution (that of its one remaining species—*Poricellaria ratoniensis*) is tropical Indo-West-Pacific (now including Australia) (Lagaaij and Cook 1973; Ryland 1974).

In general, it appears that *Nellia* never underwent quite as much diversification as either *Poricellaria* or *Vincularia*. Like those two groups it did undergo a decline from the Eocene to the Oligocene; unlike them it did not continue to extinction (*Vincularia*) or near-extinction (*Poricellaria*). Instead, the genus seems to be doing quite well, with three Recent species, one of which, *N. tenella,* has an extremely widespread distribution. Obviously, the differences in rate of differentiation and decline make *Nellia tenella* the "living fossil." It is difficult, nevertheless, to determine the reasons for its persistence, especially as most members of the group seem to have shared a rather similar ecology: cellariiform growth in shallow to moderately deep water, often on unstable substrata, in subtropical or tropical climates. *Poricellaria* and *Nellia* apparently co-occurred more often in the past than they do today; *Poricellaria* seems to have undergone greater geographic and habitat constrictions. *Poricellaria ratoniensis* is strictly tropical, while *Nellia tenella* retains a tropical/subtropical distribution. Living *Poricellaria* does seem, then, to be less able to tolerate low temperatures than *N. tenella*. It also has a much more restricted depth range (0–59 m), which may, of course, be linked to temperature tolerance. There is some indication that *Nellia tenella* may be better able to tolerate fluctuating or lowering of salinities than *Poricellaria*. On the other hand, *Poricellaria ratoniensis* is reported to survive hypersaline conditions remarkably well (to 56.7% parts per thousand according to Lagaaij and Cook 1973).

Acknowledgments: We thank JoAnn Sanner for preparing and measuring specimens, making photographic prints, and drafting Fig. 3; Susann

Braden, Walter Brown, and Mary-Jacque Mann for operating the scanning electron microscope and preparing photographic negatives; and Richard Boardman for reviewing the manuscript.

Literature

Banta, W. C. 1975. Origin and early evolution of cheilostome Bryozoa, pp. 565–582. In: Pouyet, S. (ed.), Bryozoa 1974, Proceedings of the Third Conference, International Bryozoology Association. Docum. Lab. Géol. Fac. Sci. Lyon.

Bassler, R. D. 1953. Bryozoa, p. 253. In: Moore, R. C. (ed.), Treatise on invertebrate paleontology, (G). Geol. Soc. Am. and U. Kansas Press, Lawrence, KS.

Boardman, R. S. 1981. Coloniality and the origin of post-Triassic tubular bryozoans, pp. 70–89. In: Broadhead, T. W. (ed.), Lophophorates, notes for a short course. U. Tennessee Studies in Geology, 5.

Boardman, R. S. & Cheetham, A. H. 1969. Skeletal growth, intracolony variation and evolution in Bryozoa: a review. J. Paleontol. 43:205–233.

Boardman, R. S., Cheetham, A. H. 1973. Degrees of colony dominance in stenolaemate and gymnolaemate Bryozoa, pp. 121–220. In: Boardman, R. S., Cheetham, A. H., Oliver, Jr., W. A. (eds.), Animal colonies, development and function through time. Stroudsburg, PA: Dowden, Hutchinson, and Ross.

Brood, K. 1976. Cyclostomatous Bryozoa from the coastal waters of East Africa. Zool. Scripta 5:277–300.

Brown, D. A. 1964. Fossil Bryozoa from drill holes on Eniwetak Atoll. U.S. Geol. Survey, Prof. Paper 260EE: 1113–1116.

Busk, G. 1884. Report on the Polyzoa. The Cheilostomata. Rep. Sci. Res. Voy. Challenger, Zoology 10(30):1–216.

Canu, F., Bassler, R. S. 1920. North American early Tertiary Bryozoa. U.S. Natl. Mus. Bull. 106:1–879.

Canu, F., Bassler, R. S. 1933. The bryozoan fauna of the Vincentown limesand. U.S. Natl. Mus. Bull. 165:1–108.

Cheetham, A. H. 1960. *Rimosocella,* new genus of cheilostome Bryozoa. Micropaleontology 6:287–289.

Cheetham, A. H. 1963. Late Eocene zoogeography of the eastern Gulf Coast Region. Mem. Geol. Soc. Amer. 91:1–113.

Cheetham, A. H. 1966. Cheilostomatous Polyzoa from the upper Bracklesham Beds (Eocene) of Sussex. Bull. Brit. Mus. (Nat. Hist.) Geol. 13:1–115.

Cheetham, A. H. 1968. Evolution of zooecial asymmetry and origin of poricellariid cheilostomes (Bryozoa). Att. Soc. It. Sc. Nat. e Museo Civ. St. Nat. Milano 108:185–194.

Cheetham, A. H. 1973. Study of cheilostome polymorphism using principal components analysis, pp. 385–409. In: Larwood, G. P. (ed.), Living and fossil Bryozoa. London: Academic.

Cheetham, A. H. 1975a. Preliminary report on early Eocene cheilostome bryozoans from Site 308–Leg 32, Deep Sea Drilling Project, pp. 835–851. In: Larson, R. L., Moberly, R., et al. Initial reports of the Deep Sea Drilling Project, vol. 32. Washington, DC: U.S. Govt. Printing Off.

Cheetham, A. H. 1975b. Taxonomic significance of autozooid size and shape in some early multiserial cheilostomes from the Gulf Coast, pp. 547–564. In: Pouyet, S. (ed.), Bryozoa 1974, Proceedings of the Third Conference, International Bryozoology Association. Docum Lab. Géol. Fac. Sci. Lyon.

Cheetham, A. H. 1979. pp. 1–71. In: Carney, J. N., MacFarlane, A. Geology of Tanna, Aneityum, Futuna and Aniwa. New Hebrides Govt. Geol. Survey. Reg. Rept.

Cook, P. L. 1968. Bryozoa (Polyzoa) from the coasts of tropical West Africa. Atlantide Rep. no. 10:115–262.

Harmer, S. F. 1926. The Polyzoa of the Siboga Expedition. Part 2. Cheilostomata Anasca. Siboga Exped. 28B:183–501.

Labracherie, M. 1970. Nouveaux Bryozoaires de la famille des Farciminariidae du Tertiare aquitain. Bull. Soc. Géol. Fr., ser. 7 11:630–637.

Labracherie, M. 1975. Descriptions des bryozoaires cheilostomes d'âge Eocène inférieur du Site 246 (crosiére 25, Deep Sea Drilling Project). Bull. Inst. Géol. Bassin d'Aquitaine 18:149–202.

Lagaaij, R. 1968. Fossil Bryozoa reveal long-distance sand transport along the Dutch coast. Prac. Koninkl. Nederl. Akad. van Wetenschapi, ser. B 71:31–50.

Lagaaij, R. 1969. Paleocene Bryozoa from a boring in Surinam. Geol. Mijnbouw 48:165–175.

Lagaaij, R., Cook, P. L. 1973. Some Tertiary to Recent Bryozoa, pp. 489–498. In: Hallam, A. (ed.), Atlas of paleobiogeography. Amsterdam: Elsevier.

MacGillivray, P. H. 1880. *Nellia oculata,* p. 51, pl. 49, Fig. 5. In: McCoy, F. (ed.), Prodromus of the zoology of Victoria. Melbourne, 1878–85, vol. 1, decade 5.

Marcus, E. 1939. Bryozoarios marinhos brasileiros, 3. Bol. Fac. Fil. Cienc. Letr., U. Sao Paulo, vol. 13, Zoologia 3:111–299.

D'Orbigny, A. 1851. Paléontologie Francaise, pp. 1–188 (*Nellia* on p. 29). Terrains Crétacés, V. Bryozaires.

Osburn, R. C. 1940. Bryozoa of Puerto Rico with a résumé of the West Indian fauna. Sci. Survey Puerto Rico and Virgin Is., N.Y. Acad. Sci. 16(3):321–486.

Osburn, R. C. 1950. Bryozoa of the Pacific Coast of North America. I. Cheilostomata-Anasca. Publ. Allan Hancock Pacific Exped. 14:1–269.

Pohowsky, R. A. 1973. A Jurassic cheilostome from England, pp. 447–461. In: Larwood, G. P., (ed.), Living and fossil Bryozoa. London: Academic.

Rucker, J. B. 1967. Paleoecological analysis of cheilostome Bryozoa from Venezuela-British Guiana shelf sediments. Bull. Mar. Sci. 17:787–839.

Ryland, J. S. 1974. Bryozoa in the Great Barrier Reef Province, pp. 341–348. In: Proc. 2nd Int. Coral Reef Symp., Vol. 1. Brisbane: Great Barrier Reef Committee.

Schopf, T. J. M. 1977. Patterns and themes of evolution among the Bryozoa, pp. 159–207. In: Hallam, A. (ed.), Patterns of evolution. Amsterdam: Elsevier.

Tewari, B. S., Srivastava, I. P. 1967. On some fossil Bryozoa from India. J. Geol. Soc. Incia 8:18–28.

Voigt, E. 1959. La signification stratigraphique des bryozoaires dans le Crétacé supérieur. Congrès Soc. Savantes 84:701–707.

Winston, J. E. 1981. Feeding behavior of modern bryozoans, pp. 1–21. In: Broadhead, T. W. (ed.), Lophophorates, notes for a short course, U. Tennessee Studies in Geology, 5.

32

The Cretaceous Coral *Heliopora* (Octocorallia, Coenothecalia)—a Common Indo-Pacific Reef Builder

Mitchell W. Colgan

Earth Sciences Board, University of California, Santa Cruz, CA 95064

Heliopora: a Living Fossil

The fossil record of the genus *Heliopora* begins in the Early Cretaceous Period with its oldest representative *Heliopora japonica,* and "notwithstanding its great antiquity, it differs very little from the living *Heliopora coerulea*" (Eguchi 1948). *Heliopora coerulea,* the blue coral, is a common reef former of modern Indo-Pacific reefs and the sole surviving species of the Helioporidae, the only family of the order Coenothecalia. Coenothecalia, one of six orders of the subclass Octocorallia, is unique among the Octocorallia because *Heliopora* has a trabecular skeleton of fibrous aragonite (Hill 1960). Scattered throughout the corallum are autopores (0.5 mm to 2.0 mm) surrounded by a coenenchyma of numerous smaller siphonopores (Fig. 1). The autopores and siphonopores contain the polyps and the solenium of the colony, which together somewhat mask the blue color of the corallum in a veneer of golden brown tissue. The polyp has eight soft septa, and the autopores have a varying number of short pseudosepta which emerge from the theca (Bayer 1956). The number of pseudosepta vary from 10 to 30, depending on the species.

Both the autopores and the siphonopores are tabulated, which led some authors (Gregory 1900; Bayer 1956) to suggest a link between the Coenothecalia and the tabulate corals. Furthermore, *Heliopora coerulea* resembles the Paleozoic tabulate coral *Heliolites,* because both share a similar trabecular skeleton, growth form, and pore diversity. These parallels in skeletal design were used to argue for an alliance of *Heliolites* under the family Helioporidae (Moseley 1876). Although superficially similar, the two differ in their trabecular inclination, and the design and number of the septa. These differences preclude a possible alliance between the Coenothecalia and tabulate corals (Hill 1960). Elimination of any link to the tabulate corals thus isolates *Heliopora* because it has no close living relatives and no known ancestors (Gregory 1900) that can be deduced from hard-part analysis.

Distribution of *Heliopora*

Throughout its history, *Heliopora* has had wide geographic distribution following the shallow water, reef environment of the ancient Tethyan sea and the modern Indo-Pacific (Bouillon and Houvenaghel-Crèvecoeur 1970). Since the Early Cretaceous, *Heliopora* spp. have been recognized in the fossil record and a single species from Japan, *Heliopora japonica,* is the earliest known representative of the taxon (Eguchi 1948). This record gives *Heliopora* a nearly continuous history of over 100 million years (Table 1), making it one of the oldest octocoral fossils. Although there are no further

Fig. 1. Corallum surface of *Heliopora coerulea* showing the larger autopores and numerous siphonopores (×30).

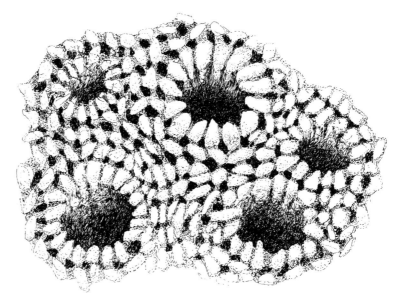

Table 1. Distribution of *Heliopora* species in time and space.

Age	Species	Location	Reference
Lower Cretaceous	*H. japonica*	Japan	Eguchi 1948
Middle to	*H. aprutina*	S. Europe	Parona et al. 1910
Upper Cretaceous	*H. bassanii*	S. Europe	Parona et al. 1910
	H. coerulea	England	Duncan 1879
	H. decipiens	S. Europe	Parona et al. 1910
	H. edwardsi	India	Stoliczka 1873
	H. edwardsi	Greece	Hackemesser 1936
	H. incrustans	Denmark	Nielsen 1917
	H. lindstromi	N. Alps	Vetters 1925
	H. macrotoma	Europe	Milne-Edwards and Haine 1860
	H. neocomiensis	Hungary	Kolosvary 1954
	H. partschi	Europe	Milne-Edwards and Haine 1860
	H. septifera	Alps	Vetters 1925
	H. somaliensis	N.E. Africa	Gregory 1900
	H. spongiosa	N. Alps	Koby 1898
Paleocene			
Eocene	*H. bellardi*	S. Europe	Sabaris 1943
	H. bennetti	Caribbean	Wells 1934
	H. boettgeri	Borneo	Gregory 1900
	H. mexicanae	Caribbean	Frost and Langenheim 1975
Oligocene	*H. bellardi*	S. Europe	Sabaris 1943
	H. boettgeri	Borneo	Gregory 1900
	H. oblite	Borneo	Felix 1921
Miocene	*H. boettgeri*	Borneo	Umbgrove 1929
	H. coerulea	Borneo	Umbgrove 1929
	H. fijiensis	Fiji	Hoffmeister 1945
	H. sparipora	Borneo	Felix 1921
Pliocene	*H. coerulea*	Indo-Pacific	Eguchi 1948

records of *Heliopora* in the region around Japan until the Pliocene on Taiwan (Hanzawa 1931), *Heliopora* spp. did flourish during the Middle to Late Cretaceous Period in Europe, India, and Africa. This time span marks the highest diversity of named species, and of these 14 species only two were found outside of Europe. One species, *Heliopora edwardsi* was recognized in India (Stoliczka 1873) and Greece (Hackemesser 1936), and among the Cretaceous *Heliopora*, it is a close ally of *Heliopora japonica* (Eguchi 1948). In Devonshire England, Cenomanian-age reef deposits contain an encrusting coral identified as *Heliopora coerulea* (Duncan 1879).

After the Cretaceous extinction, *Heliopora* seemingly disappeared for the next 15 million years from the fossil record. The absence of *Heliopora* during the Paleocene Epoch was followed by a resurgence of *Heliopora* in Eocene fossil records. The Eocene Epoch marks the second most speciose time span with four species of *Heliopora* named, and the genus had a much wider distribution than it had during the Cretaceous Period. *Heliopora* ranged from the Caribbean through Europe to the Indo-Pacific, this distribution corresponding with the shallow water borders of the Eocene Tethyan sea (Bouillon and Houvenaghel-Crèvecoeur 1970). The Caribbean *H. bennetti* and *H. mexicanae* are the only two species of the taxa found in the New World, and they disappeared from the fossil record after the Eocene Epoch. By the Oligocene there was only one European species, *H. bellardi,* (Sabaris 1943); the other species, *H. oblite* and *H. boettgeri,* were restricted to the Indo-Pacific, the province that the blue coral presently occupies.

During the Miocene Epoch, *Heliopora coerulea* was "rediscovered" in the Menkrwit Bed of Borneo (Umbgrove 1929 [*fide*] Wells 1954b). Along with *Heliopora coerulea* three other Indo-Pacific species were named from the region (Table 1). Throughout the Pliocene Epoch *Heliopora coerulea* was the only helioporid recognized in reef deposits, and was found in the central Indo-Pacific province (i.e., Indonesia, Philippines, New Guinea (Eguchi 1948), and Taiwan (Hanzawa 1931)). By the Late Pliocene Epoch, *Heliopora coerulea* reached the Marshall Islands (Wells 1954b), thereby giving a geographic range similar to its current distribution.

Ecology of *Heliopora*

The recent *Heliopora coerulea* lives within the warm, tropical waters of the Indo-Pacific between 25N and 25S latitude, and is found across nearly the entire breadth of the province from the Seychelles Islands to the Marshall Islands and the Great Barrier Reef (Bouillon and Houvenaghel-Crèvecoeur 1970). On many Indo-Pacific reefs *Heliopora coerulea* is a common shallow water coral. Its predominance in the seaward reef flats of the Marshall Island Atolls led Wells (1954a) to name the habitat the *H. coerulea* zone. The zone varies from 30.5 to 152.4 m in width and between 3.1 to 6.1 m in depth. Within the zone, *Heliopora* colonies develop into large micro-atolls between 10 and 20 ft in diameter.

Heliopora coerulea has a wide habitat range within the reef complex. On the islands of Palau, *Heliopora* thrives in the shallow water of the outer barrier islands, where the current is strong (Eguchi 1948). On Guam, *H. coerulea* is found to a depth range from less than 1 m to greater than 30 m (Colgan 1981).

Like most reef building corals, *H. coerulea* has symbiotic algae, zooxanthellae, which enhance the coral's growth (Goreau and Goreau 1959) and augment its nutrition (Muscatine 1973). *Heliopora* is able to feed actively from the water column, feed passively as a suspension feeder, and share in nutrients produced by its zooxanthellae. Therefore, it is both an autotroph and a heterotroph, and can be considered nutritionally a generalist.

Morphologic Variability

The most conspicuous feature of the living *Heliopora coerulea* is the blue color of its corallum, and the corallum color ranges widely from very pale blue to dark greyish-blue; this variation can be seen in different individuals in the same reef. In areas where *H. coerulea* abuts another coral colony, the blue pigment is often almost lost. The blue pigment is a distinguishing taxonomic characteristic of the living (Bayer 1956), but is of limited value when fossil specimens are analyzed, because the color is lost early in diagenesis. In Late Pliocene deposits on Guam, the blue corallum of *Heliopora coerulea* is already altered to a off-white similar to the other reef corals (R. Randall, personal communication). Other characters used to differentiate species vary to such a degree that there is often an overlap between the living species and many of the fossil species. An analysis of the variability of the living *H. coerulea* shows a continuity between it and the fossil species. At the generic level, both the recent and fossil *Heliopora* are structurally similar and share many characteristics.

The presence and distribution of the autopores and siphonopores shapes the appearance of the corallum surface. Species-level distinctions are often based on minor differences in the diameter of the autopore and the number of pseudosepta. Ancillary defining characteristics include the distance between the autopores (Hoffmeister 1945), the length of the pseudosepta (Gregory 1900), and sometimes the growth forms (Hoffmeister 1945).

An examination of these species-defining features in modern *Heliopora coerulea* showed these characters to be highly variable, affected both by ecology and geography. For example, five corals from Guam were sampled from depths ranging from 5 m to 25 m. My analysis showed that there was a significant within-species variation in the number of pseudosepta per calice ($F_{4,59} = 3.62$) with values ranging from 11 to 15. There was no apparent correlation between pseudosepta number and either the water depth of the habitat or the position on the corallum. *Heliopora coerulea* from the Seychelles Islands has a mean pseudosepta number of 14.9 (Bouillon and Houvenaghel-Crevecoeur 1970), and is significantly different from the pooled Guam data ($t_{144} = 12.71$).

The autopore diameter, likewise, varies. The most often quoted value for *H. coerulea* autopore diameter is 0.5 mm, with no range or variance given. The 0.5 mm value is within the range of the autopore diameters of Guam's *Heliopora*, though it appears to be on the small side, because the average diameter is 0.7 mm, with the range of 0.4 to 1.0 mm. For the Sey-

chelles Island sample, the autopore diameters of *Heliopora* range from "plus or minus 1 mm" for the upper portion of the corallum and diminished to 0.5 mm at the base (Bouillon and Houvenaghel-Crevecoeur 1970).

Among the *Heliopora* on Guam, one of the most variable characteristics is the distance between the autopores. From 20 measurements, the average distance was 3.28 mm, with a standard deviation of 2.93. The character appears to be dependent upon depth, light, and position on the corallum (pers. obs.).

The growth form of *Heliopora* is highly plastic and varies according to depth, current, and light (pers. obs.). Dana (1854) described two varieties of *H. coerulea*,—var. *tuberosa*, and var. *meandrina*—corresponding to the columnar and lobate growth forms, respectively. In addition, *H. coerulea* in shaded environments often has an encrusting growth form. The variety of growth form within the species prevents this characteristic from being useful in making species-level distinctions (Eguchi 1948).

The inherent variability of the characteristics of *Heliopora* may not have been fully appreciated by some previous workers who described and named fossil *Heliopora*, because many of the attributes particular to a fossil species often overlap with the recent coral.

Similarities Between Species of *Heliopora*

Eguchi (1948) describes the oldest *Heliopora*, the branching *H. japonica*, as having an autopore diameter between 0.75 mm and 1 mm, and between 15 and 17 pseudosepta per calice. He states "that in most respects *H. japonica* agrees fairly well with *H. coerulea* with the main distinction being in the size of the autopores and the number of septa." When compared with the specimens from Guam and Seychelles, these distinctions between the Lower Cretaceous coral and the modern coral are minor. The range of the diameter of all three groups of *Heliopora* overlaps, and this is particularly true for the Seychelles sample. Likewise, the range in the number of pseudosepta per calice overlaps for the three localities, with the highest degree of overlap being between the Seychelles *H. coerulea* and *H. japonica* (i.e., most common septa number 14 to 15, and 15 to 16, respectively). In addition to the morphologic similarities, *H. japonica* apparently occupied a niche similar to that of *H. coerulea* (Eguchi 1948).

There are several other Cretaceous species that are similar to *Heliopora japonica: H. aprutina, H. edwardsi, H. somaliensis,* and *H. urgonensis,* as well as the Recent *H. coerulea.* In turn, two species (*H. edwardsi* and *H. somaliensis*) compare closely with *H. coerulea. H. somaliensis* is characterized as having autopores ranging between 0.5 mm and 1 mm with 12 to 15 pseudosepta; both characteristics are within the range commonly seen *H. coerulea.* Or again, the encrusting *H. edwardsi* has 18 pseudosepta and was initially described as being "almost quite identical with the recent species" (Stoliczka 1873).

The recent coral finds a close ally in the Cenomanian reef deposits of Devonshire, England. An encrusting coral there was identified as *H. coerulea* because the specimens had 12 pseudosepta and "cannot be distinguished from the encrusting form of recent *Heliopora coerulea*" (Duncan 1879).

Like the Cretaceous *Heliopora,* the Paleogene species resembles the living *H. coerulea.* The New World Eocene corals, the encrusting *H. bennetti* and the branching *H. mexicanae*—have a corallum surface pattern that parallels the structure of *H. coerulea* with only modest differences in the number of pseudosepta (24, *H. bennetti* [Wells, 1934]) and the diameter of the autopores (1.1 to 1.3 mm *H. mexicanae* [Frost and Langenheim 1975]). The structures of the corallum of the Paleogene species show a continuity that links the taxa across the Eocene Tethyan sea. In the Eocene and the Oligocene deposits of southern Europe, the lobed *Heliopora bellardii* has an autopore diameter ranging from 1.0 to 1.5 mm and a pseudosepta number between 16 and 24 (Sabaris 1943); the Borneo species, *Heliopora boettgeri,* has an autopore diameter of 1 mm and 16–24 pseudosepta per calice (Felix 1921).

The Neogene *Heliopora—H. fijiensis* and *H. sparipora*—have corallum surface patterns like those in both *H. coerulea* and the Paleogene species. Like *H. boettgeri, Heliopora fijiensis*

has an encrusting growth form, an autopore diameter (0.75 to 0.9 mm), and a pseudosepta number (14 to 17) that overlap with *H. coerulea.*

In summary, then, *Heliopora* spans over 100 million years, with only slight morphologic changes. This is particularly evident when the fossil species are compared with the living *Heliopora,* because many of the variations described for the fossil species find expression in the range of characters seen in living *H. coerulea.* The similarity in morphology provides a continuity between the oldest *Heliopora* and the living *H. coerulea,* thereby making it a good candidate for consideration as a living fossil.

Acknowledgments. I would like to thank my wife Sain, for her translation and for her illustration. I deeply appreciate the editorial comments of Léo Laporte and Niles Eldredge. I also thank Lawerence Chai and Donald Colgan for their help. Contribution No. 202, University of Guam, Marine Laboratory. This report was funded in part by an Amoco grant to the University of California, Santa Cruz.

Literature

Bayer, F. 1956. Octocorallia, pp. 16–231. In: Moore, R. C. (ed.), Treatise on invertebrate paleontology, Coelenterata. Geol. Soc. Am. and U. Kansas Press, Lawrence, Ks.

Bouillon, J., Houvenaghel-Crèvecoeur, N. 1970. Étude monographique du genre *Heliopora* de Blainville. Musée Royal de L'Afrique Centrale Série in-8. Sci. Zool. 178.

Colgan, M. 1981. Long-term recovery process of a coral community after a catastrophic disturbance. U. Guam Mar. Lab. Tech Rept. 76.

Dana, J. D. 1846. Zoophytes. U.S. Explor. Exped. (Washington) V.7. 740 p., 61 pl.

Duncan, P. M. 1879. On the Upper Greensand coral fauna of Haldon, Devonshire, Geol. Soc. Lond. Quart. J. 35:94–96.

Eguchi, M. 1948. Fossil Helioporidae from Japan and the South Sea Islands. J. Paleont. 22(3):362–364.

Felix, J. 1921. Fossile Anthozoen von Borneo: Palaeontologie von Timor, Leif. 9, no. 15, p. 1–64, pls 1–6.

Frost, S. H., Langenheim, R. L. 1975. Cenozoic reef biofacies: Tertiary larger foraminifera and scleractinan coals from Chiapas Mexico. North. Illinois U. Press.

Goreau, T., Goreau, N. 1959. The physiology of skeleton formation in corals. II. Calcium deposition by hermotypic corals under various conditions in the reef. Biol. Bull. 117:239–250.

Gregory, J. 1900. Polytremacis and the ancestry of the Helioporidae. Proc. Roy. Soc. Lond. 46:291–305.

Hackemesser, M. 1936. Eine kretazeische Korallenfauna aus Mittelgriechenland und ihre palaeobiologischen Beziehungen.

Hanzawa, S. 1931. Notes on the raised coral reefs and their equivalent deposits in Taiwan (Formosa) and adjacent islets. Recent Oceanogr. Works Japan. 32:37–52.

Hill, D. 1960. Possible intermediates between Alcyonaria, Tabulata and Rugosa, and Rugosa and Hexacoralla. 21st. Int. Geol. Congr. 22:51–58.

Hoffmeister, J. 1945. Corals. In: Ladd, H. S., Hoffmeister, J. E. (eds.), Geology of Lau, Fiji. Bull. Bishop. Mus. 181:298–311.

Koby, F. 1898. Monographie des Polypiers Cratacés de la Suisse. Mém. de la Soc. Paléont. Suisse. 24.

Kolosvary, G. 1954. Les Corallaires du Cretace de la Hongrie. Jb. Hung. Geol. Anst. 42(2):64–131.

Milne-Edward, H., Haime, J. 1860. Histoire naturelle des Coralliaires: (Paris). Vol. III.

Moseley, H. 1876. On the structure and relations of the Alcyonarian *Heliopora coerulea*, with some account of the anatomy of a species of Sarcophyton. Phil. Trans. Roy. Soc. 166(1):91–132.

Muscatine, L. 1973. Nutrition of corals. In: Jones, O., Endean, R. (eds.), Biology and geology of coral reefs. 2:77–115.

Nielsen, K. 1917. *Heliopora incrustans* nov. sp. with a survey of the Octocorallia in the deposits of the Danian in Denmark. Kobenhaven. Medd. geol. 5(8):1–13.

Parona, C., Crema, C., Prever, P. 1910. La fauna Coralligena del Cretceo dei Monti d'Ocre nell'Abruzzo Aquilano. Roma mem. serv. desc. Carta geol. d'It. R. com. geol. Regno. 5(1):1–242.

Sabaris, D. 1943. Fauna coralina del Eoceno Cataln. Mem. Acad. Cience. Barcelona. 26(3):259–439.

Stoliczka, F. 1873. The corals of Anthozoa. In: Cretaceous fauna of South India. Pal. Indica. 4(8):1–59.

Umbgrove, J. H. 1929. Anthozoen van N. O. Borneo. Dienst Mijnbouw Nederlandsch-Indie Wetensch. Meded. 9:47–78.

Vetters, H. 1925. Über kretazeische Korallen und andere Fossilreste im nordalpinen Flysch. Jahrb. Geol. Bundesanstalt. Wien. 75:1–18.

Wells, J. 1934. Eocene corals. Part I. From Cuba; Part II. A new species of Madracis from Texas. Bull. Amer. Paleont. 20(70b):145–164.

Wells, J. 1954a. Recent corals of the Marshall Islands. In: Bikini and nearby atolls. U. S. Geol. Surv. Prof. Pap. 260:385–486.

Wells, J. 1954b. Fossil corals from Bikini Atoll. In: Bikini and nearby atolls. U.S. Geol. Surv. Prof. Pap. 260:609–615.

33
Simpson's Inverse: Bradytely and the Phenomenon of Living Fossils

Niles Eldredge

Department of Invertebrates, American Museum of Natural History, New York, NY 10024

The scientific problem posed by "living fossils"—loosely speaking, members of the Recent biota whose external form, at least, has changed but little since the lineage's inception—is, of course, a question of rates. Why have these creatures changed so slowly, so negligibly? In the context of Darwin's own founding conceptions, and certainly from the perspective of the modern synthesis, living fossils are something of an enigma, if not an embarrassment. Creationists have long cited the continued existence of "lower" forms of life alongside evolution's more recent and derived productions as a falsification of the very notion of evolution, but what their observation suggests is that life's diversification and the sequential addition of evolutionary novelties—"descent with modification"—is burlesqued if viewed simply as a linear process of "progress."

Yet even the most sophisticated of evolutionists have tended to see evolutionary change as inevitable, given simply the passage of vast stretches of time. For example, G. G. Simpson told us: "Organic change is so nearly universal that a state of 'evolutionary motion' is inherent in phyletic survival" (1944:149). And this feeling that evolutionary change is almost inevitable is based upon the relentless logic of adaptation through natural selection: somehow, one feels, it is inevitable that environments must change, and organisms must modify their adaptations to keep pace, or face the grim consequence of extinction. Hence the fascination for living fossils: Somehow these hoary beasts have escaped the trap. They have not kept pace with a changing world. And yet here they are, alive and well, hale and hearty—in some cases thriving, while in others just barely hanging on. *Limulus polyphemus,* it is said, is typically among the last of the metazoan species to be driven out of polluted estuaries along the eastern coast of the United States.

Thus living fossils tend to be seen as anomalies. And the theories that are available to explain their putative stagnation reflect their perceived status: Perhaps they lack the requisite genetic variability, or have not yet been subjected to strong (directional) selection pressure, or (better from the standpoint of the synthesis) perhaps they have been subjected to a straightjacket of unrelenting stabilizing selection. Stanley (1979) and Eldredge (1975, 1979) have reviewed most of these hypotheses. What they all have in common, it seems, is the viewpoint that extremely slow rates of evolution (and, for that matter, *all* problems of evolutionary rates) are strictly a matter of the *transformation of aspects of the phenotype.* This is, after all, how we detect the problem in the first place. The data are always expressed as amounts of morphologic change over an interval of time. And the synthesis sees evolution in general as a matter of the stochastic and deterministically based changes of gene frequencies within populations. Small wonder most hypotheses seeking to explain living fossils are couched squarely in such transformational terms.

Eldredge (1975, 1976, 1979; Eldredge and Cracraft 1980) and Stanley (1975, 1979) have suggested that an alternative approach, a *taxic* perspective, may shed some further light on macroevolutionary problems and, in particular, offer alternative hypotheses for the explanation of disparate rates of morphologic transformation. A taxic base begins with the assumption that both species and monophyletic taxa are spatiotemporally bounded entities. They are historical things, or in Ghiselin's (1974) and Hull's (1976) terminology, they are "individuals." Taxa are *not* classes (though taxonomic categories patently are classes; the distinction seems to have been first strongly borne home by Simpson (1963)). If, on the other hand, we choose to see taxa (including species) solely as arbitrarily defined and delineated segments of evolving lineages (as, e.g., Simpson 1961, sees them for the most part in his second major contribution to method and theory in systematics) then taxa really are classes. The importance of the distinction between seeing taxa as discrete historical entities, on the one hand, or as arbitrarily defined classes on the other, in the present context is simple enough: We can view rates of evolution solely as a transformational problem if and only if we see taxa strictly as classes (Eldredge 1979, 1982). If, on the other hand, taxa are individuals, they have origins, histories and terminations, and we might suppose there to be a spectrum of rates of origins of taxa. Eldredge and Cracraft (1980) follow conventional wisdom along these lines, and acknowledge that species exist in a dynamic sense in nature, hence have origins (speciation), whereas monophyletic taxa of rank higher than species are strictly interconnected genealogical arrays of species. Hence the "origin of taxa of higher rank" does not entail any special biological process other than speciation and the processes of genetic change that occur within populations and at the genomic level within organisms (see Eldredge 1982; Eldredge and Salthe in press; and Vrba and Eldredge in press, for discussions of the various levels of the "genealogical hierarchy"). What this means, though, is that differential rates of speciation

(and extinction) within and among genealogical clades ought to be considered *in addition to* the standard processes of drift and selection (at the population level) and whatever molecular processes pertain at the genomic level, as a full explanation of any particular problem of evolutionary rates that involves more than one species. Living fossils pose such a problem.

Briefly stated, the transformational view of the origin of species (a view most commonly found in the paleontological literature but not uncommonly broached when neotologists think about how species "behave" in geologic time) sees evolution as the accumulation of morphologic, hence genetic, change. We simply define and recognize "species" (and other taxa) on the basis of conveniently recognized "packages" of such change. It is commonly stated (e.g., Mayr 1942:153; Cain 1954:110 ff.) that were the fossil record complete, the taxonomist's job of naming and pigeon-holing species would be impossible. Gaps in the record often supply that criterion of arbitration for the delimitation of "paleospecies."

The flip side to this argument is that species themselves exist as reproductively coherent assemblages. In a 2×2 contingency table, Vrba (1980:68) nicely graphed what should be realized by all: There is no necessary correlation between amount of morphologic (and/or genetic) change and speciation. Speciation, in other words, may be accompanied by a great deal of anatomical, behavioral, and physiological change, or hardly any at all, as in sibling species. This is one way of putting the "decoupling" (Stanley 1975) between transformational and taxic rates.

Nonetheless, and in spite of the "no necessary relation" between the transformation of gene frequencies and morphologies, on the one hand, and the origin of species on the other, we might ask whether speciation—the origin of new reproductive communities from old— might not be the trigger for adaptive, transformational change, as Mayr (1954) and Eldredge (1979), to cite but two, have suggested. Empirically, the fossil record seems to show a correlation between speciation and anatomical change (Eldredge and Gould 1972; Gould and Eldredge 1977). The idea of "punctuated equilibria" does not maintain either that all evolutionary change occurs only at speciation, or that speciation always results in appreciable morphologic change. It maintains merely that speciation may not infrequently entail such detectable change, and much of the anatomical change we can see in the fossil record seems to be correlated with speciation events. The latter observation, once again, hinges on the supposition that species are real entities in nature, and that their recognition by morphologic criteria involves something more than mere arbitrary selection of a set of anatomical features that vary among a series of samples and use of these variables to define a set of morphologic classes. As Fisher (this volume) and others have indicated, being utterly sure, when dealing with fossils, that the taxic assumption is warranted, that the fossil elements (or, in many cases, even the Recent elements) of the lineage really correspond to actual species (reproductive communities) remains exceedingly difficult to test rigorously (but see Eldredge and Cracraft 1980, Chs. 3 and 6, for further discussion). What I am arguing here is simply the necessity to consider the consequences for our understanding of evolution if we variously consider taxa as classes or as individuals.

George Simpson's name stands at the head of this essay because it was largely through his efforts that a coherent study of taxonomic and transformational rates has been founded (Simpson 1944, and especially 1953 for taxonomic rates). Simpson (1944:ix) felt the fossil record had much to suggest to theorists about the precise way the various determinants of evolution worked together. (Simpson's list of such determinants included "variability, rate of mutation, character of mutations, length of generation, size of population, and natural selection" [1944:30]). But even more relevant to living fossils, Simpson devoted an entire chapter of his 1944 *Tempo and Mode in Evolution* (Ch. 4) to "Low-Rate and High-Rate Lines." As I have argued (Eldredge, in press) at length, Simpson's entire book was focused on developing the background, data, and theory for his notion of quantum evolution—essentially a transformational hypothesis of rapid shifts between adaptive zones (or peaks) that would account for the relative scarcity of annectant fossil forms between taxa of higher rank. He recognized *modes* of evolution, qualitatively distinct sorts of evolution. These were speciation (usually in-

volving a subdivision of an adaptive zone), phyletic evolution (which may involve exploration of different subzones of an adaptive zone), and quantum evolution (which Simpson felt typically involved abandoning one adaptive zone and invading another).

The point here is that Simpson based his argument for the existence of these three qualitatively distinct modes to a great degree on three qualitatively distinct classes of (transformational) evolutionary rates: "low-rate lines," "normal" rates, and "high-rate lines." In his newly introduced jargon, these were known as *bradytely, horotely,* and *tachytely,* respectively. Simpson emphasized that bradytely and tachytely are *not* merely the lower and upper tails of a normally distributed spectrum of rates; each is a distinct category of rate, with its own internal distribution—an original, if somewhat dubious, claim that helped buttress his conclusion that his three modes were likewise qualitatively distinct. I have argued (Eldredge, in press) that in acknowledging that in the very nature of things, rapidly evolving lineages are unlikely to leave much evidence in the fossil record, Simpson focused on the mirror-image problem of inordinately slow-rate lineages— partly because of their intrinsic interest, and partly because, as the opposite, inverse case of tachytely, perhaps bradytely could shed some light on tachytely. Hence "Simpson's inverse." Bradytely, whether or not qualitatively distinct from horotely and tachytely, seems tractable for study. Diversity is low, be it construed as morphologic, taxic, or both, and then as now the problem seems at least approachable in terms of gathering appropriate data of serviceable quality.

Schopf (1984) has recently reiterated his doubts that there are such things as living fossils, except in "persistence of particular traits," which he acknowledges as an interesting problem. Schopf's list of six variant definitions of living fossils is both useful and provocative:

"Depending on the author, a living fossil is: (1) A living species that has persisted over a very long interval of geologic time. (2) A living species that is morphologically and physiologically quite similar to a fossil species, as seen over long intervals of geologic time. (3) A living species that has a preponderance of primitive morphologic traits. (4) A living species that has one of the above, *and* a relict distribution. (5) A living species that was once thought to be extinct. (6) An extant clade of low taxonomic diversity whose species have one or more of the properties of (1), (2) or (3)." (Schopf, 1984, p. 272).

Only two cases presented in this book, Winston and Cheetham on the bryozoan *Nellia* and Batten on the gastropod *Neritopsis,* report using the same name for a species (in the Cretaceous and Eocene, respectively) as is used for a present-day species. The notostracan branchiopods (cited by Schopf 1984) apparently hold the record. The living species *Triops cancriformis* is also "known" from the Upper Triassic. But fossil notostracans are scrappy affairs, and only the relatively featureless carapace is known from the Triassic. The carapace, indeed, seems to have remained relatively unchanged, but the systematics of living notostracans is based on far more than carapace shape. All this means, of course, is that in terms of parts comparable between fossil and Recent specimens, no systematist can tell them apart. It is fair to conclude, I think, that no one supposes that it is the actual longevity of a single species that underlies cases of extraordinarily low-rate lines of morphologic transformation.

Thus we still must ask: Is it slow transformation, producing little anatomical diversity, that accounts for low observed taxonomic diversity in so many of these examples? Or is it the other way around, as the taxic alternative would seem to hint: Low rates of production of new taxa (i.e., speciation) set the brakes on the accumulation of much morphologic diversity? For this *is* a common pattern—Schopf's sixth alternative. And a number of cases in this book seem to conform well to this pattern.

And perhaps the most intriguing pattern to be discussed in the recent macroevolutionary literature is actually an addition to Schopf's sixth definition of living fossils. As far back as the 1880s, invertebrate paleontologist Henry Shaler Williams wrote about two contrasting sorts of fossils. Some taxa, according to Williams, are typically variable, both on single bedding planes and on up through entire sections. They tend to be found in a variety of habitats, occur over relatively wide geographic areas, and persist for long periods of time. Such species contrast with others that seem specialized, relatively invariant, and much more restricted in their distribu-

tion in space and time. Recently, William's observations have been revived and conceptually linked to the ecological notions of eurytopy and stenotypy, which some authors (e.g., Eldredge 1975, 1979; Eldredge and Cracraft 1980; Vrba 1980) have in turn linked in various sorts of ways with theories of differential rates of speciation and species survival. In short, the phenomenon of living fossils has seemed to these and other authors to be caused by, or an incidental effect of (Vrba 1980, 1983), low rates of speciation—itself a reflection of essentially eurytopic adaptations of the organisms.

The new element of this pattern, most cogently pointed out by Vrba (1980, with details provided in this volume) is the contrast afforded by sister-taxa within a monophyletic clade. Whatever the underlying causal scheme, it frequently happens that the primitive sister-group of all other members of a monophyletic group (1) are depauperate in species, (2) retain a large number of plesiomorphies, (3) are ecologically eurytopic in many physiological and behavioral attributes, and (4) show great individual species longevity with (5) broad areal and habitat distribution. All these attributes contrast with individual species within the sister-taxon, which is correspondingly (1) rich in species, (2) stenotopic (component species are relatively narrow-niched) and consequently have (3) many apomorphies (unique evolutionary specializations), and (4) tend to be relatively short-lived geologically and (5) narrowly distributed. Among the cases presented in this volume, Novacek, Fisher, Vrba, and Delson and Rosenberger all explicitly discuss this particular pattern and possible reasons for its existence. Other cases appear to fit it as well. Fisher (this volume) has gone further and added the new concept of tree symmetry/asymmetry as an approach to the analysis of bradytely; he has sharpened the focus considerably on the basic problem of testing rival hypotheses that bear on the entire problem of differential evolutionary rates.

Schopf is certainly correct that a number of somewhat different kinds of phenomena underlie our rather casual use of the expression "living fossil." Some species do have relict distributions (e.g., *Sphenodon,* unfortunately not included in this volume), while others patently do not, such as the likewise-absent *Lingula.* Some lineages are depauperate in species, such as *Limulus* and its close relatives, while others generally considered living fossils (such as the nuculoid bivalves, not included here) are relatively speciose. All sorts of combinations are possible, and many turn up in the cases presented in this book.

And, of course, some authors (e.g., Patterson, Tattersall, Schwartz) have considered both well-known and obscure putative examples of living fossils, and have found them wanting in one or more criteria of the term. This is salubrious: RNA is a synapomorphy for "all life." Thus its presence in any (subset of) living organisms is plesiomorphic. Contemplating RNA would lead us to describe all organisms as living fossils—interesting in some loose metaphysical sense, perhaps, but telling us nothing of evolutionary importance other than that some traits have persisted for long periods of time, up to and including the entire history of life. It is "overall plesiomorphy" (itself admittedly nebulous, and perhaps only interpretable in the context of comparison with relatively apomorphic sister-taxa) and its various ecological and distributional correlates that deserves special attention and the sobriquet "living fossils."

I cannot conclude, from a survey of the contributions of this volume, that any specific version of the "taxic" hypothesis of living fossils, such as the one linking plesiomorphic/apomorphic sister-taxa respectively with the ecological dichotomy of eurytopy and stenotopy, as outlined above, is abundantly corroborated. While there is nothing in these cases to support Simpson's claim that bradytely is in a class of itself and *not* merely the lower tail of a monotonically distributed spectrum of rates, there is also no reason to conclude that *all* cases of living fossils belong to low diversity, eurytopic lines with characteristically low speciation rates. It is beginning to seem that evolution is an ontologically hierarchical process (Eldredge 1982; Eldredge and Salthe in press; Vrba and Eldredge in press; Eldredge, in press). Higher-level phenomena (i.e., those involving more than one species) result from processes operating at that level (e.g., differential speciation and extinction) and from processes operating at lower levels, including genomic and ontogenetic constraints, plus the more traditionally invoked processes of drift and natural selection. The data in this book suggest that the general class of "taxic" explanations is appropriate as an additional source of processes to be considered when evaluating differential evolutionary rates. The data presented in each of the cases of this book, plus their accompanying guides to the literature, serve as a firm foundation as well as a model for a far more careful consideration of the general problem of arrested evolution, bradytely—or just plain "living fossils."

Acknowledgments: Tom Schopf kindly lent his skepticism of the entire notion of "living fossils," as well as a pre-publication copy of his manuscript. We shall miss him. And I am of course grateful to the authors of the essays in this volume who all, sooner or later, produced their contributions to this thoughtful collection.

Literature

Cain, A. J. 1954. Animal species and their evolution. New York: Harper and Bros. (reprint ed., 1960).

Eldredge, N. 1975. Survivors from the good old, old, old days. Nat. Hist. 81(10):52–59.

Eldredge, N. 1976. Differential evolutionary rates. Paleobiology 2:174–177.

Eldredge, N. 1979. Alternative approaches to evolutionary theory. In: Schwartz, J. H., Rollins, H. B. (eds.), Models and methodologies in evolutionary theory. Bull. Carnegie Mus. Nat. Hist. 13:7–19.

Eldredge, N. 1982. Phenomenological levels and evolutionary rates. Syst. Zool. 31:338–347.

Eldredge, N. In press. The integration of evolutionary theory. U. Chicago Press.

Eldredge, N., Cracraft, J. 1980. Phylogenetic patterns and the evolutionary process: method and theory in comparative biology. New York: Columbia U. Press.

Eldredge, N., Gould, S. J. 1972. Punctuated equilibria: an alternative to phyletic gradualism, pp. 82–115. In: Schopf, T. J. M. (ed.), Models in paleobiology, San Francisco: Freeman, Cooper.

Eldredge, N., Salthe, S. N. in press. Hierarchy and evolution. Oxford Surveys in Evol. Biology 1.

Ghiselin, M. T. 1974. A radical solution to the species problem. Syst. Zool. 23:536–544.

Gould, S. J., Eldredge, N. 1977. Punctuated equilibria: the tempo and mode of evolution reconsidered. Paleobiology 3:23–40.

Hull, D. L. 1976. Are species really individuals? Syst. Zool. 25:174–191.

Mayr, E. 1942. Systematics and the origin of species. New York: Columbia U. Press. (reprint ed., 1982).

Mayr, E. 1954. Change in genetic environment and evolution, pp. 188–213. In: Huxley, J., Hardy, A. C., Ford, E. B. (eds.), Evolution as a process. New York: Collier Books. (reprint ed., 1963).

Schopf, T. J. M. 1984. Rates of evolution and the notion of living fossils. Ann. Rev. Ecol. Syst. 12:245–92.

Simpson, G. G. 1944. Tempo and mode in evolution. New York: Columbia U. Press.

Simpson, G. G. 1953. The major features of evolution. New York: Columbia U. Press.

Simpson, G. G. 1961. Principles of animal taxonomy. New York: Columbia U. Press.

Simpson, G. G. 1963. The meaning of taxonomic statements, pp. 1–31. In: Washburn, S. L. (ed.), Classification and human evolution. Chicago: Aldine.

Stanley, S. M. 1975. A theory of evolution above the species level. Proc. Natl. Acad. Sci. 72:646–650.

Stanley, S. M. 1979. Macroevolution: pattern and process. San Francisco: Freeman.

Vrba, E. S. 1980. Evolution, species and fossils: how does life evolve? S. Afr. J. Sci. 76:61–84.

Vrba, E. S. 1983. Macroevolutionary trends: new perspectives on the roles of adaptation and incidental effect. Science 221:387–389.

Vrba, E. S., Eldredge, N. In Press. Individuals, hierarchies and processes: towards a more complete evolutionary theory. Paleobiology 10.

34
Does Bradytely Exist?

Steven M. Stanley

Department of Earth and Planetary Sciences, The Johns Hopkins University, Baltimore, MD 21218

To paleontologists, at least, the existence of living fossils has long seemed to represent the same puzzle as the occurrence of bradytely. Bradytely was one of three distinct tempos of evolution recognized by Simpson (1944, 1953) in his pioneering analysis of rates of evolution. Simpson coined the term *horotely* for a spectrum of rates that represented normal evolution for a particular taxon, a spectrum forming a unimodal histogram; *tachytely* was meant to be essentially synonymous with *quantum evolution,* which was exceptionally rapid change leading to a new adaptive zone; and *bradytely* was very sluggish evolution, which was alleged to produce a discrete mode beyond the "slow" end of a horotelic distribution of rates. Among the taxa alleged to be represented in this mode were ones traditionally recognized as living fossils. Simpson's concepts were developed within a gradualistic framework, and my aim in writing this article is partly to point out that the three categories of rates, including bradytely, lose their value in a punctuational framework. Simpson was explicitly considering rates of phyletic evolution, or anagenesis, which in the punctuational model of evolution plays a minor role in large-scale transformation. The problem is more fundamental, however: It is now evident that regardless of one's world view—gradualistic or punctuational—there is no empirical basis for recognizing the three types of evolutionary rates. I will consider the three tempos individu-

ally, focusing finally upon bradytely and living fossils.

Simpson presented what he termed horotelic distributions for particular higher taxa by plotting mirror images of histograms of longevities for component genera or species. The resulting histograms of rates were curiously left skewed and, in fact, the histograms were invalid because the nonlinear relationship between a variable and its reciprocal means that the mirror-image transformation was inappropriate (Stanley 1979:130). Simpson's conclusion was that for a particular higher taxon "the characteristic (modal) rate is decidedly nearer the maximum for this group than the minimum" (Simpson 1953:317). In other words, relatively rapid evolution was seen as the norm, which is a gradualistic conclusion. Also gradualistic was the use of generic longevities to plot "horotelic" distributions of phyletic (anagenetic) rates. The assumption was that the dominant way in which genera have disappeared is by pseudoextinction, or phyletic transformation: "If all genera were strictly comparable, phyletic rates of evolution would be proportional to the reciprocals of the durations of the genera in question" (Simpson 1953:32). Simpson, of course, recognized that some genera experienced true extinction, but concluded that this simply served to introduce some error, making "group rates" appear more rapid than true phyletic rates. In a punctuational framework, phyletic transformation of genera is rare, and what can be estimated from a distribution of generic longevities is a rate of extinction, not a rate of evolution.

The concept of tachytely also reflects gradualistic thinking. Tachytely was "defined as a phylogenetic phenomenon and is to be distinguished from rapid speciation or splitting" (Simpson 1953:336). Simpson noted that belief in rapid phyletic transformation does not conflict with recognition of rapidly divergent speciation events, "such as occur on invasion of new and open habitats." What the concepts of tachytely and horotely did do was to form a gradualistic model for considering rates of evolution. Well-established lineages were expected to evolve at a substantial pace, and they were expected to account for the bulk of evolutionary change through an accumulation of change over millions of years (horotelic rates) or through sudden shifts of adaptive zone (tachytelic rates). The envisioned patterns are depicted in Fig. 1.

"Bradytelic" was the adjective that Simpson applied to examples of "arrested evolution," or "lineages [that] have evolved at rates much slower than any of the horotelic distribution" (Simpson 1953:318). The purported evidence for this was the great geological longevity of certain living taxa, such as genera of bivalve mollusks and species of centric diatoms. When, for example, a survivorship curve for extinct genera of bivalves is plotted together with a similar cumulative distribution for living genera, the tail of the latter stretches well beyond that of the former (Fig. 2). Genera that contribute to the excess number in the tail of the distribution for extinct genera provided the evidence for bradytely. Some genera of the so-called bradytelic group are not yet older than 250 million years (the approximate limit of the horotelic distribution) simply because they originated within the last 250 million years. Simpson concluded that bradytelic rates were separated from horotelic rates by "a statistical discontinuity." In fact, with proper analysis this discontinuity disappears. As Raup (1975) has shown, data for longevities for living taxa cannot simply be cumulated to form a survivorship curve. Living taxa instead provide data for "time-specific" analysis that differs from the "dynamic" analysis that is appropriate for data representing extinct taxa (Kurtén 1964). Raup has further noted that any sample of extinct forms is strongly biased against long-lived taxa because for a taxon of long duration there is a relatively high probability of still being alive. By using the appropriate types of analysis for the two types of data and pooling the resulting survivorship data, one can produce a single survivorship

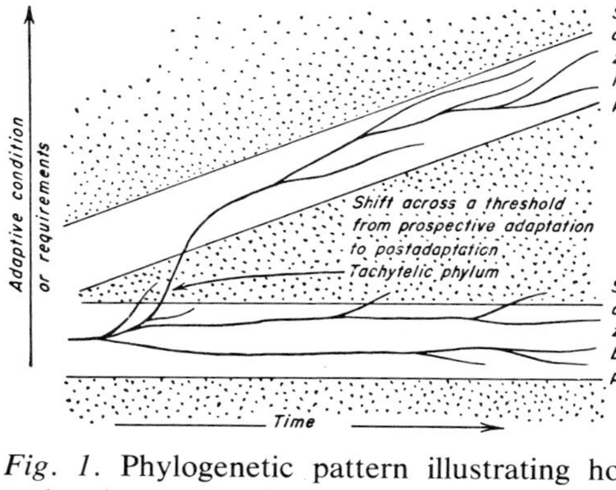

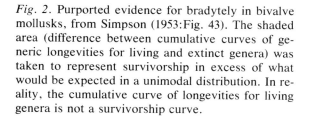

Fig. 1. Phylogenetic pattern illustrating horotely, tachytely, and bradytely as envisioned by Simpson (1953:Fig. 44). Tachytely is asserted to be unusually rapid phyletic transformation associated with a shift of adaptive zone, and bradytely is unusually slow phyletic evolution. In this gradualistic scheme, speciation does nothing more than establish new lineages that move slowly in new evolutionary directions.

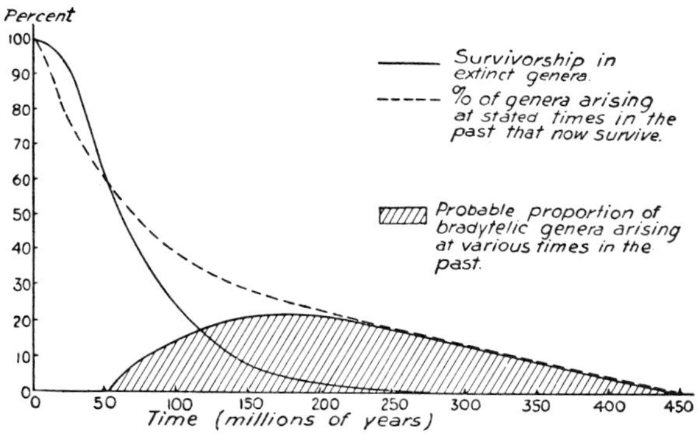

Fig. 2. Purported evidence for bradytely in bivalve mollusks, from Simpson (1953:Fig. 43). The shaded area (difference between cumulative curves of generic longevities for living and extinct genera) was taken to represent survivorship in excess of what would be expected in a unimodal distribution. In reality, the cumulative curve of longevities for living genera is not a survivorship curve.

curve, though even this curve may include errors (Raup 1975). In any event, the discrepancy between the curves in Fig. 1 has no meaning and there is no evidence of a discontinuity of rates. Thus, the alleged distinction between horotely and bradytely disappears.

In a gradualistic framework of evolution, living fossils might still be seen as resulting from unusually slow rates of evolution—ones represented in the tail of the unimodal, right-skewed distribution of rates. In a punctuational framework, the problem of living fossils takes on an entirely different aspect. The expectation here is that well-established species will normally undergo such sluggish phyletic change that this mode of change plays a subordinate role in evolution: Most evolution takes place within small populations in association with speciation events. In fact, the expectation is that even after millions of years, a surviving lineage will likely have traits very similar to those that it possessed soon after it came into being. In other words, most geologically ancient single lineages or narrow clades (ones experiencing few speciation events) that have survived to the present will constitute living fossils.

What remains to be explained in the punctuational model is why some single lineages or narrow clades have survived for so long. My view (Stanley 1979:129) is simply that a histogram of longevities for all of the single lineages or small clades within a higher taxon will normally have a right-skewed shape. Living fossil groups form the extremity of the right-hand tail. They are unusual, then, not in being strange cases of arrested evolution. They are simply champions at

warding off extinction. Still, their great longevities will normally intergrade with shorter ones for other clades of comparable size.

What I have discussed here represents what might be termed the kinetics of macroevolution, a description of evolutionary change. The dynamics, which entail causation, present more complex problems. If the punctuational view is correct and living fossils are nothing more than relatively long-lived groups, we need a set of explanations for survivorship. As many examples in the foregoing chapters suggest, a lineage or small clade may endure for a long stretch of geologic time because of great niche breadth, broad geographic distribution, protection in a small, cloistered habitat or geographic area, or other varied factors that are not all mutually exclusive. Thus, although the punctuational expectation is that living fossil groups should exist, the reasons why some groups rather than others fulfill that expectation can only be assessed on a case-by-case basis, as undertaken in many of the chapters of this book.

Literature

Kurtén, B. 1964. Population structure and paleoecology, pp. 91–106. In: Imbrie, J., Newell, N. (eds.), Approaches to paleoecology. New York: Wiley.

Raup, D. M. 1975. Taxonomic survivorship curves and Van Valen's Law. Paleobiology. 1:82–96.

Simpson, G. G. 1944. Tempo and mode in evolution. New York: Columbia U. Press.

Simpson, G. G. 1953. The major features of evolution. New York: Columbia U. Press.

Stanley, S. M. 1979. Macroevolution: pattern and process. San Francisco: Freeman.

Index